Global Urban Monitoring and Assessment

through Earth Observation

Taylor & Francis Series in Remote Sensing Applications

Series Editor
Qihao Weng
Indiana State University
Terre Haute, Indiana, U.S.A.

Taylor & Francis Series in Remote Sensing Applications

Qihao Weng, Series Editor

Global Urban Monitoring and Assessment

through Earth Observation

Edited by

Qihao Weng

CRC Press
Taylor & Francis Group
Boca Raton London New York

CRC Press is an imprint of the
Taylor & Francis Group, an **informa** business

CRC Press
Taylor & Francis Group
6000 Broken Sound Parkway NW, Suite 300
Boca Raton, FL 33487-2742

First issued in paperback 2019

© 2014 by Taylor & Francis Group, LLC
CRC Press is an imprint of Taylor & Francis Group, an Informa business

No claim to original U.S. Government works

ISBN-13: 978-1-4665-6449-7 (hbk)
ISBN-13: 978-0-367-86762-1 (pbk)

Library of Congress Cataloging-in-Publication Data

Global urban monitoring and assessment through earth observation / edited by Qihao Weng.
    pages cm -- (Remote sensing applications series ; 10)
    Includes bibliographical references and index.
    ISBN 978-1-4665-6449-7 (hardback)
    1. Urban geography--Remote sensing--International cooperation. 2. Land use, Urban--Remote sensing--International cooperation. 3. Environmental monitoring--Remote sensing--International cooperation. 4. Geographic information systems. I. Weng, Qihao.

GF125.G56 2014
910.28'7--dc23   2014001496

Visit the Taylor & Francis Web site at
http://www.taylorandfrancis.com

and the CRC Press Web site at
http://www.crcpress.com

Contents

SECTION I   Global Urban Observation: Needs and Requirements

SECTION II   Global Urban Footprint: Data Sets and Products

SECTION III   Urban Observation, Monitoring, Forecasting, and Assessment Initiatives

SECTION IV Innovative Concepts and Techniques in Urban Remote Sensing

Foreword: GEO—A Globally Integrated Approach to Urban Monitoring

Recognizing the growing need for improved Earth observations, 150 governments and leading international organizations have established the Group on Earth Observations (GEO) to collaborate and implement a Global Earth Observation System of Systems (GEOSS). Countries and organizations are sparing data from their respective Earth monitoring systems, including satellites in space and in situ instruments from terrestrial, oceanic, and atmospheric domains. They are interlinking these systems so that, together, they provide a more complete picture of Earth systems dynamics.

Universities, space agencies, and other partners are working together in the Global Urban Observation and Information Task (SB-04 of GEO), under the leadership of Professor Weng, to expand the use of Earth observations and remotely sensed data to provide information on urban environment characteristics and their change over time at various spatial scales. They are evaluating user needs and matching them with existing or planned technologies and data sets, and they are working with others in the GEO community to provide full and open access to data and services in order to expand and consolidate the network of researchers, stakeholders, and practitioners who are working for a more sustainable future in urban areas.

In fact, cities and densely populated areas are where the impact of human activities and the effects of natural forcing factors (including global climate change variability) are most directly felt by society. With half of the world's population living in cities today, urban observation and modeling is a key issue not only for the GEO but for resource managers and policy makers alike. Urban areas account for roughly 3% of the Earth's surface but host half of the global population. According to the 2011 revision of the *World Urbanization Prospects*,* in Europe, the Americas, and Oceania more than two-thirds of the population live in urban areas, whereas about one-third of the population of Asia and Africa lives in urban areas. Nonetheless, the most dramatic increase in urban population would be in these two latter continents. In Africa, urban population will increase from 414 million to over 1.2 billion by 2050, while in Asia it will soar from 1.9 billion to 3.3 billion by 2050. This fast and unprecedented dynamic is posing new challenges to governments, decision makers, and stakeholders. New settlements and the dramatic expansion of urban areas require feasible, affordable, and sustainable solutions for housing, energy,

---

* United Nations, Department of Economic and Social Affairs, Population Division, *World Urbanization Prospects, the 2011 Revision.*

and infrastructure in order to mitigate urban poverty, the expansion of slums, and a general deterioration of the urban environment.

Earth observations are key in this process since they provide a uniquely valuable vantage point for monitoring many kinds of large-scale dynamics. In situ and remotely sensed data can be provided with very little delay and can include raw data, maps, optical images, or radar images that accurately measure and track critical parameters like land use and classification, meteorological variables, heat islands phenomena, and trace gas emissions. Urban population growth is driven by a combination of factors usually related to local triggers whose long-term impacts on the environment and economy have a global effect. Growing urban populations will have a direct impact on biodiversity and ecosystem hot spots and will increase vulnerability and exposure of populations to the effects of climate change variability, including sea level rise in coastal regions, extreme weather events, more frequent or intense cyclones/hurricanes, longer dry periods and heavier rains that result in increased flooding, and large and uncontrollable outbreak and transmission of diseases and pandemics. High- and moderate-resolution imagery can be conveniently used to map and parameterize land use (including land use change), urban settlements, physical networks, wildland–urban interfaces, and urban patterns over time. In addition, deterministic or stochastic models (e.g., run-off models, meteorological models, tsunami models) can be used to design reliable and meaningful scenarios whose outputs can improve the adaptation and resilience of urban society even in a context where weather-related hazards might worsen because of climate change.

The Global Urban Observation and Information Task (SB-04) in the GEO represents a collective effort of tens of governments and organizations as well as many individuals to monitor the urban system, share and exchange data, and deliver useful information to society. Interlinking observation systems requires common standards for architecture and data sharing, but usually the architecture of an Earth observation system refers to the way in which its components are designed so that they function as a whole. Each GEOSS component, including those being contributed by Task SB-04, must be configured so that it can be linked with the other participating systems. In addition, each contributor to GEOSS subscribes to the GEO data-sharing principles, which aim to ensure the full and open exchange of data, metadata, and products. GEOSS disseminates information and analyses directly to users through its GEO Portal (http://www.geoportal.org/), a single Internet gateway to the comprehensive and near-real-time data produced by GEOSS. GEO Portal integrates diverse data sets, identifies relevant data and other portals of contributing systems, and provides access to models and other decision-support tools. GEOSS has enabled many countries to access information and thereby provide essential services to address challenges that otherwise would not yet have been met. Despite significant progress in recent years, there remain substantial gaps in ongoing national, regional, and global efforts to address these challenges. The GEO has demonstrated that it can play a key role in addressing these gaps in an effective and long-term manner through increased coordination and networking among its major stakeholders and by working together with other key international organizations.

The chapters in this book show the key accomplishments of some of the best researchers in this field and, as discussed earlier, on one of the most relevant phenomenon facing society in the future.

Barbara J. Ryan
The GEO Secretariat, Executive Director
Geneva, Switzerland

Editor

Dr. Qihao Weng is the director of the Center for Urban and Environmental Change and a professor of geography at Indiana State University, Terre Haute, Indiana. He was a visiting NASA Senior Fellow in 2008–2009. Dr. Weng is also a guest/adjunct professor at Peking University, Hong Kong Polytechnic University, Wuhan University, and Beijing Normal University and a guest research scientist at Beijing Meteorological Bureau, China.

He earned his PhD in geography from the University of Georgia in 1999. In the same year, he joined the University of Alabama as an assistant professor. Since 2001, he has been a member of the faculty in the Department of Earth and Environmental Systems at Indiana State University, where he has taught courses on remote sensing, digital image processing, remote sensing–GIS integration, GIS, and environmental modeling and has mentored 10 doctoral and 9 master's students.

Dr. Weng's research focuses on remote sensing and GIS analysis of urban ecological and environmental systems, land-use and land-cover change, environmental modeling, urbanization impacts, and human–environment interactions. He is the author of over 150 peer-reviewed journal articles and other publications and 6 books.

Dr. Weng has worked extensively with optical and thermal remote sensing data and, more recently, with LiDAR data, primarily for urban heat island study, landcover and impervious surface mapping, urban growth detection, image analysis algorithms, and the integration with socioeconomic characteristics, with financial support from US funding agencies that include NSF, NASA, USGS, USAID, NOAA, National Geographic Society, and Indiana Department of Natural Resources.

Dr. Weng was the recipient of the Robert E. Altenhofen Memorial Scholarship Award by the American Society for Photogrammetry and Remote Sensing (1999), the Best Student-Authored Paper Award by the International Geographic Information Foundation (1998), and the 2010 Erdas Award for the Best Scientific Paper in Remote Sensing by ASPRS (1st place). At Indiana State University, he received the Theodore Dreiser Distinguished Research Award in 2006 (the university's highest research honor) and was selected as a Lilly Foundation Faculty Fellow in 2005 (one of the six recipients). In May 2008, he received a prestigious NASA senior fellowship. In April 2011, Dr. Weng received the Outstanding Contributions Award in Remote Sensing sponsored by the American Association of Geographers (AAG) Remote Sensing Specialty Group.

Dr. Weng has given over 70 invited talks (including colloquia, seminars, keynote addresses, and public speeches) and has presented or copresented over 100 papers at professional conferences.

Dr. Weng is the coordinator for the GEO's SB-04, Global Urban Observation and Information Task (2012–2015). In addition, he serves as an associate editor of *ISPRS Journal of Photogrammetry and Remote Sensing* and is the series editor for both the Taylor & Francis series in remote sensing applications and the McGraw-Hill series in GIS&T. His past professional experience includes serving as the national director of the American Society for Photogrammetry and Remote Sensing (2007–2010), as chair of AAG China Geography Specialty Group (2010–2011), and as secretary of ISPRS Working Group VIII/1 (Human Settlement and Impact Analysis, 2004–2008), as well as a panel member of the US DOE's Cool Roofs Roadmap and Strategy in 2010.

Contributors

Yifang Ban
Division of Geoinformatics
Department of Urban Planning
and Environment
KTH Royal Institute of Technology
Stockholm, Sweden

Kimberly E. Baugh
Cooperative Institute for Research
in the Environmental Sciences
University of Colorado, Boulder
Boulder, Colorado

Thomas Blaschke
Department of Geoinformatics
University of Salzburg
Salzburg, Austria

Stefan Dech
German Remote Sensing Data Center
German Aerospace Center
Oberpfaffenhofen, Germany

Vivek Dey
Department of Geodesy and Geomatics
Engineering
University of New Brunswick
Fredericton, New Brunswick, Canada

Inmaculada Dopido
Hyperspectral Computing Laboratory
University of Extremadura
Badajoz, Spain

Manfred Ehlers
Institute for Geoinformatics
and Remote Sensing
University of Osnabrück
Osnabrück, Germany

Daniele Ehrlich
Global Security and Crisis
Management Unit
Institute for the Protection and
Security of the Citizen
European Commission
Joint Research Centre
Ispra, Italy

Christopher D. Elvidge
Earth Observation Group
National Geophysical Data Center
National Oceanic and Atmospheric
Administration
Boulder, Colorado

Thomas Esch
German Remote Sensing Data Center
Earth Observation Center
German Aerospace Center
Oberpfaffenhofen, Germany

Andreas Felbier
German Remote Sensing Data Center
Earth Observation Center
German Aerospace Center
Oberpfaffenhofen, Germany

Stefano Ferri
Institute for the Protection and
Security of the Citizen
European Commission
Joint Research Centre
Ispra, Italy

Yevgeniya Filippovska
Institute for Geoinformatics
and Remote Sensing
University of Osnabrück
Osnabrück, Germany

Paolo Gamba
Telecommunications and Remote
  Sensing Laboratory
and
Department of Electrical, Computer
  and Biomedical Engineering
University of Pavia
Pavia, Italy

Tilottama Ghosh
Cooperative Institute for Research
  in the Environmental Sciences
University of Colorado, Boulder
Boulder, Colorado

Huadong Guo
Institute of Remote Sensing
  and Digital Earth
Chinese Academy of Sciences
Beijing, People's Republic of China

Jianping Guo
Center for Atmosphere Watch
  and Services
Chinese Academy of Meteorological
  Sciences
Beijing, People's Republic of China

Matina Halkia
Global Security and Crisis
  Management Unit
European Commission
Joint Research Centre
Ispra, Italy

Johannes Heinzel
Global Security and Crisis
  Management Unit
Institute for the Protection and
  Security of the Citizen
European Commission
Joint Research Centre
Ispra, Italy

Peter Hofmann
Department of Geoinformatics
University of Salzburg
Salzburg, Austria

Feng-Chi Hsu
Cooperative Institute for Research
  in the Environmental Sciences
University of Colorado, Boulder
Boulder, Colorado

Hongtao Hu
Division of Geoinformatics
KTH Royal Institute of Technology
Stockholm, Sweden

Gianni Cristian Iannelli
Telecommunications and Remote
  Sensing Laboratory
and
Department of Electrical, Computer
  and Biomedical Engineering
University of Pavia
Pavia, Italy

Koki Iwao
Geological Survey of Japan
National Institute of Advanced
  Industrial Science and
  Technology
Tsukuba, Japan

Zhiben Jiang
State Key Laboratory of Remote
  Sensing Science
College of Global Change and Earth
  System Science
Beijing Normal University
Beijing, People's Republic of China

Huiran Jin
College of Environmental Science
  and Forestry
State University of New York
Syracuse, New York

Zhenyu Jin
Department of Geography
University of Utah
Salt Lake City, Utah

Martin Kada
Institute for Geoinformatics
    and Remote Sensing
University of Osnabrück
Osnabrück, Germany

Thomas Kemper
Global Security and Crisis
    Management Unit
Institute for the Protection and
    Security of the Citizen
European Commission
Joint Research Centre
Ispra, Italy

Iphigenia Keramitsoglou
Institute for Astronomy, Astrophysics,
    Space Applications and Remote
    Sensing
National Observatory of Athens
Athens, Greece

Gianni Lisini
European Center for Training and
    Research in Earthquake Engineering
Institute for Advanced Study of Pavia
Pavia, Italy

Linlin Lu
Key Laboratory of Digital Earth
    Science
Institute of Remote Sensing
    and Digital Earth
Chinese Academy of Sciences
Beijing, People's Republic of China

Xiulian Ma
Department of Sociology
University of Utah
Salt Lake City, Utah

Mattia Marconcini
German Remote Sensing Data Center
Earth Observation Center
German Aerospace Center
Oberpfaffenhofen, Germany

Hiroyuki Miyazaki
Center for Spatial Information Science
University of Tokyo
Tokyo, Japan

Giorgos Mountrakis
College of Environmental Science
    and Forestry
State University of New York
Syracuse, New York

Martino Pesaresi
Global Security and Crisis
    Management Unit
Institute for the Protection and
    Security of the Citizen
European Commission
Joint Research Centre
Ispra, Italy

Antonio Plaza
Hyperspectral Computing Laboratory
University of Extremadura
Badajoz, Spain

Dale Quattrochi
Earth Science Office
Marshall Space Flight Center
National Aeronautics and Space
    Administration
Huntsville, Alabama

Salman Quresh
Department of Geography
Humboldt University of Berlin
Berlin, Germany

and

School of Architecture
Birmingham City University
Birmingham, United Kingdom

Achim Roth
German Remote Sensing Data Center
Earth Observation Center
German Aerospace Center
Oberpfaffenhofen, Germany

Günther Sagl
GIScience/Geoinformatics
    Research Group
Department of Geography
University of Heidelberg
Heidelberg, Germany

Bahram Salehi
Department of Geodesy and Geomatics
    Engineering
University of New Brunswick
Fredericton, New Brunswick, Canada

Xiaowei Shao
Center for Spatial Information Science
University of Tokyo
Tokyo, Japan

Ryosuke Shibasaki
Center for Spatial Information Science
University of Tokyo
Tokyo, Japan

Natalia Sofina
Institute for Geoinformatics
    and Remote Sensing
University of Osnabrück
Osnabrück, Germany

Pierre Soille
Global Security and Crisis
    Management Unit
European Commission
Joint Research Centre
Ispra, Italy

Vasileios Syrris
Global Security and Crisis
    Management Unit
European Commission
Joint Research Centre
Ispra, Italy

Hannes Taubenböck
German Remote Sensing Data Center
Earth Observation Center
German Aerospace Center
Oberpfaffenhofen, Germany

Michael Timberlake
Department of Sociology
University of Utah
Salt Lake City, Utah

Qihao Weng
Center for Urban and
    Environmental Change
Department of Earth &
    Environmental Systems
Indiana State University
Terre Haute, Indiana

Michael Wiesner
German Remote Sensing Data Center
German Aerospace Center
Oberpfaffenhofen, Germany

George Xian
Earth Resources Observation
    and Science Center
U.S. Geological Survey
Sioux Falls, South Dakota

Bing Xu
School of Environment
Tsinghua University
and
College of Global Change and Earth
    System Science
Beijing Normal University
Beijing, People's Republic of China

Osama Yousif
Division of Geoinformatics
KTH Royal Institute of Technology
Stockholm, Sweden

Yun Zhang
Department of Geodesy and Geomatics
  Engineering
University of New Brunswick
Fredericton, New Brunswick, Canada

Ming Zhong
Department of Civil Engineering
University of New Brunswick
Fredericton, New Brunswick, Canada

MATLAB® Statement

MATLAB® is a registered trademark of The MathWorks, Inc. For product information, please contact:

The MathWorks, Inc.
3 Apple Hill Drive
Natick, MA 01760-2098 USA
Tel: 508-647-7000
Fax: 508-647-7001
E-mail: info@mathworks.com
Web: www.mathworks.com

1 What Is Special about Global Urban Remote Sensing?

Qihao Weng

CONTENTS

1.1 URBANIZATION, GLOBAL CHANGES, AND URBAN REMOTE SENSING

As the trend of urbanization continues worldwide, environmental problems associated with this process have become an important concern (Weng, 2001). The need for monitoring and managing urban areas is amplified by the concern over global climate changes. Although, at this time, we are not clear about how urban climate is complicated by global warming, it is important to understand the combined impacts of urbanization and global warming on urban areas so that we can better manage natural resources and develop measures for mitigation and adaptation (Grimmond, 2007). Owing to the nature of cities as complex human settlements, urban areas are more vulnerable than rural settlements to the impacts of global environmental change (CCSP, 2008). Most impact concerns, including those on health, water, and infrastructures, severe weather events, energy requirements, urban metabolism, sea level rise, economic competitiveness, opportunities and risks, and social and political structures, can be addressed by, or be better understood with, the Earth Observation (EO) technology.

Over the past decade, urban remote sensing has emerged as a new frontier in the EO technology by focusing primarily on (1) understanding the biophysical properties, patterns, and processes of urban landscapes and (2) mapping and monitoring of urban land cover and spatial extent. Driven by the societal needs and improvement in sensor technology and image processing techniques, we have recently witnessed a substantial increase in research and development, technology transfer,

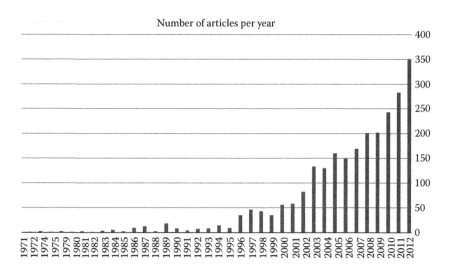

FIGURE 1.1   The number of journal articles (including review articles) on urban remote sensing derived from a Scopus search on March 27, 2013, by the author.

and engineering activities worldwide. As shown in Figure 1.1, the number of journal articles (including review articles) on urban remote sensing has been increasing rapidly since 2000. This period coincides with the advent of high spatial resolution satellite images (especially those with higher than 5 m resolution), spaceborne hyperspectral images and Lidar data, and enhanced image processing techniques such as object-based image analysis (OBIA), data mining, and data and image fusion of different sensors, wavelength regions, and spatial, spectral, and temporal resolutions (Weng, 2009). Remote sensing methods and techniques have been applied to urban areas by using all ranges of electromagnetic wavelength and close-range sensors (Weng, 2012). Table 1.1 lists most relevant peer-reviewed journals for urban remote sensing along with the most prolific authors and major research groups. It is apparent from the table that all major remote sensing journals have published articles on this subject and that researchers worldwide are interested in using remote sensing technology to study urban areas, urbanization, and associated environmental issues.

In the context of global urbanization and environmental changes and having recognized the benefits of urban imaging and mapping techniques, Group on Earth Observation (GEO), an international organization for exploiting EO technologies to support decision making, calls for the development of a global urban observation and information system. GEO decided to establish, in its 2012–2015 Work Plan, a new task of "Global Urban Observation and Information" (Weng et al., 2013). The main objectives of this task are as follows:

1. To improve the coordination of urban observations, monitoring, forecasting, and assessment initiatives worldwide
2. To produce up-to-date information on the status and development of the urban systems at different scales

TABLE 1.1

Literature Search Results Using Scopus on "Urban Remote Sensing"

Most Relevant Journals (# of Publications)	Most Prolific Authors (# of Publications on the Subject)	Major Research Groups (# of Publication Affiliated with the Organization)
International Journal of Remote Sensing (245)	Weng, Q. (39)	Chinese Academy of Sciences (57)
Photogrammetric Engineering & Remote Sensing (143)	Gamba, P. (30)	Deutsches Zentrum fur Luft- Und Raumfahrt (German Aerospace Center) (47)
	Nichol, J. (20)	
	Dell'Acqua, F. (17)	
Remote Sensing of Environment (142)	Myint, S.W. (14)	Indiana State University (44)
IEEE Transactions on Geoscience and Remote Sensing (98)	Lu, D. (13)	Beijing Normal University (43)
	Wong, M.S. (13)	NASA Goddard Space Flight Center (39)
ISPRS Journal of Photogrammetry and Remote Sensing (73)	Gong, P. (12)	
	Lo, C.P. (12)	Peking University (35)
Landscape and Urban Planning (64)	Stow, D. (12)	Arizona State University (31)
Atmospheric Environment (62)	Wang, L. (12)	Università degli Studi di Pavia (31)
Journal of the Indian Society of Remote Sensing (48)	Elvidge, C.D. (11)	Wuhan University (29)
	Shi, P. (11)	Hong Kong Polytechnic University (27)
Environmental Monitoring and Assessment (47)	Wu, C. (11)	Istanbul Teknik Üniversitesi (27)
Journal of Geophysical Research D Atmospheres (35)	Benediktsson, J.A. (10)	University of California, Santa Barbara (26)
	Canters, F. (10)	
	Roberts, D.A. (10)	European Commission Joint Research Centre, Ispra (26)
Computers Environment and Urban Systems (33)	Small, C. (10)	
	Voogt, J.A. (10)	Consiglio Nazionale delle Ricerche (National Research Council, Italy) (25)
Remote Sensing (33)	Yamaguchi, Y. (10)	
International Journal of Applied Earth Observation and Geoinformation (31)	Taubenbock, H. (9)	
	Jensen, R.R. (9)	The University of Western Ontario (24)
IEEE Geoscience and Remote Sensing Letters (29)	Li, X. (9)	University of Maryland (24)
	Bruzzone, L. (9)	Indian Institute of Remote Sensing (23)
Sensors (28)	Liu, H. (9)	UC Berkeley (23)
IEEE Journal of Selected Topics in Applied Earth Observations and Remote Sensing (27)	Zhou, W. (9)	Nanjing University (23)
	Clarke, K.C. (9)	Instituto Nacional de Pesquisas Espaciais (INPE, Brazil) (22)
	Dech, S. (9)	
Geocarto International (26)	Emeis, S. (8)	Indiana University (21)
Canadian Journal of Remote Sensing (25)	Esch, T. (8)	Indian Space Research Organization (21)
	Lisini, G. (8)	
Chinese Geographical Science (24)	Quattrochi, D.A. (8)	
Applied Geography (24)	Roth, A. (8)	
	Xian, G. (8)	
	Yeh, A.G.O. (8)	

Note: The search was conducted on March 27, 2013 (document type = article + review; language = English). A total of 2573 papers were found, with the earliest publication date in 1971.

3. To fill existing gaps in the integration of global urban land observations with
   a. Data that characterize urban ecosystems, built environment, air quality, and carbon emission
   b. Indicators of population density, environmental quality, and quality of life
   c. Patterns of human environmental and infectious diseases
4. To develop innovative concepts and techniques in support of effective and sustainable urban development

This book presents to the readers the current state of global urban monitoring, assessment, modeling, and prediction through EO and related technologies. Specifically, it introduces to the readers GEO's important international collaborative effort, the current state of global urban remote sensing, and future directions. A few selected innovative works on urban remote sensing are included as they will contribute to the timely development of innovative concepts and techniques for sustainable urban development. This book would be an excellent reference book for students, researchers, and professors in academia who conduct remote sensing research and for professionals and decision makers in government, industry, and commercial sectors who deal with urban planning, civic, environmental, and sustainability issues. It will also be suitable as a textbook or as a supplement text for undergraduate and graduate students who are interested in remote sensing, urban environment, and sustainability.

1.2   WHAT IS SPECIAL ABOUT GLOBAL URBAN REMOTE SENSING?

The vast majority of urban remote sensing studies have been conducted at local or regional scales. A Scopus search shows that there are only 243 papers on "global urban remote sensing," as compared to 2573 articles on "urban remote sensing." Global urban remote sensing differs from studies conducted at local and regional scales mainly in the following aspects:

A top-down approach: Most previous urban studies have focused on one locality and once only. Even though those methods are effective for spatial studies, the results of characterization, analysis, and modeling may not prove effective for spatiotemporal analyses. The limitations include difficulties in characterizing several satellite images from various times of the same region or several images of the same time for different regions (Rajasekar and Weng, 2009). The transferability of a method from one study site to another is always a concern. In contrast, a global study must design a scheme that can be applied to all localities, thus possessing the feature of a top-down approach. However, the top-down approach may have its own limitations, such as uneven classification and accuracy across different localities (Xian and Homer, 2010).

The multiscaling issue: The majority of urban phenomena are scale dependent, which means that urban patterns change with the scale of observation. In reality, very few geographical phenomena are scale independent, that is, the patterns do not change across scales (Cao and Lam, 1992). Urban landscape processes appear to be hierarchical in pattern and structure. Therefore, a study of the relationship between the patterns at different levels in the hierarchy may help in obtaining a better understanding of the scale and resolution problem in urban areas (Weng et al., 2004) and in finding the optimal scale for examining the relationship between urban landscape pattern and process (Frohn, 1998; Liu and Weng, 2009). For example, Oke (2008) suggested that there were often several linked urban heat islands (UHIs) in one city and that each was distinguished from the other mainly by scales imposed by the biophysical structure of a given city and the structure of the urban atmosphere. Each UHI, therefore, requires measurement arrays appropriate to the scale (Oke, 2008). Most previous literature failed to account for this multiscaling concern in the UHI studies (e.g., Streutker, 2002; Imhoff et al., 2010). Weng et al. (2011) were the first to examine multiple UHIs of various sizes (spatial extent) using the same sets of ASTER imagery. By characterizing the urban thermal landscape at the microscale, the mesoscale, and the regional scale, and by further linking UHIs with biophysical parameters, they were able to analyze the UHI phenomenon in Indianapolis, United States, as a scale-dependent process.

Use of coarse spatial resolution imagery: The relationship between the geographical scale of a study area and the spatial resolution of a remote sensing image must be carefully examined (Quattrochi and Goodchild, 1997). For mapping at the continental or global scale, coarse spatial resolution data are usually employed. Gamba and Herold (2009) assessed eight major research efforts in global urban extent mapping and found that most maps were produced at the spatial resolution of 1–2 km. When using coarse-resolution images, a threshold has to be defined with respect to what constitutes a built-up/impervious pixel (Lu et al., 2008; Schneider et al., 2010). Reliable impervious surface data that derive from medium-resolution imagery are helpful for validating and predicting urban/built-up extent at the coarse-resolution level (Lu et al., 2008). Similarly, fine spatial resolution imagery, such as IKONOS and QuickBird data, has been used for purposes of sample training and/or validation, where medium-resolution imagery is the main data source for a project. Such a nested multiscale approach needs to consider the mixed pixel problem. The presence of mixed pixels has been recognized as a major problem affecting the effective use of satellite imagery in urban studies (Lu and Weng, 2004). The mixed pixel problem results from the fact that the observational scale (i.e., spatial resolution) fails to correspond to the spatial characteristics of the target (Mather, 1999). Different approaches have been developed to handle mixed pixels (e.g., Wang, 1990; Ridd, 1995; Foody, 1999), which should also be applicable for studies at the global scale.

Globally consistent mapping system: Several global urban area maps have been created using coarse- or medium-resolution satellite imagery (Gamba and Herold, 2009; Potere et al., 2009). These maps yield very different levels of accuracy with estimate of the total global urban land varying from 0.27×10^6 to 3.52×10^6 km^2 (Potere et al., 2009). The mapping procedures are often found effective for certain regions but not for others. It appears that the mapping accuracy is higher for Europe, North America, and Latin America whereas it is lower for Asia and Africa for nearly all the existing global urban maps (Potere et al., 2009). Therefore, it is important to develop a consistent mapping system that is applicable to all world regions and to improve global coverage and data accuracy of urban observation through integrating satellite imagery from multiple platforms/sensors with in situ data. Equally important is to define requirements for global urban monitoring and assessment in terms of data products and expectations for data validation, archiving, update, and sharing. These objectives will be addressed by GEO SB-04 team during the period 2012–2015 (Weng et al., 2013). A fundamental issue that must be addressed is the definition of urban areas, which will affect the way we process satellite imagery. To support the development of a globally consistent urban mapping, a grid-based mapping system that integrates an algorithm of mosaicking for the cities in the world, an algorithm of urban area mapping, and a ground truthing may be necessary (Miyazaki et al., 2013).

Global perspective on sustainable urban development: Urban areas represent the centers of economy, society, culture, and policy. As a result, most of the current and future ecological, economic, and societal challenges are either directly or indirectly related to human activities in or around settlements. A global urban morphological database, as a further step of global urban mapping, will provide a good tool for urban climate modeling to better understand the impacts of global climate change on urban areas. Furthermore, urban analyses at the global scale that link satellite observation data products with socioeconomic, demographic, and in situ data will improve knowledge on urban environments and ecosystems, air quality and carbon emission, population density, quality of life, and human environmental and infectious diseases. These types of analyses offer irreplaceable, important insights into future urban development in the world, which is impossible to derive from local or regional studies. The above two objectives will also be pursued by the GEO SB-04 team in the next few years (Weng et al., 2013).

1.3   SYNOPSIS OF THE BOOK

This book consists of 4 sections and 18 chapters. Section I deals with the needs and requirements of global urban observation and assessment, and the priority actions of GEO SB-04 between 2012 and 2015 and beyond; Section II introduces four international efforts at mapping global urban footprint from Germany, the United States, Japan, and the Joint Research Centre of European Commission; Section III selects and presents a few important initiatives on urban observation, monitoring, forecasting, and assessment from different countries; and finally, Section IV develops innovative concepts and techniques in support of effective urban sensing and sustainable urban development.

Section I starts with an introduction to GEO's Global Urban Observation and Information Task, its origin, rationale, and objectives, and its key activities between 2012 and 2015. Following the introduction, Chapter 2 describes in detail the GEO's urban supersites initiative through which eight cities, including Los Angeles, Atlanta, Mexico City, Athens, Istanbul, Sao Paolo, Beijing, and Hong Kong, are selected as test sites for its developing methodology. Through this partnership, the participants will generate an agreed set of protocols to share and visualize the intermediate and final global information products and other resources and to foster joint studies and dissemination of results. In Chapter 3, Gamba and his colleagues propose a framework to exploit EO-derived information in urban studies. The need for multiple spatial scales and multiple information about human settlements requires a multiscale and multifeature data fusion approach to be included in a specific application.

Examples related to land use detection and risk assessment at the built-up area level are discussed to provide a clear view of the advantages and drawbacks of the proposed framework. Also included in Section I is Chapter 4. A writing team of 17 contributors were chosen to write by category about the satellite sensors for urban mapping, assessing, and monitoring. These categories include optical sensors (fine-, medium-, and coarse-resolution sensors), thermal IR sensors (medium- and coarse-resolution), SAR (fine-, medium-, and coarse-resolution sensors), and night-light sensors. For each type of sensor, the most updated information is provided with respect to

- Date, month, and year a sensor starts to acquire image data
- Data characteristics (spatial, spectral, temporal, and radiometric resolution)
- Satellite orbital characteristics
- How the satellite orbital and image data characteristics of a particular sensor satisfy the needs for urban mapping, assessing, and monitoring
- Current practice by conducting a short literature review
- Key limitations
- Future perspective

Section II introduces initiatives directed by co-leads or contributors of GEO SB-04 to global urban footprint mapping. Chapter 5 presents with the Global Urban Footprint (GUF) project conducted by the German Aerospace Center (DLR). The German TanDEM-X mission collects a global coverage of VHR imagery that can be used to map settlement patterns worldwide in a unique spatial detail. Based on a fully automated, operational image analysis procedure, the DLR uses approximately 180,000 TanDEM-X images to generate a worldwide inventory of human settlements—the GUF layer. The free-access version of the GUF shows a spatial resolution of 75 m whereas the commercial product will have the spatial resolution of about 12 m. The GUF data layer offers valuable information for an analysis of the worldwide urbanization pattern. In Chapter 6, Elvidge and his colleagues assess national trends in satellite-observed lighting between 1992 and 2012. Analysis of a time series of global annual satellite maps of nighttime lights spanning 21 years and six satellites reveals several distinct patterns linked to population changes, economic development, and improvements in lighting efficiency. The results indicate that there are national-level differences in the behavior of nighttime lights over time. Seven categories of national lighting trends have been defined: (1) rapid growth, (2) moderate growth, (3) population-centric lighting, (4) economic-centric lighting, (5) stable lighting, (6) erratic lighting, and (7) antipole lighting. Recognition of these patterns may lead to improved spatial modeling of socioeconomic processes based on satellite-observed nighttime lights. In Chapter 7, Miyazaki and his colleagues present the methodology for the development of a global built-up area map at 15 m resolution using ASTER images and existing GIS data. The methodology includes three components: ground truth data for urban sites, an automated algorithm of mosaic operation for cities around the world, and an automated algorithm of built-up area mapping. These components were implemented in a consistent system with grid computing. A built-up area map has been developed for 3,374 cities around the globe using 11,802 scenes of ASTER data. Chapter 8 investigates the possibility of producing fine-scale geo-information layers using high-resolution or very-high-resolution satellite images and covering large regions such as at global, regional, or national scales. This chapter reports about the methodological choices made during the design of the Global Human Settlement Layer (GHSL) production and summarizes the main results of the experiment conducted by the Joint Research Center (JRC) of the European Commission during 2012; this experiment involved 24.3 million square kilometers of test areas spread over four continents, automatically mapped to the image data collected by a variety of optical satellite and airborne sensors with spatial resolutions ranging from 0.5 to 10 m.

Section III exemplifies a few international efforts on the observation, monitoring, forecasting, and assessment of urban areas and other types of human settlements. Taubenböck et al. in Chapter 9 present an application-oriented approach using multi-temporal remote sensing data to monitor the spatiotemporal dynamics of megacities, using four Chinese megacities—Beijing, Shanghai, Guangzhou, and Shenzhen—as case studies. Object-oriented and pixel-based classification image analysis techniques are applied to multitemporal Landsat as well as to TerraSAR-X and TanDEM-X data to delineate urbanized areas of the megacities at different dates. Assessment of urban growth patterns using spatial metrics are then performed to find out their similarities and differences. Chapter 10 addresses the special situation of refugees and internally displaced persons (IDPs) living in camps, sometimes the size of a large city. After defining the differences of refugees and IDPs, it provides examples of a variety of camp situations. The methods for mapping and monitoring refugee/IDP camps with EO data and for estimating the population in those camps are proposed and examined. In Chapter 11, Lu et al. present a methodology for mapping built-up areas in China at fine spatial resolution with CBERS-2B data and to assess the GHSL products generated using the image data. The CBERS-2B panchromatic imagery at 2.36 m resolution, covering mainly the eastern part of China, is processed to generate GHSL products. A benchmark experiment was conducted between GHSL at three scales and the MOD500 urban layer with Chinese Landuse2000 data as the reference data. GHSL at every scale ranks better than MODIS data. The GHSL at 50 m resolution was found to yield the highest accuracy for estimating the population of villages, whereas the GHSL at 200 m resolution showed a high degree of agreement with the reference data in both cities and villages. In Chapter 12, Xu and her colleagues assess the climatological and geographical factors responsible for the global pandemic of influenza A (H1N1) in 2009. A time series of global risk maps are generated to predict environmental exposure based on modeling Merra data including daily temperature, precipitation, and absolute humidity. These maps reveal clear seasonal changes in environmental risks over various parts of the world and provide information for developing early warning signs. From the indicators that define human travel flow in a global travel network, the authors find that a relatively small number of countries or cities could account for most of the outbreak cases worldwide. These findings confirm that the climatological and geographical factors have significant impacts on the global transmission of the virus. The last chapter of Section III, Chapter 13, showcases a methodology for monitoring urban thermal environment, based on a study in Athens, Greece, from May until September 2009. The goal is to downscale the quarter-hour geostationary land surface temperate

(MSG-SEVIRI LST data of 3–5 km pixel size) images to 1 km spatial resolution—characteristics currently not available with any EO mission—and then to extract the surface UHI patterns so as to study their diurnal variability.

Chapters selected for Section IV illustrate innovative concepts, methods, and techniques in urban remote sensing in recent years. In Chapter 14, Sagl and Blaschke argue that multiple coordinated views of spatiotemporal data provide unprecedented opportunities for geographic analysis in times of "big data" and that different types of data generation enable an integrated sensing. They analyze the intersection between machine-generated (satellite imagery, weather stations) and user-generated (social media, mobile phone data) data, and suggest that GIS is at the core of integrated urban sensing, especially in urban monitoring studies. They further demonstrate that GIS-based integrated urban sensing enables analyses, forecasts, and visualizations of a variety of spatial components of socioeconomic phenomena, including people, urban commodities, information flows, human interaction with urban commodities, as well as the relationship between networks of human interaction and natural environments. Chapter 15 focuses on OBIA for urban studies. With the advent of very high spatial resolution satellite images, automatic extraction of urban land cover information from them has soared. The authors of this chapter review the techniques of OBIA with a major focus on image segmentation and intend to provide fundamental knowledge of OBIA in the context of urban remote sensing. They argue that OBIA can be used to quickly identify land cover parcels and features of fast-growing urban areas in spite of many technical challenges and problems. In Chapter 16, the OBIA technique is further employed for detecting informal settlements. In Chapter 17, Ehlers and his colleagues develop a methodology for rapid detection and visualization of changes in areas of crisis or catastrophes, especially of buildings and infrastructure. Because standard methods for automated change detection failed, several new methods have been developed and tested. These methods are based on frequency analysis, segmentation, and texture parameters and can also be combined with a decision tree approach. In comparison to five standard change detection methods, their new combined approach, called combined edge segment texture (CEST), shows superior results. If available, GIS and/or 3D information from stereo or shadow analysis can be further included in this change detection algorithm. The last chapter of Section IV, Chapter 18, provides a detailed review on the fusion of SAR and optical data for urban land cover mapping and change detection with two case studies. The fusion of ENVISAT ASAR and a single-date MERIS data with low spatial resolution offer the potential to distinguish between several classes, including the separation of forest from low-density built-up areas, parks and roads, and so forth, which would be impossible with SAR images alone. Similar improvement can be seen in the accuracy of change detection where combined SAR and optical data outperform the SAR or the optical data alone.

ACKNOWLEDGMENTS

I wish to extend my gratitude to all the contributors for their help in this endeavor. Without the continued support of many colleagues of GEO SB-04, this book would have hardly been possible. From its conceptualization to final production,

Drs. Paolo Gamba, Thomas Esch, Dale Quattrochi, George Xian, Bing Xu, Martino Pesaresi, Yifan Ban, Iphigenia Keramitsoglou, and Xiaoli Ding, among others, offered important insights and provided generous support. My gratitude further goes to Barbara J. Ryan, director of GEO Secretariat, who was kind enough to write the foreword to this book. Dr. Francesco Gaetani at the GEO Secretariat was always an information source and a great support for the SB-04 task. Drs. Jinlong Fan and Guicai Li at the National Satellite Meteorological Center, Beijing, advocated having an urban task in GEO. I offer my deepest appreciation to all the reviewers, who have taken precious time from their busy schedules to review the chapters submitted to this book. Finally, I am indebted to my family for their enduring love and support. It is my hope that the publication of this book will provide stimulation to students, professors, and researchers to produce more in-depth works on urban remote sensing at various spatial scales, in particular at the global and continental scales. The information on global data products, methods, techniques, and modeling results discussed in this book will help to address issues of urban environmental characteristics and expansion as well as the concerns over climate change, human health and welfare, and sustainable urban development. It becomes apparent that the realization of many societal benefits of EO technology requires international collaboration and cooperation.

The reviewers of the chapters for this book are listed below in alphabetical order: Demetre Argialas, Benjamin Bechtel, Jón Atli Benediktsson, Guillermo Castilla, Alex de Sherbinin, Pinliang Dong, Thomas Esch, Dell'Acqua Fabio, Paolo Gamba, Peter Hofmann, Xuefei Hu, Thomas Kemper, Minho Kim, Yongmei Lu, Xin Miao, Zina Mitraka, Hiroyuki Miyazaki, Janet E. Nichol, Junmei Tang, Hannes Taubenboeck, Tobias Ullmann, Changshan Wu, George Xian, Zhixiao Xie, Klemen Zaksek, Hongsheng Zhang, Qingling Zhang, Weiqi Zhou, Yuyu Zhou, and Zhe Zhu.

REFERENCES

Cao, C. and Lam, N. S.-N. 1992. Understanding the scale and resolution effects in remote sensing. In: D. A. Quattrochi and M. F. Goodchild, eds., *Scale in Remote Sensing and GIS*. Boca Raton, FL: CRC Press, pp. 57–72.

Foody, G. M. 1999. Image classification with a neural network: From completely-crisp to fully-fuzzy situation. In: P. M. Atkinson and N. J. Tate, eds., *Advances in Remote Sensing and GIS Analysis*. New York: John Wiley & Sons, pp. 17–37.

Frohn, R. C. 1998. *Remote Sensing for Landscape Ecology: New Metric Indicators for Monitoring, Modelling, and Assessment of Ecosystems*. Boca Raton, FL: Lewis Publishers.

Gamba, P. and Herold, M. 2009. *Global Mapping of Human Settlements: Experiences, Datasets, and Prospects*. Boca Raton, FL: CRC Press.

Grimmond, S. 2007. Urbanization and global environmental changes: Local effects of urban warming. *Geographical Journal*, 173(1), 83–88.

Imhoff, M. L., Zhang, P., Wolfe, R. E., and Bounoua, L. 2010. Remote sensing of the urban heat island effect across biomes in the continental USA. *Remote Sensing of Environment*, 114(3), 504–513.

Liu, H. and Weng, Q. 2009. Scaling-up effect on the relationship between landscape pattern and land surface temperature. *Photogrammetric Engineering & Remote Sensing*, 75(3), 291–304.

Lu, D., Tian, H., Zhou, G., and Ge, H. 2008. Regional mapping of human settlements in southeastern China with multisensor remotely sensed data. *Remote Sensing of Environment*, 112(9), 3668–3679.

Lu, D. and Weng, Q. 2004. Spectral mixture analysis of the urban landscape in Indianapolis with Landsat ETM+ imagery. *Photogrammetric Engineering & Remote Sensing*, 70, 1053–1062.

Mather, P. M. 1999. Land cover classification revisited. In: P. M. Atkinson and N. J. Tate, eds., *Advances in Remote Sensing and GIS*. New York: John Wiley & Sons, pp. 7–16.

Miyazaki, H., Shao, X., Iwao, K., and Shibasaki, R. 2013. An automated method for global urban area mapping by integrating ASTER satellite images and GIS data. *IEEE Journal of Selected Topics in Applied Earth Observations and Remote Sensing*, 6(2), 1004–1019.

Oke, T. R. 2008. The continuing quest to understand urban heat islands. Presented at the *Second Workshop on Earth Observation for Urban Planning Management*, Kowloon, Hong Kong, May 20–21, 2008.

Potere, D., Schneider, A., Angel, S., and Civco, D. L. 2009. Mapping urban areas on a global scale: Which of the eight maps now available is more accurate? *International Journal of Remote Sensing*, 30(24), 6531–6558.

Quattrochi, D. A. and Goodchild, M. F. 1997. *Scale in Remote Sensing and GIS*. New York: Lewis Publishers.

Rajasekar, U. and Weng, Q. 2009. Spatio-temporal modeling and analysis of urban heat islands by using Landsat TM and ETM+ imagery. *International Journal of Remote Sensing*, 30(13), 3531–3548.

Ridd, M. K. 1995. Exploring a V-I-S (vegetation-impervious surface-soil) model for urban ecosystem analysis through remote sensing: Comparative anatomy for cities. *International Journal of Remote Sensing*, 16(12), 2165–2185.

Schneider, A., Friedl, M. A., and Potere, D. 2010. Mapping global urban areas using MODIS 500-m data: New methods and datasets based on 'urban ecoregions'. *Remote Sensing of Environment*, 114, 1733–1746.

Streutker, D. R. 2002. A remote sensing study of the urban heat island of Houston, Texas. *International Journal of Remote Sensing*, 23(13), 2595–2608.

U.S. Climate Change Science Program (CCSP). 2008. Analyses of the effects of global change on human health, welfare, and human systems. A Report by the U.S. Climate Change Science Program and the Subcommittee on Global Change Research [Gamble, J.L. (ed.), K.L. Ebi, F.G. Sussman, T.J. Wilbanks]. Washington, DC: U.S. Environmental Protection Agency.

Wang, F. 1990. Fuzzy supervised classification of remote sensing images. *IEEE Transactions on Geoscience and Remote Sensing*, 28(2), 194–201.

Weng, Q. 2001. A remote sensing—GIS evaluation of urban expansion and its impact on surface temperature in the Zhujiang Delta, China. *International Journal of Remote Sensing*, 22(10), 1999–2014.

Weng, Q. 2009. *Remote Sensing and GIS Integration: Theories, Methods, and Applications*. New York: McGraw-Hill Professional, p. 397.

Weng, Q. 2012. Remote sensing of impervious surfaces in the urban areas: Requirements, methods, and trends. *Remote Sensing of Environment*, 117(2), 34–49.

Weng, Q., Esch, T., Gamba, P., Quattrochi, D. A., and Xian, G. 2013. Global urban observation and information: GEO's effort to address the impacts of human settlements. In: Q. Weng, ed., *Global Urban Monitoring and Assessment through Earth Observation*. Boca Raton, FL: CRC Press/Taylor & Francis, pp. 15–34.

Weng, Q., Lu, D., and Schubring, J. 2004. Estimation of land surface temperature-vegetation abundance relationship for urban heat island studies. *Remote Sensing of Environment*, 89(4), 467–483.

Weng, Q., Rajasekar, U., and Hu, X. 2011. Modeling urban heat islands with multi-temporal ASTER images. *IEEE Transactions on Geosciences and Remote Sensing*, 49(10), 4080–4089.

Xian, G. and Homer, C. 2010. Updating the 2001 National Land Cover Database impervious surface products to 2006 using Landsat imagery change detection methods. *Remote Sensing of Environment*, 114(2), 1676–1686.

Section I

Global Urban Observation

Needs and Requirements

2 Global Urban Observation and Information

GEO's Effort to Address the Impacts of Human Settlements

Qihao Weng, Thomas Esch, Paolo Gamba,
Dale Quattrochi, and George Xian

CONTENTS

2.1 SCIENTIFIC MOTIVATION AND BACKGROUND OF THE SB-04

The twenty-first century is the first "urban century" according to the United Nations Development Programme. The focus on cities reflects awareness of the growing percentage of the world population that live in urban areas. In environmental terms, as has been pointed out at the UN Conference on Human Settlement, cities and towns are the original producers of many of the global environmental problems related to waste disposal and air and water pollution. There is a rapidly growing need for technologies that will enable monitoring of the world's natural resources and urban assets and managing exposure to natural and man-made risks (Weng and Quattrochi, 2006a).

This need is driven by continued urbanization and global climate change. Although currently urbanized land only covers approximately 2% of global land area, more than half of the world population (3.3 billion people) lives in the urban environment. By 2030, urbanized areas will expand to provide homes for 81% of the world population, according to the United Nations, and most of the population growth will be in developing countries. The number of megacities, which is defined as cities with population of over 10 million, will increase to 100 by 2025 (as compared to 25 today). Thus, there is a critical need to understand urban areas to help improve and foster the environmental and human sustainability of cities around the world (Weng and Quattrochi, 2006a).

Urban environmental problems have become unprecedentedly important in the twenty-first century. This is not a simple consequence of ever-increasing urban population and land, but this is because urbanization generates one of the most profound examples of human modification of the Earth (Weng, 2011). Alteration of the landscape through urbanization involves the transformation of the radiative, thermal, moisture, and aerodynamic characteristics of the Earth's surface (Weng, 2011). As humans alter the character of the natural landscape in the urbanization process, they impact the exchange of heat and moisture between land surface and lower atmosphere, create urban heat island (UHI) phenomenon, and influence the local, mesoscale, and even larger scale climate (Weng, 2011). A city is a human-central ecosystem. Cities are the most complex of all human settlements and differ from rural settlements in a number of ways (Weng and Yang, 2003). A significant distinction is that cities have greater size and functional complexity. Economic and societal values are stressed whereas the ecological value is often ignored (Weng and Yang, 2003). Because cities are, by nature, complex human settlements, urban ecosystems are more complex, dynamic, and vulnerable to the impacts of global climate change (Weng and Yang, 2003). The US Climate Change Science Program (CCSP, 2008) defines one of its five goals to be "understanding the sensitivity and adaptability of different natural and managed ecosystems and human systems to climate and related global changes." Although vulnerabilities of settlements to impacts of climate change vary regionally, they generally include some or many of the following impact concerns: health, water, and infrastructures, severe weather events, energy requirements, urban metabolism, sea level rise, economic competitiveness, opportunities and risks, and social and political structures. CCSP (2008) further recommends that research on climate change effects on human settlements in the United States be given a much higher priority to provide better metropolitan-area scale decision making. Earth observation (EO) technology, in conjunction with in situ data collection, has been used to observe, monitor, measure, and model many of the components that comprise urban environmental systems and ecosystem cycles for decades (Weng, 2012). There are a number of satellite remote sensing systems capable of imaging urban areas to the detail needed for global assessment of urban ecosystems. Most of these satellite systems are at the level of coarse resolution (larger than 100 m in spatial resolution) and medium resolution (10–100 m). Fewer satellite systems of high resolution (less than 10 m) can provide data for global monitoring and assessment. Several nations, including the United States, countries of the European Union, China, Japan, and India, are developing very capable satellite

TABLE 2.1
GEO-Defined Societal Benefits and Earth Observation Objectives for Global Urban Observation

Societal Benefit Area	Global Earth Observation Requirements Relating to Urban Areas
Disasters	Human infrastructure, population density, urban extent/sprawl
Health	UHI and air quality, population density, land cover
Energy	Land use and land cover, urban extent/sprawl
Climate	Land cover, urban extent/sprawl
Water	Land use, industrial water demand, population density
Ecosystems	Population density, urban extent/sprawl
Agriculture	Land cover, population density, urban extent/sprawl

Source: Modified after Herold, M., Some recommendations for global efforts in urban monitoring and assessments from remote sensing. In: P. Gamba and M. Herold (eds.), *Global Mapping of Human Settlement*, CRC Press, Boca Raton, FL, 2009, pp. 11–23.

systems, which are suited for establishment of a global urban observatory, to be launched in the next decade. Therefore, international collaboration is needed to produce consistent global maps of human settlements from various data sources, to validate data products, and to provide harmonized information through a common land cover classification system for urban areas.

The Group on Earth Observation (GEO) calls for strengthening the cooperation and coordination among global observing systems and research programs for integrated global observations. The GEO has further developed a framework for establishing a Global Earth Observation System of Systems (GEOSS). In the GEOSS Implementation Plan of 2005 (Group on Earth Observations, 2005), global urban observation and mapping are discussed in several circumstances. Table 2.1 outlines the benefits of global urban imaging and mapping. Some observation requirements are directed to multiple areas of the societal benefit. Urban extent, land cover, and land use maps are not yet available globally but could become available in the next 2–10 years. The GEOSS plan emphasizes the need for the integration of relevant existing observation systems and an integrative analysis of the EO mapping products. Gaps exist in the integration of global urban land observations with data that characterize urban ecosystems, built environment, air quality, and carbon emission, and with the indicators of population density, environmental quality, quality of life, and the patterns of human environmental and infectious diseases. Despite the relevance emphasized in the GEOSS Implementation Plan and a specific GEO task dealing with global land cover, the GEO plan as of 2011 lacks activities relating to urban ecosystems and environmental issues.

In April 2010, Dr. Jinlong Fan from the GEO Secretariat contacted Dr. Qihao Weng, inquiring about his interest in developing a global urban observation task within the framework of GEOSS. Dr. Weng further gained the support of the US GEO representative in principle by addressing technical issues that the representative raised. Through further efforts of Dr. George Xian and Dr. Dale Quattrochi, an urban observation task was endorsed by the United States Geological Survey (USGS) and National Aeronautics and Space Administration (NASA). In the meantime, Dr. Thomas Esch from German Aerospace Center (DLR) indicated to Dr. Weng that he shared the same vision, and further convinced the German GEO delegation to support an urban task. Dr. Paolo Gamba secured a support letter from IEEE Geoscience and Remote Sensing Society, which is a participating organization of GEO. When Dr. Fan finished his term at GEO Secretariat and went back to Beijing, he continued to converse with the China GEO delegation in order to secure its support for an urban task.

In a letter to the GEO Secretariat on August 29, 2011, Dr. Weng outlined the main reasons why a global urban task—the coordination of urban observations, monitoring, assessment, and modeling initiative worldwide—should be considered as an independent task. Several international colleagues, in addition to the authors of this chapter, provided insightful comments and suggestions. The rationales are listed here:

- A global urban observation system contributes to all nine societal benefit areas of GEO (Table 2.1).
- A global urban observation task relates to many other tasks listed in Version 1 of GEO 2012–2015 Work Plan and, therefore, shows a clear "cross-cutting" characteristic.
- In the previous GEO work plans, there were no tasks addressing "human presence" or "human settlements." A unique urban observation task would draw attention to all aspects of human impacts. The twenty-first century is the first urban century in human history.
- EO technologies have evolved rapidly. Urban remote sensing will be the next frontier in EO technologies.

We emphasize that urban areas represent centers of economy, society, culture, and policy. As a result, most of the current and future ecological, economic, and societal challenges are either directly or indirectly related to human activities in or around settlements. Urban areas should be addressed as a complex of ecological, physical, economic, and social processes occurring at multiple scales. Urbanization is one of the most profound examples of human modification of the Earth's surface. An urban observation system has to address the interactions between human settlements and physical environments as a whole and not only a specific aspect of human activity.

In December 2011, the GEO approved 26 tasks and related components. A new task called "Global Urban Observation and Information" was included in its 2012–2015 Work Plan as one of the tasks in the category of Information for Societal Benefits and was listed as SB-04.

2.2   OBJECTIVES AND KEY ACTIVITIES, 2012–2015

The objectives of SB-04 are as follows:

- Improve the overall coordination of urban observations, monitoring, forecasting, and assessment initiatives worldwide.
- Support the development of global urban observation and analysis systems.

- Produce up-to-date information on the status and development of the urban system—from local to global scale.
- Fill gaps in the integration of global urban observations with (1) data characterizing urban ecosystems, built environment, air quality, and carbon emission; (2) indicators of population density, environmental quality, and quality of life; and (3) patterns of human, environmental, and infectious diseases.
- Develop innovative concepts and techniques in support of effective urban sensing and sustainable urban development.

Key activities during the period 2012–2015 include the following:

- Improve global coverage and data accuracy of urban observing systems through integrating satellite data observed from multiple platforms/sensors with different resolutions, with in situ data.
- Define requirements for global urban monitoring and assessment in terms of data products and expectations for data validation, archiving, updating, and sharing.
- Develop a global urban observing network under the umbrella of GEOSS, establishing regional alliances and encouraging the establishment of a program office.
- Create a global urban morphological database for urban monitoring/ assessment and climate modeling to better understand the impacts of global climate change on urban areas.
- Conduct global urban analyses, including time series for assessing the development of megacities (e.g., urban sprawl) and a worldwide inventory of human settlements based on satellite data.
- Conduct urban analyses linking EO products to socioeconomic and environmental quality data to improve knowledge of urban ecosystems.
- Conduct surveys to assess the magnitude and dynamics of the urban heat island effect, particularly for cities in developing countries, and identify environmental impacts on megacities.

2.3 URBAN SUPERSITES INITIATIVE

The SB-04 team has selected eight cities as the supersites: Los Angeles, Atlanta, Mexico City, Athens, Istanbul, São Paulo, Beijing, and Hong Kong. Further, we have identified the following as the benefits of the initiative:

- Identifying and exploiting synergies of resources in R&D as well as applications and benefits
- Expanding the impact of SB-04 developments by specifically addressing selected cities and interested user communities
- Facilitating joint studies and publications and broadening the knowledge on global urban remote sensing
- Enhancing interactions among existing SB-04 contributors and encouraging new partners

- Optimizing the presentation of SB-04's product portfolio to potential stakeholders
- Establishing potential synergies between urban supersites and geohazard supersites

We have so far developed a detailed work plan for the initiative by incorporating contributions from all co-leads and key contributors (Figure 2.1).

The long-term goal of the urban supersites initiative is to establish an urban supersites partnership. Through the partnership, the participants will generate an agreed set of protocols that we will use to share and visualize the intermediate and final global information products, allowing systematic interlaced analysis tasks between the available image-derived information sources (including reference data) and to foster joint studies and dissemination of results. Collective urban sensing is considered to be a promising direction that warrants further investigations.

2.3.1 OBJECTIVES OF THE GLOBAL URBAN SUPERSITES INITIATIVE

Five main objectives are identified for the urban supersites initiative:

1. To provide globally distributed data (EO and derived products), to establish an urban data repository, and to develop standards for specification and validation
2. To estimate urban extent and associated changes
3. To assess risks associated with natural disasters, air and water qualities, and health hazards caused by vector- and animal-borne diseases
4. To derive urban biophysical parameters for characterizing urban land surface–atmosphere interactions (e.g., temperature, emissivity, albedo, vegetation cover) and for climate change mitigation and adaptation
5. To augment and enhance analysis techniques and methods that illustrate the causes and effects of urbanization at local, regional, and global scales

Each task is treated as a work package (WP). A group of team members will work to define the subobjectives of each WP, the requirements in terms of data products, and the expectations for data validation, archiving, updating and sharing; to identify the past, current, and future remote sensing datasets and models; and, finally, to set up the timeline and milestones for implementation.

2.3.2 WP1: GLOBAL HUMAN SETTLEMENTS DATA

Dynamic global urbanization and the constant growth of the settlement area that accompanies this phenomenon are some of the most pressing challenges to sustainable development. The human habitat, with the built-up area as its basic physical manifestation, is the driver for the continuing revolution of society, culture, economy, and policy. Hence, the urban extent and settlements pattern as well as their spatiotemporal development are key indicators for assessing the impacts of urbanization on central sectors such as wealth, health, climate, biodiversity, energy, and water.

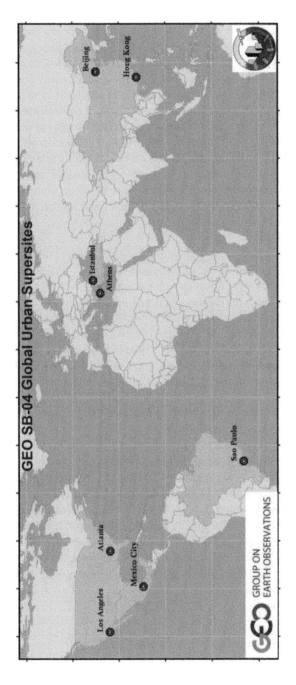

FIGURE 2.1 Locations of the urban supersites.

Spaceborne EO has been established as an effective technique for the worldwide monitoring of human settlements. Currently, there are several EO-derived or EO-supported global human settlement layers (GHSL) that are mostly generated on the basis of medium-resolution (MR) optical EO imagery (Gamba and Herold, 2009; Potere et al., 2009). In this context, the MODIS 500 data set with a spatial resolution of 463 m (Schneider et al., 2010) and the GlobCover layer with a spatial resolution of 309 m (ESA, 2009) are considered to be the geometrically and thematically most accurate global data sets on the location and extent of human settlements. However, the spatial resolution of the current GHSL does not facilitate the detection of small settlements (e.g., villages in rural areas) and the analysis of related phenomena such as peri-urbanization. Therefore, the overall objective of WP1 is to coordinate the generation and provision of spatially and thematically more detailed GHSL by identifying and assessing innovative processing techniques and exploring new EO data sets for the mapping of settlement extents based on high resolution (HR) and very high resolution (VHR) EO imagery. To achieve this goal, the WP1 includes the following tasks:

- Assessment of appropriate and innovative data sources and methods for mapping urban extent at the global scale, especially remote sensing data that can provide both regional and global scale details
- Mapping and characterization of global settlements pattern in unprecedented spatial detail based on remote sensing imagery, including data of established systems such as MODIS, Landsat, IRS, and SPOT and recent satellites such as Suomi NPP Visible Infrared Imaging Radiometer Suite (VIIRS, nighttime lights) or TerraSAR-X/TanDEM-X (SAR data)
- Systematic analysis of the spatiotemporal development of megacities over the last 40 years
- Identification and examination of the hot spots of rapid urban growths in the last few decades around the world

Considering the expertise and activities of the SB-04 participants, the listed tasks will in particular include the examination, application, and comparison of the various new data sets and analysis techniques. Regarding new sensor systems, the data recorded by the VIIRS of the Suomi NPP satellite is of particular interest since it allows for the derivation of new nightlights data. Moreover, new automatic image analysis frameworks for the global delineation of built-up areas based on HR/VHR optical imagery presented by Pesaresi et al. (2011) and Miyazaki et al. (2013) will be taken into consideration. In addition, novel approaches toward the derivation of HR GHSL from SAR data collected by ENVISAT ASAR (Gamba and Lisini, 2012) and TerraSAR-X/TanDEM-X (Esch et al., 2012) are applied.

2.3.3   WP2: URBAN EXTENT AND ASSOCIATED CHANGE

Studies on urban growth monitoring, natural resources management, transportation development, and environmental impact assessment require information about urban extents and associated changes. Despite the growing importance of urban land cover and land use in regional to global scale system change analyses, it remains difficult to determine urban extents and changes around the world because of lack of information and inconsistencies in urban definition across administrative boundaries (Maktav et al., 2005; Schneider et al., 2009). Remote sensing is a good solution to quantify urban extents and associated changes in both regional and global scales (Small, 2005). The spatial extent of urban areas has been mapped successfully for a long time using remote sensing data. These efforts include using either hard classification methods along with medium-resolution remote sensing data or using continuous variable models by treating the urban landscape as a continuum such as percentage of impervious surface (Ridd, 1995; Civco et al., 2002; Xian et al., 2008). Data about urban extent and associate structure change can serve as a surrogate to monitor urban extent and infrastructure change and to assess changes in the urban environment. However, a consistent collection of data for the global urban extent and change information is still unavailable.

The objective of WP2 "Estimate urban extent and associated change" of GEO SB-04 is to estimate urban extent and associated urban land cover changes in both regional and global scales by using existing and new satellite remote sensing data. WP2 will focus on evaluating the existing and new remote sensing data that are able to estimate urban extents and their changes at both regional and global scales. The WP2 tasks will conduct urban extent delineation in the global scale and perform urban land cover change analysis for the most rapidly growing cities and regions in the world. Implementing the current and new satellite data, the most optimal and practical approach, and research projects around the world for urban land quantification is the overall objective of WP2. To achieve this goal, WP2 includes the following tasks:

- Assess appropriate data sources and methods for mapping urban extent at the global scale, especially remote sensing data that can provide both regional and global scale details.
- Highlight hot spots that experienced most rapid growths in both population and urban land cover in the last decade around the world.
- Characterize and quantify urban extents by using remote sensing data, including nighttime light, MOIDS, Landsat imagery and SAR data.
- Examine urban land cover change patterns for the most rapidly growing regions around the world.

The WP2 participants will work closely with several groups that are conducting urban land cover mapping for either regional or global scale. In the United States, the USGS National Land Cover Database (NLCD) project will complete NLCD 2011 in 2014. The dataset will be used to highlight urban land cover change in the entire United States. The Global Land Cover Mapping data gathered by China will be evaluated and suitable portions selected for quantifying the urban extent and change.

2.3.4 WP3: RISKS IN URBAN AREAS

The use of EO data for risk analysis has undergone a sharp increase in the last few years, thanks to the possibility of exploiting data sets at the global level to extract consistent information to map assets exposed to risk (Jaiswal et al., 2013) and to understand their vulnerability (Polli et al., 2009). EO data are currently used, for instance, to spatially disaggregate available information about population at a coarser level via census tracks (Linard and Tatem, 2012) and provide a suitable input to health risk models. The high correlation between human settlements and population location as well as the high concentration of production activities and transport/energy infrastructures within and immediately outside urban areas make urban remote sensing a valuable tool to extract information about exposure assets in multiple scales. However, current activities in this area are lacking: there is no global effort to select the minimum set of urban features common to multiple hazard models that can be extracted from EO data. This is a necessary step to define which information is to be collected using remote sensing, possibly via semiautomatic approaches. In line with this requirement, the efforts in the TOoLs for Open Multirisk assessment using EO data (TOLOMEO) project (TOLOMEO, 2013) show that the tools for exploiting urban remote sensing data for exposure and vulnerability mapping are still to be designed and evaluated on a global scale. It is true, however, that global data layers are currently available, and the challenge is more on the interpretation and the data fusion side, as well as on the definition of the requirements for mapping products that can be obtained via EO data and help in the routines of risk assessment.

According to this scenario, the objectives of the SB-04 working program on "risks in urban areas" can be delineated as follows:

The first aim is to survey existing projects to link EO data and risk computation at the global level and to assess the usefulness of the global layers to be developed within SB-04. In fact, human settlement layers, such as those developed by DLR (Taubenbock et al., 2011) and ESA (Herold et al., 2008) or the University of Pavia (Gamba and Lisini, 2013), lead to the possibility to spatially disaggregate human settlements into buildings or blocks. Accordingly, these results allow an unprecedented level of detailed population/infrastructure mapping on a global scale. The usefulness of these data, however, has still to be proved, and can be estimated to be very different for different risks. For instance, earthquake risk requires information about structural elements that can be obtained (and even so, only partially) using VHR data (Polli et al., 2009), and the global layers mentioned earlier are not effective for building vulnerability estimate. They are, however, effective for population mapping and thus for economic and toll loss computation after a seismic event, and this is true also for many other hazards. Many of the most advanced global risk calculators (Jaiswal et al., 2011; UN-ISDR, 2011) use population as proxy to economic data and will benefit from exploiting the SB-04 data sets.

The second aim, connected to the previous one, of this WP is to dictate guidelines to further develop existing data sets into more valuable products Specifically, the idea of looking for additional information (building spatial density, building height, impervious surface percentage, urban land use maps, etc.) at the global level starting from the first SB-04 products is challenging but very interesting. In this sense, the urban supersite initiative is a welcome addition to the GEO task plan. One or more test sites will be considered to implement and validate the set of newly proposed implementations. The idea is, however, to provide guidelines to validate future improvements of the algorithms and software for new product development rather than to implement a complete validation procedure.

To achieve these aims, the work in the WP will be grouped into tasks. After a collection of existing technical papers are examined, the best practices in the use of EO data in risk-related activities for urban area will be identified, and the definition of the requirements for global/regional information products extracted from EO data for risk-related activities will be detailed. As an example, different urban land use zones according to specific risk-related legends can be explored. There are preliminary works, for instance, in the case of "urban climate zones" (Gamba et al., 2012) that show that it is possible to define a methodology working on different urban areas in different parts of the world to extract areas that have homogeneous behavior with respect to urban micrometeorological models.

The final step in the development of this research is the implementation of a pilot analysis aimed at extracting one or more of these products in one or more test sites, and *the* validation of the results of the pilot analysis using existing data, ground survey, and any other approach that the SB-04 partners can implement. The City of São Paulo in Brazil will be considered for this final step, thanks to the availability of optical and radar data set as well as existing GIS data for urban extents for the whole state of São Paulo.

2.3.5 WP4: Urban Biophysical Parameters

The diversity of urban areas is not only evident in the socioeconomic makeup of cities, but it extends to the biophysical environment in which urban areas preside. The interactions with the biophysical environment that is impacted by and impacts urban areas constitute a feedback mechanism, which can be complex. Within the purview of the urban ecosystem, there are essentially five primary "nodes" that interact with each other: the urban atmosphere, the urban hydrosphere, the urban biosphere, the urban lithosphere (e.g., soils and bedrock), and the "urban fabric," which is comprised of the components that make up the urban landscape (e.g., buildings, roads, houses) (Douglas, 1983; Quattrochi, 2006). The reciprocal actions between these nodes establish the biophysical environment, which, outside of the urban lithosphere (except for such events as earthquakes), are highly reactive with each other. The impact of humans on the urban biophysical environment is paramount and can enhance the effects created by the interactions of the four primary nodes extant of the lithosphere. Moreover, climate change can modify or magnify these effects whereby they are synergistic in impact. For example, it is anticipated that climate change will affect precipitation amount and the intensity of precipitation events; thus, there may be too little precipitation resulting in long periods of droughty conditions, or there may be an increase in heavy precipitation over short time spans as a result of convective storm activity. Climate change, therefore, will impact the urban atmosphere, hydrosphere, and biosphere, and the interactions that occur between these urban ecosystem nodes, and, ultimately, will affect the urban fabric, human health, and the socioeconomic structure of cities (Rosenzweig et al., 2011a,b). Hence, mitigation or adaptation strategies to climate change must be implemented for dealing with the resulting impacts within the overall perspective of the urban ecosystem (Seto and Shepherd, 2009).

Within the scope of observing, measuring, and modeling both the interactions of urban biophysical parameters and potential impacts that climate change will have on the urban ecosystem, and subsequently on the humans that live in urban areas (Chrysoulakis et al., 2013), EO data have been proven to be an invaluable resource. The synoptic, near-real-time, and multispectral, multispatial, and multitemporal characteristics of EO data offer an unparalleled advantage in collecting data that can be used to measure and model human–biophysical interactions across the urban landscape, and between the urban surface and the lower atmosphere (Quattrochi et al., 2009). As a consequence, EO data are imperative for developing a better understanding of how humans affect the urban biosphere and vice-versa and for identifying and monitoring indicators of climate change and the ensuing impacts on urban areas (Ridd and Hipple, 2006; Weng and Quattrochi, 2006a). From this construct, WP4 has four objectives that, in context, will provide insight for understanding how the urban ecosystem operates, how humans interact with the urban biosphere parameters, what the potential impacts of climate change will be on urban areas, and what mitigation or adaptation strategies can be developed and implemented to lessen climate change impacts both on the biophysical and human environments.

The first WP4 objective is to survey the scientific literature for information from studies that have used EO data to derive urban biophysical parameters and characteristics and those that have used EO data to analyze the drivers of urban biosphysical interactions. Foundations for this literature survey have been published by Quattrochi and Luvall (2004), Weng and Quattrochi (2006a,b), and Quattrochi et al. (2009). Because of the complexity of these interactions, it is necessary to segment specific biophysical parameters to efficiently and effectively target analysis of drivers of interactions where EO data can be most useful within a reasonable span of time, for example, goals that are achievable within a 3–5-year period. From this perspective, WP4 will focus on drivers of urban land–atmosphere interactions such as those related to land surface temperature, emissivity of urban surfaces, albedo, and vegetation cover (e.g., evapotranspiration) where EO data have proven to be useful in providing quantitative information for measuring and modeling urban land–atmosphere interactions (e.g., Weng et al., 2006; Weng, 2009a). One example of this is the application of EO data for analysis of the UHI effect, which has been ongoing since the introduction of thermal infrared detecting instruments on EO satellite missions since the mid-1980s with the introduction of the Landsat Thematic Mapper sensor (Quattrochi et al., 2003; Weng, 2009a).

Objectives two and three of WP4 focus on assessing the potential impacts of climate change on urban areas and to evaluate the usefulness of EO data for the analysis of these impacts (Seto and Shepherd, 2009; Imhoff et al., 2010). A key aspect of these objectives is to define indicators of climate change, which can provide meaningful, authoritative, climate-relevant measures about the status, rates, and trends of key physical, ecological, and societal variables and values to inform decisions on management, research, and education at local to regional scales. In turn, these indicators can be used to identify climate-related conditions and impacts to help develop effective mitigation and adaptation measures. A foundation for the development and expression of climate change indicators is currently being defined and described by the US National Climate Assessment (NCA, 2012). Examples of specific indicators that are applicable to urban biophysical parameters are heat (air quality, ozone, and particulate matter), the UHI, land cover and change, ecological health, and wildland/urban interface issues (NCA, 2012).

A fourth objective of WP4 is to investigate how multitemporal and multispectral EO data can be integrated into modeling schemes to assist in accomplishing the tasks related to the three previous WP4 objectives. "Top-down" and "bottom-up" scaled models are required to understand land surface–atmosphere interactions and the implications of climate change on these interactions. Here, the power of EO data can come to bear via research on the evaluation of land surface thermal parameters across urban areas at multispatial scales, through the collection of EO data over cities at different temporal periods and from the analysis of multispectral and hyper-spectral data over thermal infrared, visible, and shortwave infrared (VSWIR) wave-lengths (Sobrino et al., 2013; Zakšek and Oštir, 2013). Few studies of urban thermal properties and the UHI have been conducted using EO data collected from different satellite platforms with different spatial resolutions (Pu et al., 2006; Buyantuyev et al., 2010). More such studies are needed to quantify and model the variability or invariability of specific land surface thermal characteristics, such as the emissivity of urban surfaces, that can be integrated into a robust spatial modeling framework (i.e., top-down or bottom-up) (Nichol, 2009; Mitraka et al., 2013). EO data collected over cities at daily, weekly, or monthly temporal periods for both daytime and night-time are also required to better understand and quantify the thermal dynamics of the urban land surface in a temporally systematic and systemic manner (Nichol, 2005). Although nascent at the present time, there are future EO satellite missions that offer great promise for attaining these temporal data to provide a diurnal time-sequenced modeling structure of urban thermal dynamics for megacities and smaller cities around the world (see Weng et al., 2006; Xiao et al., 2008; Zhou and Wang, 2011; Xiong et al., 2012 for the rationale to support this objective). Multispectral and hypserspectral EO data obtained from different sensors in varying orbits and with spatial/temporal configurations are required to observe and quantify urban surface reflectance properties that comprise the urban biophysical realm, thereby providing information that can be used to develop spectral libraries of urban surface albedo and emissivity, which are either not well defined or nonexistent at present (Roberts et al., 2012; Zhu et al., 2013). The information contained in these libraries can be compared and verified against each other for input into spectral models of urban land surface "building blocks"—those surface components that are ubiquitous to and comprise the urban landscape.

2.3.6   WP5: Urban Analysis Methods

The impacts of urbanization may be at the local, regional, or global scale. Associated with a rapid urbanization worldwide is an exacerbation in environmental issues and health problems. EO data can be used to detect and measure changes in urban growth patterns and to elucidate how urbanization impacts the environment and human health. Remote sensing data, with their advantages in spectral, spatial, and temporal resolutions, have demonstrated their power in providing information about physical characteristics of urban areas, including the size, shape, and rate of change and have been widely used for mapping and monitoring of urban biophysical features (Jensen and Cowen, 1999; Weng, 2012). Some examples of use of remote sensing images in the urban areas include providing land cover/use data and biophysical attributes (Haack et al., 2002; Weng et al., 2006; Weng and Hu, 2008), extracting and updating transportation network (Harvey et al., 2004; Song and Civco, 2004) and buildings (Lee et al., 2003; Miliaresis and Kokkas, 2007), and detecting urban expansion (Yeh and Li, 1997; Weng, 2002). The EO data products, modeling results, methods, and techniques over the past decade have provided substantial support for sustainable urban development worldwide.

Furthermore, EO data can be used to assist in taking into account multiple factors affecting human health such as those contributing to environmental health hazards and contagious and infectious diseases. EO data, in combination with other data sources, can provide geospatial information on environmental conditions for understanding distributions of water-borne diseases, air quality, soil, and vegetation as they influence community health and livestock. For example, remote sensing and GIS technologies, in combination with biological, ecological, and statistical methods, have been extensively applied in West Nile Virus (WNV) epidemiology studies globally (Ruiz et al., 2004, 2007; Liu et al., 2008; Pan et al., 2008). Finally, remote sensing, GIS, and census data have been integrated to estimate population and residential density (Harvey, 2002; Li and Weng, 2005), to appraise socioeconomic conditions (Thomson, 2000; Hall et al., 2001), and to evaluate urban environmental quality (Nichol and Wong, 2006; Liang and Weng, 2011) and the quality of life in the cities (Lo and Faber, 1997; Li and Weng, 2007).

The aims of WP5 are set to address the concerns over urbanization and environmental impacts using EO data and technology and to improve urban remote sensing methods and techniques for better global observation and monitoring of the urbanization pattern and process. Specific objectives include

- Enhancing urban analysis by developing innovative methods and techniques in support of effective urban sensing and sustainable urban development
- Conducting urban analyses by linking EO products with socioeconomic and in situ data to improve knowledge of urban environments and ecosystems
- Conducting urban analyses at local, regional, and global scales to improve the understanding of the patterns, processes, and consequences of urbanization in different geographical and socioeconomic settings

To meet these objectives, the WP participants will focus on the following tasks between 2012 and 2015. A literature survey will be conducted of existing satellite sensors for urban analysis and monitoring, examining current practices, and addressing key limitations and future perspectives of remote sensor technology. Urban landscape processes appear to be hierarchical in both pattern and structure. A study of the relationship between the patterns at different levels in the hierarchy may help in obtaining a better understanding of the scale problem (Cao and Lam, 1997) and in finding the optimal scale for examining the relationship between urbanization pattern and process (Liu and Weng, 2009). The use of data from various satellite sensors may result in different research results because they usually have different spatial resolutions. Therefore, it is important to examine the changes in the spatial configuration of any landscape pattern that is the result of using different spatial resolutions of satellite imagery. We thus intend to reexamine the scale issue in urban observation and analysis.

New frontiers in EO technology since 1999—such as VHR, hyperspectral, Lidar sensing, and their synergy with existing technologies—and advances in remote sensing imaging science, such as object-oriented image analysis, artificial neural network, data fusion, and data mining, are changing the image information we obtain and the way we handle the image processing. Both aspects will reshape the scope and contents of urban remote sensing. Reflection on these recent changes in EO/remote sensing technology and implications for urban analysis become imperative for this WP. The temptation to take advantage of the opportunity of combining ever-increasing computational power, Internet and modern telecommunication technologies, sensor webs, more plentiful and capable digital data, volunteered geographic information, and more advanced data processing algorithms has resulted in a new round of attention to the integration of remote sensing, GIS, and GPS for environmental, resources, and urban studies (Weng, 2009b). Collective urban sensing (Blaschke et al., 2011) and people as sensors (Hay et al., 2011) are two examples pushing through this direction in order to develop innovative methods for observing, monitoring, and analyzing urban systems. Through the collaboration of the SB-04 team members, we will develop various approaches to a synergy of datasets and techniques for urban analysis in one or more supersites.

2.3.7 DATA REPOSITORY

We will gather and generate essential datasets for the supersites and establish a data repository for these EO data and derived products. The repository will serve the urban supersites initiative and function as an information hub for disseminating important study results. Selected datasets and results, providing invaluable information and reference datasets for decision makers in urban planning, environmental management, human health, energy, and sustainability will be accessible to the general public too. This tool can also be employed as an education source for students, researchers, and professionals worldwide. At the global scale, data users may include UN-Habitat, World Bank, urban climate modelers, epidemiologists, and the like.

Raw data to be collected may include VHR SAR (TerraSAR-X), Landsat data, VHR Optical data (e.g., IKONOS, QuickBird, WorldView imagery), VIIRS/Suomi NPP, and so forth. Thematic layers, which may include urban extent, urban extent change maps, essential environmental variables (land surface temperature, emissivity, albedo, vegetation cover, impervious surface), and urban morphology (built-up structures, average distance between built-up structures, classification) will be generated.

Within GEO, hot spot areas in terms of urban phenomena addressed by SB-04 and expected benefits for broad range of SB-04 products will provide interactions and leveraging with SB-02 (Global Land Cover), SB-03 (Global Forest

Observation), SB-05 (Impact Assessment of Human Activities), DI-01 (Informing Risk Management and Disaster Reduction), HE-01 (Tools and Information for Health Decision Making), HE-02 (Tracking Pollutants), EN-01 (Energy and Geo-Resources Management), CL-01 (Climate Information for Adaptation), WA-01 (Integrated Water Information including floods and droughts), and EC-01 (Global Ecosystem Monitoring). Collaboration with SB-02, SB-05, HE-01, HE-02, and CL-01 is high priority of SB-04.

REFERENCES

Blaschke, T., G. J. Hay, Q. Weng, and B. Resch. 2011. Collective sensing: Integrating geospatial technologies to understand urban systems—An overview. *Remote Sensing*, 3(8), 1743–1776; doi:10.3390/rs3081743.

Buyantuyev, A., J. Wu, and C. Gries. 2010. Multiscale analysis of the urbanization of the Phoenix metropolitan landscape of USA: Time, space, and thematic resolution. *Landscape and Urban Planning*, 94, 206–217.

Cao, C. and N. S.-N. Lam. 1997. Understanding the scale and resolution effects in remote sensing. In D.A. Quattrochi and M.F. Goodchild (eds.), *Scale in Remote Sensing and GIS*. CRC Press, Boca Raton, FL, pp. 57–72.

Chrysoulakis, N., M. Lopes, R. San Jose, C. S. B. Grimmond, M. B. Jones, V. Magliulo, J. E. M. Klostermann et al. 2013. Sustainable urban metabolism as a link between bio-physical sciences and urban planning: The BRIDGE project. *Landscape and Urban Planning*, 112, 100–117.

Civco, D., J. Hurd, E. H. Wilson, C. L. Arnold, and M. P. Prisloe, Jr. 2002. Quantifying and describing urbanizing landscapes in the northeast United States. *Photogrammetric Engineering & Remote Sensing*, 68(10), 1083–1090.

Douglas, I. 1983. *The Urban Environment*. Edward Arnold, Baltimore, MD, 229pp.

Esch, T., H. Taubenböck, A. Roth, W. Heldens, A. Felbier, M. Thiel, M. Schmidt, A. Müller, and S. Dech. 2012. TanDEM-X mission—New perspectives for the inventory and monitoring of global settlement patterns. *Journal of Applied Remote Sensing*, 6(1), 1–21.

European Space Agency (ESA). 2009. GLOBCOVER 2009: Products description and validation report. Available at http://due.esrin.esa.int/globcover/LandCover2009/GLOBCOVER2009_Validation_Report_2.2.pdf (accessed on March 18, 2013).

Gamba, P. and M. Herold. 2009. *Global Mapping of Human Settlement—Experiences, Datasets, and Prospects*. CRC Press, Boca Raton, FL.

Gamba, P. and G. Lisini. 2013. Fast and efficient urban extent extraction using ASAR Wide Swath Mode data. *IEEE Journal of Selected Topics in Applied Earth Observation and Remote Sensing*, 6, 2184–2195.

Gamba, P., G. Lisini, P. Liu, P. Du, and H. Lin. 2012. Urban climate zone detection and discrimination using object-based analysis of VHR scenes. *Proceedings of GEOBIA 2012*, Rio de Janeiro, Brazil, May 2012, pp. 70–74.

Group on Earth Observations. 2005. The Global Earth Observation System of Systems (GEOSS) 10-year implementation plan (as adopted February 16, 2005). Available at http://www.earthobservations.org/documents/10-Year%20Implementation%20Plan.pdf (last accessed January 19, 2014).

Haack, B. N., E. K. Solomon, M. A. Bechdol, and N. D. Herold. 2002. Radar and optical data comparison/integration for urban delineation: A case study. *Photogrammetric Engineering & Remote Sensing*, 68, 1289–1296.

Hall, G. B., N. W. Malcolm, and J. M. Piwowar. 2001. Integration of remote sensing and GIS to detect pockets of urban poverty: The case of Rosario, Argentina. *Transactions in GIS*, 5, 235–253.

Harvey, J. T. 2002. Estimation census district population from satellite imagery: Some approaches and limitations. *International Journal of Remote Sensing*, 23, 2071–2095.

Harvey, W., J. C. McGlone, D. M. McKeown, and J. M. Irvine. 2004. User-centric evaluation of semi-automated road network extraction. *Photogrammetric Engineering & Remote Sensing*, 70, 1353–1364.

Hay, G. J., C. Kyle, B. Hemachandran, G. Chen, M. M. Rahman, T. S. Fung, and J. L. Arvai. 2011. Geospatial technologies to improve urban energy efficiency. *Remote Sensing*, 3(7), 1380–1405.

Herold, M. 2009. Some recommendations for global efforts in urban monitoring and assessments from remote sensing. In: P. Gamba and M. Herold (eds.), *Global Mapping of Human Settlement*. CRC Press, Boca Raton, FL, pp. 11–23.

Herold, M., C. E. Woodcock, T. R. Loveland, J. Townshend, M. Brady, C. Steenmans, and C. Schmullius. 2008. Land-cover observations as part of a Global Earth Observation System of Systems (GEOSS): Progress, activities, and prospects. *IEEE Systems Journal*, 2(3), 414–423.

Imhoff, M. L., P. Zhang, R. E. Wolfe, and L. Bounoua. 2010. Remote sensing of the urban heat island effect across biomes in the continental USA. *Remote Sensing of Environment*, 114, 504–513.

Jaiswal, K., F. Dell'Acqua, and P. Gamba. 2013. Spatial aspects of building and population exposure data and their implications for global earthquake exposure modeling. *Natural Hazards*, 68, 1291–1309; doi: 10.1007/s11069-012-0241-2.

Jaiswal, K., D. Wald, P. Earle, K. Porter, and M. Hearne. 2011. Earthquake casualty models within the USGS prompt assessment of global earthquakes for response (PAGER) system, Chapter 6. In: R. Spence, E. So, and C. Scawthorn (eds.), *Human Casualties in Earthquakes: Progress in Modelling and Mitigation*. Springer, Berlin, Germany.

Jensen, J. R. and D. C. Cowen. 1999. Remote sensing of urban/suburban infrastructure and socioeconomic attributes. *Photogrammetric Engineering & Remote Sensing*, 65, 611–622.

Lee, D. S., J. Shan, and J. S. Bethel. 2003. Class-guided building extraction from IKONOS imagery. *Photogrammetric Engineering & Remote Sensing*, 69(2), 143–150.

Li, G. and Q. Weng. 2005. Using Landsat ETM+ imagery to measure population density in Indianapolis, Indiana, USA. *Photogrammetric Engineering & Remote Sensing*, 71, 947–958.

Li, G. and Q. Weng. 2007. Measuring the quality of life in city of Indianapolis by integration of remote sensing and census data. *International Journal of Remote Sensing*, 28(2), 249–267.

Liang, B. and Q. Weng. 2011. Assessing urban environmental quality change of Indianapolis, United States, by the remote sensing and GIS integration. *IEEE Journal of Selected Topics in Applied Earth Observations & Remote Sensing*, 4(1), 43–55.

Linard, C. and A. J. Tatem. 2012. Large-scale spatial population databases in infectious disease research. *International Journal of Health Geographics*, 11, 7.

Liu, H. and Q. Weng. 2009. Scaling-up effect on the relationship between landscape pattern and land surface temperature. *Photogrammetric Engineering & Remote Sensing*, 75(3), 291–304.

Liu, H., Q. Weng, and D. Gaines. 2008. Multi-temporal analysis of the relationship between WNV dissemination and environmental variables in Indianapolis, USA. *International Journal of Health Geographics*, 7, 66; doi: 10.1186/1476-072X-7-66.

Lo, C. P. and B. J. Faber. 1997. Integration of Landsat Thematic Mapper and census data for quality of life assessment. *Remote Sensing of Environment*, 62, 143–157.

Maktav, D., F. S. Erbek. and C. Jürgens. 2005. Remote sensing of urban areas. *International Journal of Remote Sensing*, 26(4), 655–659.

Miliaresis, G. and N. Kokkas. 2007. Segmentation and object-based classification for the extraction of the building class from LiDAR DEMs. *Computers and Geosciences*, 33(8), 1076–1087.

Mitraka, Z., N. Chrysoulakis, Y. Kamarianakis, P. Partsinevelos, and A. Tsouchlaraki. 2013. Improving the estimation of urban surface emissivity based on sub-pixel classification of high resolution satellite imagery. *Remote Sensing of Environment*, 117, 125–134.

Miyazaki, H., X. Shao, K. Iwao, and R. Shibasaki. 2013. An automated method for global urban area mapping by integrating ASTER satellite images and GIS data. *IEEE Journal of Selected Topics in Applied Earth Observations and Remote Sensing*, 99, 1–16.

NCA. 2012. *Climate Change Impacts and Responses: NCA Report Series, 5C, Societal Indicators for the National Climate Assessment*. US Global Change Research Program, Washington, DC, 121pp. http://downloads.usgcrp.gov/NCA/Activities/Societal_Indicators_FINAL.pdf (last accessed on April 20, 2013).

Nichol, J. 2005. Remote sensing of urban heat islands by day and night. *Photogrammetric Engineering and Remote Sensing*, 71, 613–621.

Nichol, J. 2009. An emissivity modulation method for spatial enhancement of thermal satellite images of urban heat island analysis. *Photogrammetric Engineering and Remote Sensing*, 75, 547–556.

Nichol, J. E. and M. S. Wong. 2006. Assessment of urban environmental quality in a subtropical city using multispectral satellite images. *Environment and Planning B: Planning and Design*, 33, 39–58.

Pan, L. L., L. X. Qin, S. X. Yang, and J. P. Shuai. 2008. A neural network-based method for risk factor analysis of West Nile virus. *Risk Analysis*, 28(2), 487–496.

Pesaresi, M., D. Ehrlich, I. Caravaggi, M. Kauffmann, and C. Louvrier. 2011. Towards global automatic built-up area recognition using optical VHR imagery. *IEEE Journal of Selected Topics in Applied Earth Observations*, 4(4), 923–934.

Polli, D., F. Dell'Acqua, and P. Gamba. 2009. First steps towards a framework for Earth Observation (EO)-based seismic vulnerability evaluation. *Environmental Semeiotics*, 2(1), 16–20.

Potere, D., A. Schneider, S. Angel, and D. L. Civco. 2009. Mapping urban areas on a global scale: Which of the eight maps now available is more accurate? *International Journal of Remote Sensing*, 30(24), 6531–6558.

Pu, R., P. Gong, R. Mkichishita, and T. Sasqawa. 2006. Assessment of multi-resolution and multisensory data for urban surface temperature retrieval. *Remote Sensing of Environment*, 104, 211–225.

Quattrochi, D. A. 2006. Environmental dynamics of human settlements. In: M.R. Ridd and J.D. Hipple (eds.), *Remote Sensing of Human Settlements, Manual of Remote Sensing*, 3rd edn., Vol. 5, Section 10.3. American Society for Photogrammetry and Remote Sensing, Bethesda, MD, pp. 564–577.

Quattrochi, D. A. and J. C. Luvall. 2004. *Thermal Remote Sensing in Land Surface Processes*. CRC Press, Boca Raton, FL, 440pp.

Quattrochi, D. A., A. Prakash, M. Eneva, R. Wright, D. K. Hall, M. Anderson, W. P. Kustas, R. G. Allen, T. Pagano, and M. T. Coolbaugh. 2009. Thermal remote sensing: Theory, sensors, and applications. In: M.W. Jackson (ed.), *Earth Observing Platforms & Sensors, Manual of Remote Sensing*, 3rd edn., Vol. 1.1. American Society for Photogrammetry and Remote Sensing, Bethesda, MD, pp. 107–187.

Quattrochi, D. A., S. J. Walsh, J. R. Jensen, and M. K. Ridd. 2003. Remote sensing. In: G.L. Gaile and C.J. Wilmott (eds.), *Geography in American at the Dawn of the 21st Century*. Oxford University Press, New York, pp. 376–416.

Ridd, M. K. 1995. Exploring a V-I-S (vegetation-impervious surface-soil) model for urban ecosystem analysis through remote-sensing-comparative anatomy for cities. *International Journal of Remote Sensing*, 16(12), 2165–2185.

Ridd, M. K. and J. D. Hipple. 2006. *Remote Sensing of Human Settlements, Manual of Remote Sensing*, 3rd edn., Vol. 5. American Society for Photogrammetry and Remote Sensing, Bethesda, MD, 659pp.

Roberts, D. A., D. A. Quattrochi, G. C. Hulley, S. J. Hook, and R. O. Green. 2012. Synergies between VSWIR and TIR data for the urban environment: An evaluation of the potential for the Hysperspectral Infrared Imager (HyspIRI) Decadal Survey mission. *Remote Sensing of Environment*, 117, 83–101.

Rosenzweig, C., W. D. Solecki, S. A. Hammer, and S. Mehrotra (eds.). 2011a. *Climate Change and Cities*. 1st edn. Cambridge University Press, Cambridge, U.K. Cambridge Books Online, January 19, 2014. Available at http://dx.doi.org/10.1017/CBO9780511783142.

Rosenzweig, C., W. D. Solecki, S. A. Hammer, and S. Mehrotra. 2011b. Urban climate change in context. In: C. Rosenzweig, W.D. Solecki, S.A. Hammer, and S. Mehrotra (eds.), *Climate Change and Cities: First Assessment Report of the Urban Climate Change Research Network*. Cambridge University Press, Cambridge, U.K., pp. 3–11.

Ruiz, M. O., C. Tedesco, T. J. McTighe, C. Austin, and U. Kitron. 2004. Environmental and social determinants of human risk during a West Nile virus outbreak in the greater Chicago area, 2002. *International Journal of Health Geographics*, 3, 8–18.

Ruiz, M. O., E. D. Walker, E. S. Foster, L. D. Haramis, and U. D. Kitron. 2007. Association of west Nile virus illness and urban landscapes in Chicago and Detroit. *International Journal of Health Geographics*, 6(10), 1–11.

Schneider, A., M. A. Friedl, and D. Potere. 2009. A new map of global urban extent from MODIS satellite data. *Environmental Research Letter*, 4; doi:10.1088/1748–9326/4/044003.

Schneider, A., M. A. Friedl, and D. Potere. 2010. Mapping global urban areas using MODIS 500-m data: New methods and datasets based on urban ecoregions. *Remote Sensing of Environment*, 114, 1733–1746.

Seto, K. and J. M. Shepherd. 2009. Global urban land-use trends and climate change. *Current Opinion in Environmental Sustainability*, 1, 89–95.

Small, C. 2005. A global analysis of urban reflectance. *International Journal of Remote Sensing*, 26(4), 661–681.

Sobrino, J. R., R. Oltra-Carrió, G. Sòria, R. Bianchi, and M. Paganini. 2013. Impact of spatial resolution and satellite overpass time on evaluation of the surface urban heat island effects. *Remote Sensing of Environment*, 117, 50–56.

Song, M. and D. Civco. 2004. Road extraction using SVM and image segmentation. *Photogrammetric Engineering & Remote Sensing*, 70, 1365–1372.

Taubenbock, H., T. Esch, A. Felbier, A. Roth, and S. Dech. 2011. Pattern-based accuracy assessment of an urban footprint classification using TerraSAR-X data. *IEEE Geoscience and Remote Sensing Letters*, 8(2), 278–282.

Thomson, C. N. 2000. Remote sensing/GIS integration to identify potential low-income housing sites. *Cities*, 17, 97–109.

TOLOMEO. 2013. Available at http://tolomeofp7.unipv.it (last accessed on March 15, 2013).

UN-ISDR. 2011. Global assessment or risk report (GARR11). Available at http://www.preventionweb.net/english/hyogo/gar/2011/en/home/index.html. (last accessed on March 15, 2013).

US Climate Change Science Program (CCSP). 2008. Analyses of the effects of global change on human health, welfare, and human systems. A report by the US Climate Change Science Program and the Subcommittee on Global Change Research [Gamble, J.L. (ed.), K.L. Ebi, F.G. Sussman, T.J. Wilbanks]. US Environmental Protection Agency, Washington, DC.

Weng, Q. 2002. Land use change analysis in the Zhujiang Delta of China using satellite remote sensing, GIS, and stochastic modeling. *Journal of Environmental Management*, 64, 273–284.

Weng, Q. 2009a. Thermal infrared remote sensing for urban climate and environmental studies: Methods, applications, and trends. *ISPRS Journal of Photogrammetry and Remote Sensing*, 64, 335–344.

Weng, Q. 2009b. *Remote Sensing and GIS Integration: Theories, Methods, and Applications*. McGraw-Hill Professional, New York, p. 397.

Weng, Q. 2011. Remote sensing of urban biophysical environment. In: Q. Weng (ed.), *Advances in Environmental Remote Sensing: Sensors, Algorithms, and Applications.* CRC Press, Boca Raton, FL, Chapter 20, pp. 513–533.

Weng, Q. 2012. Remote sensing of impervious surfaces in the urban areas: Requirements, methods, and trends. *Remote Sensing of Environment*, 117(2), 34–49.

Weng, Q. and X. Hu. 2008. Medium spatial resolution satellite imagery for estimating and mapping urban impervious surfaces using LSMA and ANN. *IEEE Transactions on Geosciences and Remote Sensing*, 46(8), 2397–2406.

Weng, Q., D. Lu, and B. Liang. 2006. Urban surface biophysical descriptors and land surface temperature variations. *Photogrammetric Engineering & Remote Sensing*, 72(11), 1275–1286.

Weng, Q. and D. A. Quattrochi. 2006a. An introduction to urban remote sensing. In: Q. Weng and D.A. Quattrochi (eds.), *Urban Remote Sensing*. Boca Raton, FL: CRC/Taylor & Francis, pp. xiii–xvi.

Weng, Q. and D. A. Quattrochi. 2006b. Thermal remote sensing of urban areas: An introduction to the special issue. *Remote Sensing of Environment*, 104(2), 119–122.

Weng, Q. and S. Yang. 2003. An approach to evaluation of sustainability for Guangzhou's urban ecosystem. *International Journal of Sustainable Development and World Ecology*, 10(1), 69–81.

Xian, G., M. Crane, and C. McMahon. 2008. Quantifying multi-temporal urban development characteristics in Las Vegas from Landsat and ASTER data. *Photogrammetric Engineering & Remote Sensing*, 74(4), 473–481.

Xiao, R., Q. Weng, Z. Ouyang, W. Li, E. W. Schlenke, and Z. Zhang. 2008. Land surface temperature variations and major factors in Beijing, China. *Photogrammetric Engineering & Remote Sensing*, 74, 451–461.

Xiong, Y., S. Huang, F. Chen, H. Ye, C. Wang, and C. Zhu. 2012. The impacts of rapid urbanization on thermal environment: A remote sensing study of Guangzhou, South China. *Remote Sensing*, 4, 2033–2056; doi: 10.3390/rs4072033.

Yeh, A. G. O. and X. Li. 1997. An integrated remote sensing and GIS approach in the monitoring and evaluation of rapid urban growth for sustainable development in the Pearl River Delta, China. *International Planning Studies*, 2, 193–210.

Zakšek, J. and K. Oštir. 2012. Downscaling land surface temperature for urban heat island diurnal cycle analysis. *Remote Sensing of Environment*, 117, 114–124.

Zhou, X. and Y.-C. Wang. 2011. Dynamics of land surface temperature in response to land-use/cover change. *Geographical Research*, 49, 23–36.

Zhu, Z., C. E. Woodcock, J. Rogan, and J. Kellndorfer. 2013. Assessment of spectral, polarimetric, temporal, and spatial dimensions for urban and peri-urban land cover classification using Landsat and SAR data. *Remote Sensing of Environment*, 117, 72–82.

3 EO Data Processing and Interpretation for Human Settlement Characterization
A Really Global Challenge

Paolo Gamba, Gianni Lisini, Gianni Cristian Iannelli,
Inmaculada Dopido, and Antonio Plaza

CONTENTS

3.1 INTRODUCTION

The need for increasingly accurate models of the complex interactions between mankind and the environment calls for more precise monitoring of many different areas, from forests to oceans, from inland waters to urban areas. Specifically, human settlements appear to be the focus of a number of issues such as desertification and pollution as well as water, energy, and waste management. Since most of the population nowadays lives in urban areas, threats to human lives, such as diseases and man-made and natural disasters, are increasingly perceived as the causes of social and economic losses in urban areas. Using urban areas and some of their specific features as essential input information, scientists and researchers have developed models for climate change [1], earthquake risk [2], disease spread [3], and many others. To achieve this aim, global analyses, including more information than just the knowledge of urban area locations, are mandatory. And yet the latest global data sets on urban areas are still incomplete or limited to the aforementioned sources of information. These involve spatial scales that are not completely useful to address intra-urban activities.

The task of extracting and managing urban datasets at different scales by using Earth Observation (EO) data is one of the great challenges of remote sensing data interpretation. However, many of the most promising techniques that have been proposed have been applied to small or limited data sets so far; this was originally due to the limited amount of available data, but currently it is related more to the complexity of designing data analysis procedures for multiple sensors at multiple spatial and spectral resolutions. Moreover, the finest spatial resolution available from EO sensors does not fit the requirements of all urban studies, and VHR data, with all their details, may be less suited to tasks like urban land use mapping. Finally, from a global perspective, issues come from the huge amount of data available as well as the need for an efficient methodology to extract useful information with a consistent approach in different geographical areas.

This chapter initially provides an overview of existing methodologies to address some of these issues as well as the challenges related to the implementation/realization of these methodologies. To explain the sort of "high-level" message included in these pages, a few examples from existing research by the authors are included.

The remainder of the chapter is organized as follows. Section 3.2 describes a general procedure to derive scalable models from EO data and urban features. Section 3.3 discusses several challenges still open for research in the aforementioned procedure. Section 3.4 concludes with some remarks.

3.2   EO DATA AND URBAN AREAS

This section provides an overview of available mechanisms to derive relevant, scalable models from EO data using features collected in urban environments. This includes a description of success cases and processing chains under different application scenarios. Remaining challenges are outlined in Section 3.3.

3.2.1   FROM EO DATA TO URBAN FEATURES

The standard approach to EO data exploitation in urban area characterization is to extract man-made features and use them as basic elements. By exploiting and combining these elements, more detailed analyses can be obtained. This is the case, for instance, in urban extent delineation—starting from elementary spatial patterns [4], road network extraction, grouping road candidates [5], or building two-dimensional and three-dimensional "builtscape" characterization to clustering 3D primitives [6]. The challenge in this type of approach is to design a sufficiently flexible algorithm, able to adapt to the multiple and different ways that these features may appear in EO data (particularly, in different geographical areas), and to recognize significant clusters as hints or proxies to urban elements. However, once the information is available, it may be exploited for a number of different applications, thus providing inputs to multiple models.

The main limitation of this processing framework is that the basic urban/artificial features are extracted at a given scale so that their use for multiple scale models is not always immediate. For example, after 2D building information is extracted, a thermal model for each building may be considered, but a thermal model for a whole city would require some sort of reprocessing and clustering of the extracted data. Similarly, and as another example, road network extraction may be used for traffic modeling, mobility management, or noise pollution monitoring and prevention. To be useful, however, a road network extraction approach should be fast, efficient, and precise enough in addition to providing an output as scalable as possible.

To improve over existing approaches and propose a unitary framework, the methodology discussed in [7] may be considered. Specifically, the idea is to include in the information extraction from an urban scene many different features corresponding to multiple scales. These features can be used either to logically and spatially cluster them into elements at a lower (coarser) scale or to infer other features at a higher (finer) scale. If each feature extraction algorithm is designed within such a framework, it can incorporate enough flexibility to combine different features at different scales and thus simultaneously obtain more information. Similarly, this framework may be used to design techniques that can be tailored to work in multiple geographical areas.

3.2.2 FROM EO DATA TO SCALABLE MODELS

A different and recently considered way to exploit EO data is related to the direct extraction of model-related features, less generic and more connected to the local/regional (and, eventually, global) models required by current studies. With respect to climate change, for instance, the thermal behavior of urban areas has been actively investigated [8] but without a strong link to global climate models. Similarly, risk analysis is a very important topic related to the impact of natural disasters that may occur in and around human settlements [9]. As shown in the aforementioned examples, on a city level (or more detailed scale), atmospheric circulation analyses, involving urban meteorological models as well as risk computations including physical and social vulnerability, have already been considered for one or more cities. They still need to be tuned and validated on a global scale as opposed to a case by case approach. As a preliminary step in this direction, there is a need to characterize every urban area at a global scale, according to land use/land cover typologies, which are peculiar to different environmental or risk models. This is the case of "urban climate zones" (areas with the same microclimatic behavior) for urban meteorology [10] or "uniformly built dwellings" (areas with buildings that have the same structural typology) for earthquake vulnerability [23].

Specifically, urban climate zones represent a comprehensive classification system for characterizing the urban environment with respect to urban meteorology, as reported in [10,11]. The same classification scheme was applied to a different environment in [12]. These works introduced the concept of "thermal climate zones" or "local climate zones," defined as regions with relatively uniform surface–air temperature distribution across different horizontal scales [10]. These climate zones can be differentiated by means of multiple characteristics from the urban 2D and 3D landscapes such as the built surface fraction, the building height-to-width ratio, the sky view factor (percentage of sky visible from the ground), the height of roughness elements, the anthropogenic heat flux, and the surface thermal admittance.

Most of these characteristics, ultimately connected to physical characteristics of the urban objects, can be extracted from remote sensing data.

The methodology reported here is based on two different processing chains. The first one is devoted to the extraction of spatially homogeneous urban areas within the scene, which may be labeled as "block." The idea is that these blocks may be then assigned, using the second part of the procedure, to one of the urban climate zone classes by considering a suitable combination of spatial and spectral indexes.

The first processing chain can be subdivided into two subsequent steps. The first one is the identification of the human settlement (as opposed to all the other land use classes in the area). To achieve this, we use the PanTex index proposed in [13], which proved to be effective to extract human settlement extents starting from panchromatic images at 2.5 m spatial resolution. The area identified as human settlement is further segmented into a homogenous zone using a spanning tree reduction scheme [14]. Alternatively, a more complex approach based on a combination of geometrical features into closed boundaries can also be used [15].

The employed processing chain, aiming at classifying each homogenous area into an urban climate zone class, is based on the joint analysis of a few indexes that, we feel, may capture most of the features listed in the introduction. We assume that a multiscale version of the same index used for urban area detection may be useful as it helps in enhancing spatial patterns at multiple geographical scales. To obtain a multiscale PanTex, the same textural feature (contrast) used to build the original index is now computed with different lag distances (which is equivalent to assuming a different spatial resolution of the data). Additionally, the original image and the results of an edge extraction technique (implemented using a Sobel filter) are included to insert spectral and edge density information, respectively.

Using these indexes, a decision tree classifier is designed using training data and is eventually applied to the whole data set. The decision tree structure used to label the segmented blocks and assign them to the different climate zones is obtained by a detailed analysis of a small sample of the blocks in the first test case described in the next section. Although this approach is apparently biased by a specific city structure and location, the same rules apparently work in different locations, as also discussed in the following paragraphs. The main rationale is that these rules refer to spatial indexes, which in turn describe quantitatively the spatial structure of the different parts of a town.

The decision tree, tuned with empirical tests, accepts as inputs three images:

1. The original image (OR)
2. The PanTex filter output with a kernel of 5 × 5 pixels applied at the full scale data (P1)
3. The PanTex filter output applied to a subsampled data set at 5 m/pixel (P5)

Experimental results were obtained on August 12, 2008 from a scene by the ALOS PRISM sensor with a spatial resolution of 2.5 m and depicting a portion of the town of Xuzhou in the Jiangsu province, People's Republic of China. The challenge, as highlighted in the previous section, was to use 2D data without spectral information to obtain spatial indexes allowing an analysis of different zones and their classification into thermal climate zones. Results depended on both segmentation and accuracy.

An incorrect segmentation may result in less precise classification of the urban blocks as the spatial indexes used by the decision tree are averaged for each block. However, the rules defined for the decision tree are more important because they allow assigning each block to a climate zone, once its spatial boundaries have been individuated by the segmentation step. For this reason and because the segmented urban image can be obtained by various means—for instance, by using available geographic information system (GIS) layers for a town—the evaluation described in the following paragraph will focus mostly on the second step of the procedure, without paying much attention to the approach used to achieve a correct segmentation.

For the test area, only five urban climate zones were considered, that is, those that were present in the scene settlements: "open set mid rise," "compact low rise," "open set low rise," "dispersed low rise," and "extensive low rise." By applying the aforementioned procedure to a first set of urban blocks, shown in Figure 3.1 together with the corresponding color legend, a relatively high overall accuracy at the object level is achieved (81%).

(a)

(b)

Typology	Color
Open set mid rise	Red
Compact low rise	Blue
Open set low rise	Green
Dispersed low rise	Yellow
Extensive low rise	Purple

(c)

FIGURE 3.1 **(See color insert.)** Experimental results for the urban climate zone extraction in a small subsample of the Xuzhou (People's Republic of China) scene: (a) the urban climate zone map to be compared with (b) a ground truth obtained by visual classification and superimposed on the original data set. Classes are identified by colors, according to the legends displayed in (c).

(a)   (b)

FIGURE 3.2   (See color insert.) Climate zone extraction for the town of Xuzhou, People's Republic of China: (a) final results of the proposed segmentation and classification procedure and (b) detailed ground truth obtained by manually delineating and labeling individual blocks.

In the previous example, the blocks were obtained manually as our focus was on the definition of the decision rules to be applied for block labeling. The same approach was however applied to the whole urban area, after performing an automatic segmentation, and the results are depicted in Figure 3.2a. These results should be compared with the detailed ground truth in Figure 3.2b. Although the color patterns appear visually similar, the overall accuracy at the block level is about 51% if computed regardless of the block size. Overall accuracy at the pixel level instead reaches 63%. The worst discrimination is achieved between the "open set low rise" and "dispersed low rise" classes. This may be due to the fact that the two typologies are very similar considering only two texture scales. Another option considering multiple spatial scales would consist of using differential attribute profiles [16].

One important consideration on these numbers is that the ground truth maps were not obtained by a meteorologist but by a remote sensing specialist using the panchromatic band only. Accordingly, we do not expect that the ground truth would be 100% accurate. In other words, a better validation procedure (including feedback from local experts) may be required.

To further illustrate the aforementioned observations, Figure 3.3 shows a panchromatic and the corresponding pansharpened images of a small portion of the area. The two additional images in this figure correspond to two different visual assessments of the urban climate zones made by two different experts and using the same color legend as in Figure 3.1. It is clear that the panchromatic image does not allow an easy discrimination between the classes, "open set mid rise" and "open set low rise." The color image may provide hints for discrimination, but these are connected to an a priori knowledge of building typologies and, thus not easily generalizable.

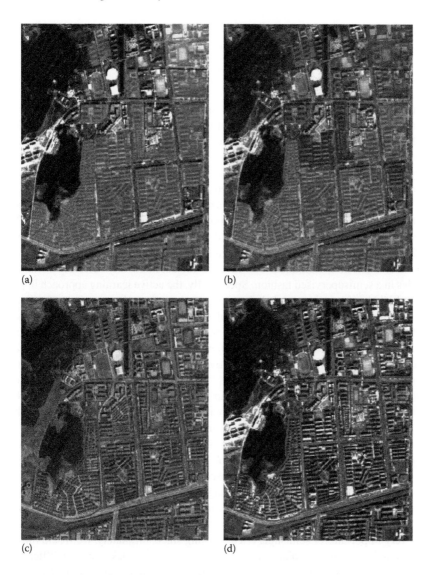

(a)   (b)

(c)   (d)

FIGURE 3.3 **(See color insert.)** Example of problematic assignment of urban blocks to urban climate zones: (a, b) two different visual interpretations by remote sensing experts, to be compared with the block borders superimposed on (c) the panchromatic and (d) the color image of the same area.

3.3 OPEN GLOBAL CHALLENGES

The few examples discussed in the previous section show that there are plenty of challenges still open for research. With the huge amount of EO data sets available and the need to consider existing information, many of these challenges can be grouped under two separate needs (1) to select and (2) to fuse the more relevant bits of information. Accordingly, and following the usual sequence of steps, feature extraction–feature selection–feature fusion (but considering the need for multiple scales and a model-based output), urban remote sensing currently faces very interesting challenges.

3.3.1 FINDING THE RIGHT BIT

Feature extraction and selection is a very important part of EO data exploitation to globally characterize urban areas. As mentioned in [7], a very promising methodology for the selection of urban scene features is the use of active learning approaches, which allow exploiting both the spectral and the spatial information in urban areas, thus enabling determination of the context that better characterizes a specific location.

Following the methodology introduced in [17], for instance, it is possible to develop a novel approach to perform semisupervised classification of urban hyperspectral images by exploiting the information retrieved with spectral unmixing. This is because many pixels in remotely sensed images are "mixed," that is, given by a combination of different substances that reside at the subpixel level. Within this framework, active learning techniques can be used for automatically selecting unlabeled samples in a semisupervised fashion. Specifically, the active learning approach in [17] selects highly informative unlabeled training samples in order to enlarge the initial (possibly very limited) set of labeled samples and perform semisupervised classification based on the information provided by well-established discriminative classifiers.

The proposed approach consists, therefore, of three main ingredients: semisupervised learning, spectral unmixing, and active learning.

1. For the semisupervised part of our approach, the multinomial logistic regression (MLR) classifier [18] provides probabilistic outputs, which play an essential role in our active learning process. Furthermore, a sparsity-inducing prior is added to the regressors to obtain sparse estimates. As a result, most of the components of the regressors are zero. This allows controlling the complexity of the proposed techniques and their generalization capacity. Finally, we use LORSAL algorithm [19] to learn the MLR classifier as it is able to learn the posterior class distributions directly and deal with the high dimensionality of hyperspectral data in a very effective way. This is very important for semisupervised learning since, ultimately, we would like to include as many unlabeled samples as possible, a task which is difficult for normal algorithms from the viewpoint of computational complexity.

2. The unmixing strategies considered in the second step include those attempting to consider spatial information within the extraction procedure. The first one is the fully constrained linear spectral unmixing (FCLSU), which first assumes that labeled samples are made up of spectrally pure constituents (endmembers) and then calculates their abundances and provides a set of fractional abundance maps (one per labeled class). An alternative approach is mixture tuned matched filtering (MTMF), which also assumes that the labeled samples are made up of spectrally pure constituents (endmembers) but then calculates their abundances by means of the MTMF method, which is a hybrid between target detection and unmixing,

thus providing a set of fractional abundance maps (one per labeled class) without the need to know the full set of endmembers in the data.

3. The third ingredient of our proposed method consists of using active learning to improve the selection of unlabeled samples for semisupervised learning. In our proposed strategy, the candidate set for the active learning process (based on the available labeled and unlabeled samples) is inferred using spatial information (specifically, by applying a first-order spatial neighborhood on available samples) so that high confidence can be expected in the class labels of the obtained candidate set. This is similar to human interaction in supervised active learning, where the class labels are known and given by an expert. In a second step, we run active learning to select the most informative samples from the candidate set. This is similar to the machine interaction level in supervised active learning, where in both cases the goal is to find the samples with higher uncertainty. Due to the fact that we use a discriminative classifier (MLR) and spectral unmixing techniques, active learning algorithms, which focus on the boundaries between the classes (which are often dominated by mixed pixels), are preferred. This way, we can combine the properties of the probabilistic MLR classifier and spectral unmixing concepts to find the most suitable (complex) unlabeled samples for improving the classification results through the selected active learning strategy. It should be noted that many active learning techniques are available in the literature [20]. In this work, we use the well-known breaking ties (BT) [21] to evaluate the proposed approach. This algorithm finds the samples minimizing the distance between the first two most probable classes.

Results for the hyperspectral ROSIS Pavia dataset (13 m spatial resolution, 610 × 340 pixels, 103 spectral bands, 9 ground-truth classes [22]) are shown in Figure 3.4. The use of BT alone leads to a mapping result (see Figure 3.4b) with an overall accuracy of 75.5%, definitely larger than the one achievable considering a standard supervised approach (63.6%). The joint use of unmixing information further improves this result, reaching an accuracy value of 79.3% in case FCLSU is used (see Figure 3.4c) while MTMF has a slightly worse performance (79.1%).

3.3.2 Fusing Spaceborne, Airborne, and Ground Data

The process of using EO data to characterize urban areas cannot avoid the fact that, in urban areas, much information will increasingly be collected and stored. Accordingly, the challenge is to include and combine the relevant existing information with the EO extracted features to obtain the multiscale model input mandatory for global models. In doing so, spaceborne remotely sensed data should somehow be "fused" with available spaceborne data, GIS layers, as well as ancillary information collected on the ground (e.g., by means of sensors or sensor networks). As an example of this procedure, we provide here a quick introduction to the approach developed to map exposure within the Global Exposure Database for the Global Earthquake Model (GED4GEM) project [23].

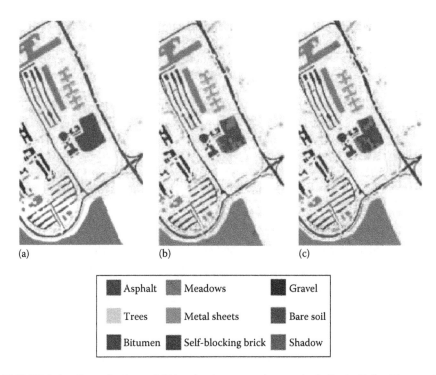

(a)　　　　　　　　　　(b)　　　　　　　　　　(c)

■ Asphalt	■ Meadows	■ Gravel
■ Trees	■ Metal sheets	■ Bare soil
■ Bitumen	■ Self-blocking brick	■ Shadow

FIGURE 3.4 **(See color insert.)** Urban land use mapping results in Pavia, Italy: (a) ground truth for the hyperspectral ROSIS data sets, (b) mapping results using semisupervised classification but no unmixing information, and (c) mapping results jointly considering a semisupervised technique and unmixing information.

Obtaining building exposure is typically a problem related to the combination of information from multiple spatial scales. Indeed, a map of all buildings is currently out of reach using EO data only, either because we do not have a full geographical coverage or because the methods to extract building features (useful for identifying vulnerability) from EO data require inputs from multiple sensors, and this is not feasible for wide areas [9]. One possible solution for a globally suitable and manageable approach is to combine available EO information with ancillary data at different scales.

The idea is graphically represented in Figure 3.5, and includes the use of globally available EO data at coarse-resolution or existing maps to select the built-up areas to focus on, VHR EO data to extract building counts, GIS and census/survey data available from local/international databases to extract dwelling/building fractions according to building typologies, and a good deal of a priori knowledge to logically connect all these features.

The part of the methodology relying on EO data can be implemented according to a procedure for urban spatial pattern recognition. Specifically, the artificial composition of built-up structures and gaps among them (mostly, but not only, roads) is a general and globally valid assumption in all urban areas and usually results in a higher local contrast within these areas than in any natural environments. Accordingly, an option to

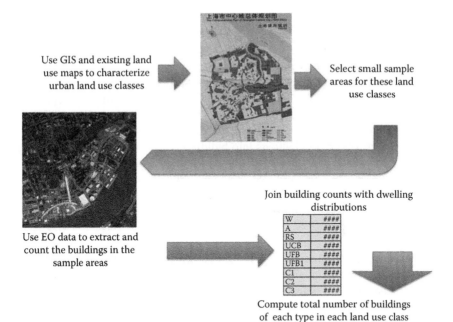

Use GIS and existing land use maps to characterize urban land use classes

Select small sample areas for these land use classes

Use EO data to extract and count the buildings in the sample areas

Join building counts with dwelling distributions

W	####
A	####
RS	####
UCB	####
UFB	####
UFB1	####
C1	####
C2	####
C3	####

Compute total number of buildings of each type in each land use class

FIGURE 3.5 Graphical representation of the proposed procedure to obtain building exposure data and earthquake risk vulnerability data exploiting EO images, ancillary (GIS/map) information, and available ancillary data.

detect urban areas, widely used on EO analysis and already proved reliable at the global level for both optical (panchromatic) and SAR data, is the use of the textural features [13,24,25]. Here, we specifically refer to the *range* textural feature [26], computed starting from the occurrence matrix and defined as the difference between the maximum and the minimum value of the reflectance in a 5 × 5 pixel kernel moving over the image. After range extraction, a few postprocessing steps are performed, as shown in Figure 3.6.

As shown in Figure 3.5, another step in the exposure mapping procedure proposed in GED4GEM is building count extraction from VHR data, either for the whole area or for some samples. This step can be then performed in many

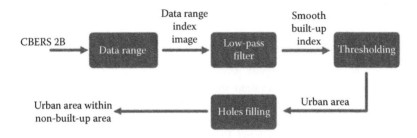

CBERS 2B → Data range → Data range index image → Low-pass filter → Smooth built-up index → Thresholding

Urban area within non-built-up area ← Holes filling ← Urban area

FIGURE 3.6 Data processing chain aimed at urban extent extraction from VHR panchromatic/ SAR data.

different ways, according to the trade-off between accuracy and adaptivity to different geographical areas and building styles. A first option, as suggested in [24], is to classify the multispectral image available for the area of interest as delineated by the previous step and then segment the classification map into significant elements with the areas assigned into building classes. A second (less precise) option is to have a very rough estimate of the building counts by means of the same approach described in Figure 3.6 but this time, looking for the peaks of the *range* index [25]. The choice between these two options depends on the accuracy of the ancillary data (e.g., the dwelling distributions into building typologies) and their statistical significance. For instance, if the dwelling distribution is available at the country level, there is no point in precisely extracting building counts at a spatial resolution of a few meters.

3.4   CONCLUSIONS

This chapter briefly touches the huge range of challenges and opportunities brought by the use of EO data to monitor urban areas all around the world. These challenges are both in the data processing domain as well as in the domain of theoretical information extraction. Specifically, it has been made clear by a few examples that the use of data from satellites may improve our knowledge of urban areas and provide invaluable inputs to models that attempt to capture the interaction between artificial and natural environments. In many situations, these inputs must be at multiple scales, and EO-related information must be combined with data from other sources, according to the final application under study.

Moreover, the need to design algorithms and information extraction procedures that are valid at the global scale includes the necessity to address settlements and artificial elements of the landscape with very different spatial and spectral properties in different parts of the world. Distilling what is common to all these properties is one of the most serious issues to be considered.

Notwithstanding the many open issues highlighted by this work, this chapter includes a few examples of successful application of the multiple scale and multiple feature framework recently proposed in [7] for EO data information extraction in urban areas. Moreover, the efforts currently channeled through the Global Earth Observation working plan for 2011–2015 and, especially, the task related to the design and management of a Global Urban Observatory will help to provide some answers to the burning questions that are still open.

ACKNOWLEDGMENTS

The authors acknowledge support provided by the TOoLs for Open Multi-risk assessment using Earth Observation data (TOLOMEO) project, funded by the European Research Agency under the International Research Staff Exchange Scheme (Contract no. 269115) for some of the research efforts proposed in this chapter. Similarly, the support of the Global Earthquake Model initiative for risk-related analyses is gratefully acknowledged.

REFERENCES

1. E. Bartholome and A.S. Belward, GLC2000: A new approach to global land cover mapping from Earth observation data, *International Journal of Remote Sensing*, 26(9), 1959–1977, 2005.

2. A. Smolka, GEM—The Global Earthquake Model, *EGU General Assembly 2009*, Vienna, Austria, April 19–24, 2009, p. 4104.

3. C. Linard and A.J. Tatem, Large-scale spatial population databases in infectious disease research, *International Journal of Health Geographics*, 11, 7, 2012.

4. P. Gamba and M. Stasolla, Spatial indexes for the extraction of formal and informal human settlements from high resolution SAR images, *IEEE Journal of Selected Topics in Applied Earth Observation and Remote Sensing*, 1(2), 98–106, June 2008.

5. M. Negri, P. Gamba, G. Lisini, and F. Tupin, Junction-aware extraction and regularization of urban road networks in high resolution SAR images, *IEEE Transactions on Geoscience and Remote Sensing*, 44(10), 2962–2971, October 2006.

6. Q.-Y. Zhou and U. Neumann, Complete residential urban area reconstruction from dense aerial LiDAR point clouds, *Graphical Models*, 75(3), 118–125, available online October 11, 2012. doi:10.1016/j.gmod.2012.09.001.

7. P. Gamba, Human settlements: A global challenge for EO data processing and interpretation, *Proceedings of IEEE*, 101(3), 570–581, March 2013.

8. Q. Weng, H. Liu, B. Liang, and D. Lu, The spatial variations of urban land surface temperatures: Pertinent factors, zoning effect, and seasonal variability, *IEEE Journal of Selected Topics in Applied Earth Observations and Remote Sensing*, 1(2), 154–166, 2008.

9. D. Polli, F. Dell'Acqua, and P. Gamba, First steps towards a framework for earth observation (EO)-based seismic vulnerability evaluation, *Environmental Semeiotics*, 2(1), 16–20, 2009.

10. I. Stewart and T. Oke, Classifying urban climate field sites by 'local climate zones': The case of Nagano, Japan, *Proceedings of the Seventh International Conference on Urban Climate*, Yokohama, Japan, 2009. Available online at http://www.ide.titech.ac.jp/~icuc7/extended_abstracts/pdf/385055-1-090515165722-002.pdf (last accessed in November 2013).

11. I. Stewart, Newly developed 'thermal climate zones' for defining and measuring urban heat island magnitude in the canopy layer, *Proceedings of the Eighth Symposium on Urban Environment*, Phoenix, AZ, 2009. Available online at http://ams.confex.com/ams/pdfpapers/150476.pdf (last accessed in November 2013).

12. I.C. Nduka and A.I. Abdulhamed, Classifying urban climate field sites by 'thermal climate zones' the case of Onitsha metropolis, *Research Journal of Environmental and Earth Sciences*, 3(2), 75–80, 2011.

13. M. Pesaresi, A. Gerhardinger, and F. Kayitakire, A robust built-up area presence index by anisotropic rotation-invariant textural measure, *IEEE Journal of Selected Topics in Earth Observations and Applied Remote Sensing*, 1(3), 180–192, 2008.

14. P.R. Marpu, I. Niemeyer, and R. Gloaguen, Unsupervised image segmentation by identifying natural clusters, *Proceedings of the IEEE International Geoscience and Remote Sensing Symposium, IGARSS'07*, Barcelona, Spain, July 23–27, 2007, pp. 1903–1904.

15. G. Lisini, F. Dell'Acqua, P. Gamba, and W. Thompkinson, Image interpretation through problem segmentation for very high resolution data, *Proceedings of the IEEE International Geoscience and Remote Sensing Symposium, IGARSS'05*, Seoul, South Korea, July 2005, pp. 5634–5637.

16. M. Dalla Mura, J.A. Benediktsson, B. Waske, and L. Bruzzone, Morphological attribute profiles for the analysis of very high resolution images, *IEEE Transactions on Geoscience and Remote Sensing*, 48(10), 3747–3762, 2010.

17. A. Plaza, I. Dopido, J. Li, and P. Gamba, Semi-supervised classification of hyperspectral data using spectral unmixing concepts, *Proceedings of the Tyrrhenian Workshop 2012 on Advances in Radar and Remote Sensing*, Naples, Italy, September 2012, unformatted CD-ROM.

18. D. Boehning, Multinomial logistic regression algorithm, *Annals of the Institute of Statistical Mathematics*, 44, 197–200, 1992.

19. J. Bioucas-Dias and M. Figueiredo, Logistic regression via variable splitting and augmented Lagrangian tools, Instituto Superior Tecnico, TULisbon, Portugal, Technical Report, 2009.

20. D. Tuia and G. Camps-Valls, Urban image classification with semisupervised multiscale cluster kernels, *IEEE Journal of Selected Topics in Applied Earth Observations and Remote Sensing*, 4(1), 65–74, March 2011.

21. T. Luo, K. Kramer, D.B. Goldgof, S. Samson, A. Remsen, T. Hopkins, and D. Cohn, Active learning to recognize multiple types of plankton, *Journal of Machine Learning Research*, 6, 589–613, 2005.

22. P. Gamba, A collection of data for urban area characterization, *Proceedings of the IEEE International Geoscience and Remote Sensing Symposium, IGARSS'04*, Anchorage, AK, September 2004, Vol. I, pp. 69–72.

23. P. Gamba, K. Jaiswal, D. Cavalca, C. Huyck, and H. Crowley, The GED4GEM project: Development of a global exposure database for the Global Earthquake Model initiative, *Proceedings of the 15th World Conference on Earthquake Engineering*, Lisbon, Portugal, September 2012, unformatted CD-ROM.

24. T. Esch, H. Taubenböck, A. Roth, W. Heldens, A. Felbier, M. Thiel, M. Schmidt, A. Müller, and S. Dech, TanDEM-X mission-new perspectives for the inventory and monitoring of global settlement patterns, *Journal of Applied Remote Sensing*, 6(1), 1–21, 2012.

25. P. Gamba, M. Aldrighi, M. Stasolla, and E. Sirtori, A detailed comparison between two fast approaches to urban extent extraction in VHR SAR images, *Proceedings of JURSE 2009*, Shanghai, China, May 22–24, 2009, unformatted CD-ROM.

26. F. Dell'Acqua, P. Gamba, and G. Lisini, Rapid mapping of high resolution SAR scenes, *ISPRS Journal of Photogrammetry and Remote Sensing*, 64(5), 482–489, September 2009.

4 Urban Observing Sensors

Qihao Weng, Paolo Gamba, Giorgos Mountrakis,
Martino Pesaresi, Linlin Lu, Thomas Kemper,
Johannes Heinzel, George Xian, Huiran Jin,
Hiroyuki Miyazaki, Bing Xu, Salman Quresh,
Iphigenia Keramitsoglou, Yifang Ban, Thomas Esch,
Achim Roth, and Christopher D. Elvidge

CONTENTS

4.1 INTRODUCTION

Urban land cover (ULC) has a considerable impact on local, regional, and global environmental change, and has significant ecological, biophysical, social, and climatic effects (Seto and Shepherd, 2009; DeFries et al., 2010). These effects are further amplified by the temporal duration of urban changes that tend to last for decades and are often irreversible. Optical sensors on board various satellite platforms play a significant role in urban monitoring and assessment. Two representative examples are indicative of the importance of optical sensors. First, since 2009 after USGS made the Landsat archive freely available, a 60-fold increase was observed in data downloads (NASA, 2013). Second, in the last decade, there has been a strong interest from the commercial sector to launch satellite optical sensors. This interest is clearly driven by the constantly increasing demand for such products from governmental, military, nonprofit, and commercial sectors.

Remote sensing thermal infrared (TIR) data have been widely used to retrieve land surface temperature (LST) (Quattrochi and Luvall, 1999; Weng et al., 2004). A series of satellite and airborne sensors, such as HCMM, Landsat TM/ETM+, AVHRR, ASTER, TIMS, have been developed to collect TIR data from the Earth's surface. In addition to LST measurements, these TIR sensors may also be utilized to obtain emissivity data of different surfaces with varied resolutions and accuracies. LST and emissivity data have been used in urban climate and environmental studies, mainly for analyzing LST patterns and their relationship with surface characteristics, assessing urban heat island (UHI), and relating LSTs with surface energy fluxes for characterizing landscape properties, patterns, and processes (Quattrochi and Luvall, 1999). Remotely sensed TIR data are a unique source of information to define surface heat islands, which are related to canopy layer heat islands. In situ data (in particular, permanent meteorological station data) offer high temporal resolution and long-term coverage but lack spatial details. Moving observations overcome this limitation to some extent but do not provide a synchronized view over a city. Only remotely sensed TIR data can provide a continuous and simultaneous view of a whole city, which is of prime importance for detailed investigation of urban surface temperature. Generally speaking, the application of TIR data has been limited in urban surface energy modeling (Voogt and Oke, 2003). Previous works have focused on the methods for estimating variables related to energy driving forces, soil moisture availability, and vegetation–soil interaction from satellite remote sensing data, but little has been done to estimate surface atmospheric parameters (Schmugge et al., 1998). These parameters are measured in the traditional way in the network of meteorological stations or in situ field measurements.

Traditional urban remote sensing studies did not make use of synthetic aperture radar (SAR) data, mainly because of issues in their interpretation. SAR sensors are active imaging systems that use runtime length and intensity of a transmitted microwave pulse for generating a consistent image. The appearance of objects and surfaces in radar images is dominated by geometric properties (imaging and object geometry, surface roughness) rather than by their chemical or biophysical characteristics (as in the case of optical data). In the geometrically highly structured urban landscape, the complex interaction of the radar pulse and the small-scale urban features leads to certain ambiguities in the received signal. Hence, especially for very high resolution (VHR) SAR, urban imagery lacks clarity. For example, there are distortions and shadow regions, which limit the capability for certain applications such as the exact delineation of buildings or other urban infrastructural elements. Moreover, the appearance of identical urban spots or objects might differ significantly depending on the imaging geometry of the data acquisition. Nevertheless, the basic phenomena affecting backscattering from man-made structures have been extensively discussed (Guida et al., 2008), with a focus not only on clearly defining mapping limitations but also on discovering very important applications, such as differential interferometry and persistent scatterers (Ferretti et al., 2001). The three-dimensional (3D) capabilities of SAR systems have also been analyzed with respect to urban areas to quantify flood risk, and their all-weather data availability is invaluable in managing catastrophic events, both at local and global scales. Therefore, notwithstanding the issues highlighted earlier, after the seminal paper by Henderson and Xia in 2001, summarizing the relatively few achievements at that point, urban remote sensing using SAR has been flourishing, with applications from urban extent extraction (Gamba et al., 2011) to detailed ULC mapping (Hu and Ban, 2012), urban change detection (Bovolo and Bruzzone, 2005), and 3D building characterization (Soergel et al., 2009), including road network extraction (Hedman et al., 2010) and damage detection at both the block (Dell'Acqua et al., 2011) and the building levels (Bovolo et al., 2012).

4.2 OPTICAL SENSORS

4.2.1 COARSE SPATIAL RESOLUTION OPTICAL SENSORS

Regional, continental, and global changes in urban land cover/use have been monitored using optical data with coarse spatial resolution (>100 m), such as NOAA advanced very high resolution radiometer (AVHRR) and terra moderate resolution imaging spectroradiometer (MODIS). The AVHRR sensor was first launched by the satellite TIROS-N in November 1978 and then by the NOAA series, which started with NOAA-6 in June 1979 and continued with NOAA-7 through NOAA-19 between 1981 and 2009 (NOAA, 2013). All satellite series launched before 2001 have ended their missions while NOAA-15 through -19 are still in operation. The MODIS sensor on board the Terra satellite was launched in December 1999 as part of NASA's Earth Observing System. It still acquires images although the life expectancy of Terra was designed for 6 years. AVHRR has five spectral bands targeting the wavelengths of red, NIR, and three TIR bands, with primary use of cloud, snow, ice, vegetation, cloud and surface temperature mapping, and land/water interface and hot target monitoring. MODIS has 36 spectral bands ranging from wavelengths of visible to shortwave and TIR, with primary use of land, cloud, vegetation, sediment, cloud and surface temperature mapping, chlorophyll, atmospheric properties, cloud fraction and height derivation, and forest fire and volcano monitoring. AVHRR acquires images of the entire Earth twice a day with a spatial resolution of approximately 1.1 km at the satellite nadir, while MODIS covers the entire surface of the Earth every 1–2 days with a spatial resolution of 250 m (bands 1–2), 500 m (bands 3–7), and 1 km (bands 8–36). Radiometric resolution is 10 bits for the AVHRR data and 12 bits for the MODIS data. The NOAA series are sun-synchronous, polar-orbiting satellites at 830–870 km above Earth, having 2500 km in swath width, while MODIS on board sun-synchronous, near-polar orbiting satellite acquires images at an altitude of 705 km at 10:30 a.m. local time in descending node (Terra) or 1:30 p.m. in ascending node (Aqua), and a swath width of 2330 km (NASA, 2013; NOAA, 2013).

It has been a challenge to apply the coarse-resolution remotely sensed data for urban observation and monitoring due to its limited spatial resolution (Schneider et al., 2003), yet it has proven useful for climatic studies due to its high temporal frequency and large spatial coverage (Gallo et al., 1993; Jonsson, 2004; Stathopoulou and Cartalis, 2009). Therefore, the effectiveness of urban studies has been dependent upon the fusion of AVHRR data with either finer spatial resolution images such as Landsat TM (Stathopoulou et al., 2004) or continuously observed meteorological (ground) data (Bengang and Shu, 2000; Ji and Peters, 2004; Stathopoulou et al., 2006). A major advancement for using coarse-resolution data in urban observation was initiated by the launch of NASA's Terra platform and specifically MODIS. The improved spectral resolution allows monitoring ecosystem processes across multiple scales (Stefanov and Netzband, 2005) and has resulted in a range of applications, including urban land use/land cover changes (Netzband and Stefanov, 2004; Clark et al., 2012), heat island studies (Schwarz et al., 2011), and vegetation phenology (Zhang et al., 2003). Another main application area has been in air quality monitoring to assess aerosol optical depth over urban areas (Hutchison et al., 2005) and to investigate particulate matter in aerosols during transboundary events, and again in combination with ground-based data (Engel-Cox et al., 2004; Alam et al., 2010, 2011).

Data fusion techniques can be applied to fusing coarse spatial resolution imagery with either finer spatial or spectral resolution remotely sensed data or both depending on the study objective. Alternative finer spatial resolution data are not limited to Landsat (Xu et al., 2006; Michishita et al., 2012a), ASTER (Xu et al., 2004), IKONOS (Xu et al., 2003), or aerial photographs (Wu et al., 2006). Finer spectral resolution data such as EO-1 Hyperion imagery with more than 200 bands (Xu and Gong, 2008) can also be combined with the coarse-resolution data in similar ways using spectral fusion models (Xu and Gong, 2007). Data fusion algorithms usually work on surface reflectance and NDVI that requires a smoothing procedure before the MODIS NDVI time series can be reconstructed for various applications (Jin and Xu, 2013). MODIS data that have been used to perform classification for urban areas resulted in confusion between urban and barren areas (Schneider et al., 2003). Few studies have attempted to develop algorithms that can optimize the potential of MODIS for urban area mapping (Schneider et al., 2009). Most of the studies that use MODIS data necessarily consider a combination with other image data or ancillary information. For example, Langer et al. (2007) fused MODIS and Landsat to monitor land cover changes, whereas Kasimu and Tateishi (2008) combined MODIS with population statistics and meteorological data for urban area mapping across the globe. It is concluded that although coarse-resolution data is advantageous because of frequent temporal acquisition and large spatial coverage for rapid large-scale observation and monitoring of urban areas, its ability to produce accurate information independently is limited. Therefore, these data sets produce robust results if combined with other data sources using appropriate fusion algorithms (Michishita et al., 2012b,c).

4.2.2 Medium Spatial Resolution Optical Sensors

There is strong demand for historical and current ULC information over large geographic areas. The Landsat Thematic Mapper (TM) and Enhanced Thematic Mapper Plus (ETM+) sensors have spatial and spectral characteristics that are well suited for characterizing terrestrial ecosystem features, including the highly heterogeneous features of ULC. Landsat TM and ETM+ sensors that have spatial resolution of 30 m for visible, near-IR, and shortwave infrared (SWIR) bands provide consistent and repetitive observations that are suitable for monitoring dynamics of ULC. In addition, TM and ETM+ data have been systematically acquired for large portions of the globe since the launch of Landsat 5 in 1984, and thus a rich archive is available for analysis. These data sets have been widely used to monitor ULC change at local and regional scales (Small, 2003; Maktav et al., 2005; Potere et al., 2009; Weng and Lu, 2009).

Urban extents and structures cannot be clearly determined by using discrete classification methods along with medium-resolution remote sensing data, in part because of highly heterogeneous features of ULC. Most urban areas, especially in single-house development areas, exhibit subpixel characteristics that mix impervious surface with other land covers (e.g., grass) in medium-resolution satellite imagery (Lu and Weng, 2004). However, by treating the urban landscape as a continuum such as percent impervious surface (PIS) while using modeling techniques to extract urban characteristics, the continuous field estimate of PIS derived from satellite data can serve as a surrogate to determine urban extent and infrastructure and to assess changes in the urban environment (Powell et al., 2007; Xian et al., 2008). The USGS National Land Cover Database (NLCD) has produced land cover and impervious surface products by using Landsat as the primary data source, and the PIS product has been used to assess the extent of urban development and associated ecological effects in the conterminous United States (Imhoff, 2010; Xian and Homer, 2010; Xian et al., 2011, 2012). Figure 4.1 shows the distribution of impervious surface in the conterminous United States in 2006. The figure further provides details of impervious surface change from 2001 to 2006 in two metropolitan areas: Los Angeles, California, and Atlanta, Georgia. The spatial patterns and new growths of ULC between the two times are displayed. The Landsat data continuity mission (LDCM) that was launched on February 11, 2013, will ensure the continued acquisition of Landsat-like data. LDCM will continue to provide valuable medium-resolution data and imagery that will be consistent with current standard Landsat data products.

Like Landsat, the SPOT (Système Pour l'Observation de la Terre) program initiated by the French government in the 1970s has been designed to provide long-term data continuity with successive improvements in sensor performance. SPOT-5 is a current popular choice for medium-resolution sensors. Launched on May 3, 2002, in addition to other sensors, SPOT-5 carries two high-resolution geometric (HRG) instruments with increased spatial resolution (compared to its predecessors) of 2.5 or 5 m in the panchromatic (0.48–0.71 µm); 10 m in the green (0.50–0.59 µm), red (0.61–0.68 µm), and near-IR (0.78–0.89 µm); and 20 m in the mid-IR (1.58–1.75 µm) bands. Images have an 8-bit radiometric resolution. The satellite flies in a sun-synchronous orbit with an altitude of 822 km, an inclination of 98.7°, and a 26-day repeat cycle. SPOT-5, due to its increased spatial resolution compared to Landsat (especially in the panchromatic band), has been extensively applied for ULC classification (e.g., Zhang et al., 2003) and building/settlement extraction in urban sprawl areas (e.g., Durieux et al., 2008; Rhinane et al., 2011). Meanwhile, SPOT-5 multispectral data at 10 m resolution were also employed in studies of suburban mapping and urban land use change detection (e.g., Deng et al., 2009; Yang and Wang, 2012). The SPOT-5 imagery was also used for urban road mapping (Couloigner et al., 1998). The latest addition to the SPOT family is SPOT-6, which was launched on September 9, 2012. SPOT-6 has an increased potential for urban-related applications due to the even higher spatial resolution.

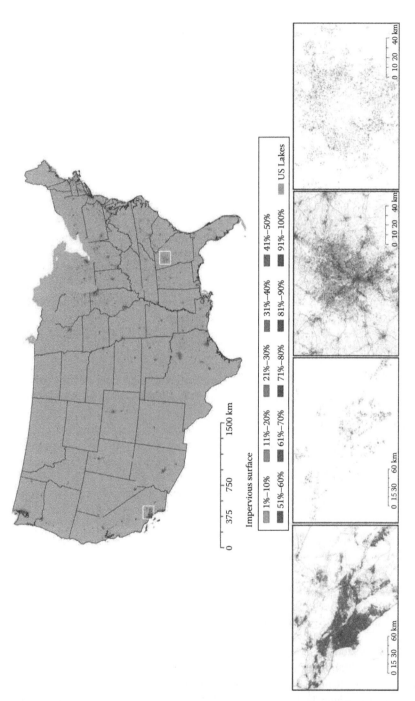

FIGURE 4.1 **(See color insert.)** Impervious surface over the conterminous United States in 2006. The lower panels from left to right are the maps of 2001 impervious surface and 2006 new impervious surface in Los Angeles, California, and those in Atlanta, Georgia.

Another commonly used medium-resolution satellite sensor for urban studies is the advanced spaceborne thermal emission and reflection (ASTER) radiometer, which is being flown on the Terra platform since December 1999. ASTER consists of three instrument subsystems: visible and near-infrared (VNIR) with three spectral bands and stereoscopic band of 15 m resolution, SWIR with six spectral bands of 30 m resolution, and TIR with five spectral bands of 90 m resolution. Ground track repeat cycle is 16 days though observations are operated on demand. The 15 m spatial resolution of VNIR sensor makes ASTER data valuable in extracting urban objects (Small, 2005) and mapping impervious surface (Weng and Hu, 2008; Orenstein et al., 2011) and ULC (Zhu and Blumberg, 2002; Lu and Weng, 2006). The SWIR detectors are not functioning since April 2008 due to anomaly of SWIR detector temperatures.

Like other optical sensing systems, Landsat, SPOT, and ASTER have their limitations. They are highly restricted by weather conditions, such as clouds, haze, snow, and ice covers. Some approaches have been introduced to reduce these limitations. For example, the synergistic use of SPOT-5 multispectral imagery and SAR remote sensing data has been proposed to map impervious surfaces at the subpixel level, and notable improvements were achieved in comparison to using SPOT imagery exclusively (Leinenkugel et al., 2011). Similarly, the combination of ASTER data with other data sources can be a key technique for extending applications of ASTER data to urban areas around the world (e.g., Miyazaki et al., in press), where a large city is usually not captured within a single ASTER scene. Contextual analysis has also been demonstrated to be useful to enhance the classification process using medium-resolution satellite data (Luo and Mountrakis, 2010).

Another issue significantly affecting sensor popularity is data availability and cost. In ASTER's case, the major limitation is the on-demand observation schedule, which limits spatial coverage. In SPOT's case, the issue of data cost is prominent as the French government has not yet matched the free-of-charge policy for Landsat scenes. Decisions by the French government on data distribution policies will significantly affect the future popularity of SPOT sensors.

4.2.3   HIGH SPATIAL RESOLUTION OPTICAL SENSORS

A wide range of high resolution (HR) and VHR spatial sensors are available from governmental and commercial consortiums. Figure 4.2 lists known satellite platforms and sensors collecting optical (passive) image data with a spatial resolution equal to or finer than 10 m. The list includes more than 50 different platforms and sensors active for 2013 or planned for 2014. For spatial resolutions of 1 m or higher, only panchromatic sensors are currently available. Multispectral data are available only at 2 m pixel size or larger. Multispectral sensors with improved spatial resolution are planned for 2014 with GeoEye-2 and WorldView-3 satellites.

Due to their fine spatial resolution, HR/VHR input image data have been used for recognition and characterization of all basic components of human settlements, such as built-up structures or buildings (Shettigara et al., 1995; Lin and Nevatia, 1998;

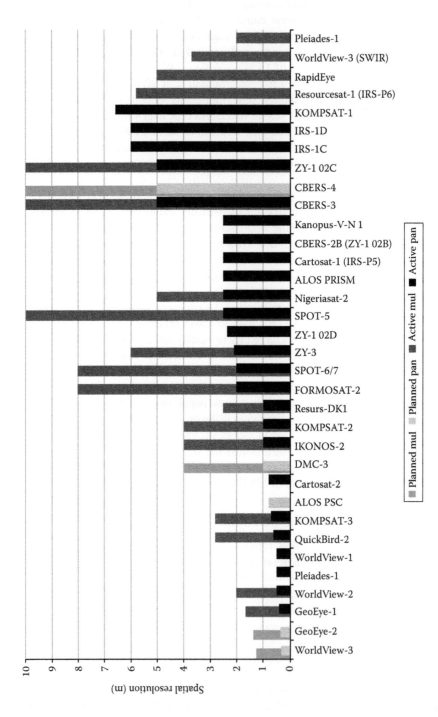

FIGURE 4.2 (See color insert.) Pixel size (meters) of current and planned high spatial resolution satellite sensors.

Benediktsson et al., 2003; Unsalan and Boyer, 2004; Khoshelham et al., 2010; Sirmacek and Unsalan, 2011), roads (Zhu et al., 2005; Chaudhuri et al., 2012), and open spaces, including city squares, public and private gardens and parks, walking areas, parking lots, and the like. In particular, urban open spaces have mostly been addressed by analyzing urban vegetated areas (Nichol and Lee, 2005; Nichol and Wong, 2007), including by detection of individual tree crowns in urban areas (Ouyang et al., 2011; Ardila et al., 2012). At the maximum, HR/VHR resolution imagery allows for civilian usage, even for detection of targets having a smaller dimension than the standard settlement components. Such examples of urban analysis may include detection of cars and other vehicles (Gerhardinger et al., 2005; Leitloff et al., 2010), including the analysis of their direction and velocity (Pesaresi et al., 2008), and monitoring of human crowds in open spaces (Sirmacek and Reinartz, 2011; Schmidt and Hinz, 2011). Furthermore, VHR image data have been critical in the detection and monitoring of built-up structures that may be functional for disaster and crisis management operations. In particular, detection of damages in urban areas has taken place in earthquake and tsunami postdisaster damage assessment (Pesaresi et al., 2007; Chesnel et al., 2008; Ouzounis et al., 2011; Lu et al., 2012; Parape et al. 2012) and in postconflict damage and reconstruction assessment (Pagot and Pesaresi, 2008; Gueguen et al., 2009). Finally, the use of VHR image data has been demonstrated for the monitoring and analysis of informal and temporary settlements, which are usually not included in the standard land use/land cover classification schemes. In particular, slum and poor urban areas (Kit et al., 2012; Kohli et al., 2012) and refugee and internally displaced people (IDP) camps (Giada et al., 2003; Jenerowicz et al., 2011; Pesaresi and Gerhardinger, 2011) are special cases of temporary human settlements relevant in crisis management operations.

From the methodological point of view, three general image-processing approaches have been used to process satellite HR/VHR input data for the analysis of human settlements: (1) 2D monocular image-derived features and classification, (2) 3D processing of stereo pairs and derived features, and (3) multisource information fusion. Each approach has its own advantages and disadvantages. The "information fusion" approach has the key advantage of the possibility of combining the best results of 2D and 3D processing approaches and usually to increase accuracy and effectiveness (Baltsavias et al., 1995; Haala and Hahan, 1995). Moreover, the inclusion of external data sources such as digital cartography, cadastral data, socioeconomic surveys, and even social media can improve the automatic information extraction process. Despite the existence of various approaches, it is worth noting a methodological constant, that is, the increased importance of structural (texture, shape) and contextual (spatial relations) image descriptors in the inferential models for processing HR/VHR image data, when compared with the models using moderate-resolution image data. This is due to the fact that as spatial resolution improves, capturing sufficient energy to register an acceptable signal-to-noise ratio becomes more challenging, leading to limited spectral separability of urban targets, especially with shorter wave bands. This decrease of spectral separability encourages the inclusion of structural and contextual image descriptors in the image information extraction models. The importance of structural and contextual HR/VHR image analysis is also amplified by the fact that urban classification often includes more or less explicitly spatial and contextual criteria in order to discriminate the relevant urban information, such as typically local densities of specific features, land cover heterogeneity measures, spatial pattern characteristics, and the sizes of built-up structures.

The key limitations concerning satellite VHR image data exploitation involve the following: commercial and confidentiality issues, high data volume, intrinsic spatial inconsistency, and limited spectral, temporal, and multitemporal archives. VHR image data are intrinsically spatially inconsistent: even accurate processing of stereo pairs cannot reach subpixel RMSE positional error, assuming a pixel size of 0.5 m. Because of the capacity to collect off-nadir image data from VHR platforms, the apparent displacement of image pixels increases further due to panoramic and parallax distortions. Unfortunately, these effects are more evident in above-ground urban targets as in the case of rooftops of buildings that are some of the key entities collected in remote sensing urban studies. In practice, these facts lead to an expected apparent displacement of the rooftops in the order of several tenths of pixels, assuming 0.5 m spatial resolution, tall buildings, and usual off-nadir data collection ranges. This fact has direct bearing in increasing the complexity of reference data collection and in decreasing the expected accuracy and repeatability of the image information retrieval tasks, especially in the frame of monitoring activities. In general, VHR multispectral sensors collect less number of bands than low- or medium-resolution sensors. This has a direct impact when applying inferential models based on spectral reflectance criteria. Moreover, image data input with spatial details of 1 m or more are available only in the panchromatic mode, which in VHR platforms is usually by summing VNIR bands. This has a direct impact when applying multispectral analysis to meter- and submeter-resolution input data. The majority of the available VHR platforms declare a nominal revisiting time in the range of 1–5 days. In some areas, because of the high probability of cloud cover, this may lead to several weeks (or even months) of unavailability of VHR data. VHR platforms are tasked only for specific commercial/governmental requests. Therefore, except for some places, usually no consistent multitemporal archived data is available for a specific area of interest, leading to a radical decrease in the multitemporal analysis capacity using VHR data.

Future perspectives include the increase of available spatial resolution in both panchromatic and multispectral sensors in the WorldView and GeoEye platforms. They reach 0.3 and 1.3 m, respectively, in the pan and multispectral modes. It is still unclear how and under which constraints these new data will be available for scientific use and for the general public. International commercial and security issues are directly proportional to the advances of the sensor and platform technologies, including the increasing spatial and spectral resolution and the increasing absolute pointing accuracy of the platforms. The de facto standard set by the US government limiting the pixel size of satellite sensors to 0.5 m for nonmilitary applications could be potentially revised in order to make the new image data available. The list of entities or users having access to VHR image data may also change accordingly. As a general trend, we can observe that legal and licensing barriers in both the input data and image-derived information products are becoming more influential as technology advances.

4.3 TIR SENSORS

4.3.1 Coarse-Resolution TIR Sensors

Several meteorological satellite missions have on board coarse spatial resolution TIR sensors and have by now acquired a considerable global archive of LST images over the last 40 years. According to their orbit, these are divided into two distinct groups, namely, geostationary missions (e.g., MSG-SEVIRI viewing Europe and Africa, GOES over America, Kalpana over India, Fengyun viewing China, and MTSAT observing East Asia) and low Earth orbiters like NOAA and Metop AVHRR. In the latter group, we may include Terra and Aqua MODIS, due to the similar TIR band characteristics and products, although Terra and Aqua are not weather satellites. These missions have been providing continuous monitoring of LST distribution at the spatial resolution ranging from 3 to 5 km for geostationary platforms to 1.1 km at nadir for low Earth orbiters. In most cases, service providers (e.g., NASA, ESA, EUMETSAT) distribute LST images as standard data products. The coarse spatial resolution of geostationary TIR imagery has prohibited their extensive use for urban studies; yet recently, scientific interest in these sensors has been revived as computational methods for sharpening these imagery to 1 km (Zakšek and Oštir, 2012; Keramitsoglou et al., 2013) or better (Bechtel et al., 2012) have become available. A clear advantage of coarse-resolution sensors is their temporal resolution. The temporal measurement frequency of polar orbiting satellite systems at ~850 km is approximately two times per day, yet ordinarily a few acquisitions are available daily from similar sensors on board different platforms (see Table 4.1). The geostationary orbit TIR sensors provide images of the Earth's disk from 36,000 km every 15–30 min, making them a unique means for capturing the diurnal variability of surface UHIs. The specific details of coarse-resolution TIR sensors are presented in Table 4.1.

LST from multispectral TIR imagery can be retrieved (Schmugge et al., 1998) either using a radiative transfer equation to correct the at-sensor radiance to surface radiance or by applying the split-window technique for sea surfaces to land surfaces, assuming that the emissivity in the channels used for the split window is similar (Dash et al., 2002). Land surface brightness temperatures are then calculated as a linear combination of the two channels. Jiménez-Muñoz and Sobrino (2008) provide a complete set of split-window coefficients that can be used to retrieve LST from TIR sensors on board the most popular coarse-resolution remote sensing satellites. Past studies of SUHI have been conducted primarily by using AVHRR or MODIS data (Kidder and Wu, 1987; Balling and Brazell, 1988; Roth et al. 1989; Gallo et al., 1993; Stathopoulou et al., 2004; Hung et al., 2006; Peng et al., 2012). Keramitsoglou et al. (2012) concluded that the spatial resolution of 1 km offered now by MODIS and AVHRR, and until April 2012 also by AATSR, is adequate for large-area urban temperature mapping and for observing the differences between daytime and nighttime patterns, although not acquired at the best overpass time to observe SUHI (Sobrino et al., 2011). Streutker (2002, 2003) used AVHRR data to quantify the SUHI of Houston, Texas, assuming an ellipsoid footprint to derive the SUHI parameters of intensity, spatial extent, orientation, and central location. Hung et al. (2006) adopted this method to measure the spatial extents and magnitudes of the

TABLE 4.1

Coarse-Resolution TIR Sensors Currently Operational (October 2012)

Instrument Short Name	Instrument Name Full	Instrument Agencies	Missions	Orbit	Spatial Resolution	Swath Width	Temporal Resolution	TIR Bands
AVHRR/3	Advanced very high resolution radiometer/3	NOAA	NOAA-15–19, Metop-B, Metop-A	Sun-synchronous at 705 km, inclination 98.6°–98.8°	1.1 km at nadir	~3000 km	Depending on the number of operational NOAA and Metop platforms; two images per platform	TIR: 10.3–11.3 μm, 11.5–12.5 μm
MODIS	Moderate-resolution imaging spectroradiometer	NASA	Terra, Aqua	Sun-synchronous at ~850 km, inclination 98.2°	1 km at nadir	2330 km	Four images daily, two daytime and two nighttime	VIS-TIR: 36 bands in the range 0.4–14.4 μm
SEVIRI	Spinning-enhanced visible and infrared imager	EUMETSAT (ESA)	Meteosat second generation	Geostationary at 36,000 km viewing Europe and Africa	3–5 km	Full Earth disk	15 min	IR9.7 = 9.52–9.8 μm, IR10.8 = 10.3–11.3 μm, IR12.0 = 11.5–12.5 μm, IR13.4 = 12.9–13.9 μm
Imager	Imager	NOAA	GOES-12, GOES-14, GOES-15, GOES-13	Geostationary at 36,000 km viewing America	10 km	Full Earth disk	15 min	IR: 4 channels: 3.9 μm, 6.7 μm, 10.7 μm, and 13.3 μm
IVISSR (FY-2)	Improved multispectral visible and infrared scan radiometer (five channels)	NRSCC (NSMC-CMA, CNSA, CAST)	Fengyun-2	Geostationary at 36,000 km viewing China	5 km	Full Earth disk	15 min	VIS-TIR: 0.5–12.5 μm (five channels)

Source: Data compiled from the *CEOS Handbook.* http://database.eohandbook.com/ (last accessed January 19, 2014).

SUHIs for eight megacities in Asia using both daytime and nighttime MODIS data acquired over the period 2001–2003. Rajasekar and Weng (2009a) applied a non-parametric model by using fast Fourier transformation (FFT) to MODIS imagery for characterization of the SUHI over space, so as to derive SUHI magnitude and other parameters. Keramitsoglou et al. (2011) applied an object-based image analysis procedure to extract urban thermal patterns to more than 3000 MODIS images acquired from May until September from the years 2000 to 2009 for the Greater Athens Area, Greece, revealing the qualitative and quantitative characteristics of Athens' SUHI retaining the original LST values, thus circumventing modeling.

Regarding the near future of sensors and satellite platforms, a number of relevant projects are under way. The European Space Agency (ESA) Sentinel-3 satellite is planned for launch from 2014, offering a sea and land surface temperature radiometer (SLSTR) with a 1 km resolution in the thermal channels and a daily revisit time. The geostationary GOES-R satellite is due in 2015, with a 2 km resolution in the thermal channels from a new advanced baseline imager (ABI). The National Polar-orbiting Operational Environmental Satellite System (NPOESS) is due for launch in 2016, designed to replace NASA's Aqua, Terra, and Aura satellites and offering the visible and infrared imagery radiometer suite (VIIRS) sensor for LST. Coupled with these large "traditional" missions, in the future there is likely to be an increase in "small satellites" (Sandau et al., 2010) that enable relatively quick and inexpensive missions, which could, for example, help to observe dynamic surface temperature patterns.

4.3.2 Medium-Resolution TIR Sensors

Currently, only a few spaceborne sensors with global imaging capacity can deliver medium-resolution TIR data required to address urban LST heterogeneity and to assess the UHI effect (Weng, 2009). The TM sensor on board Landsat 5 has been acquiring images of the Earth nearly continuously from July 16, 1982, to the present, with a single TIR band of 120 m resolution, and is thus long overdue. Figure 4.3 shows the mean annual surface temperature based on the ATC (annual temperature cycle)-modeled LST values of all available 115 Landsat-5 TM scenes (less than 30% cloud cover) between 2000 and 2010 in Los Angeles. Another TIR sensor that has global imaging capacity is with Landsat 7 ETM+ since April 15, 1999. The ETM+ provides an enhanced TIR band of 60 m resolution. Unfortunately, the scan-line-corrector on board Landsat 7 started malfunctioning after May 31, 2003, which caused a loss of approximately 25% of the data, mostly located between scan lines toward the scene edges. Although some gap-filling remedy methods can recover some of the data lost, the gap-filled data cannot match the quality of the original data. In addition, ASTER sensor flown on the Terra satellite collects five TIR bands with a ground resolution of 90 m. These multispectral infrared measurements can be converted into LST and emissivity products by using the ASTER temperature/emissivity separation algorithm (Gillespie et al., 1998). LST values calculated using this algorithm are expected to have an absolute accuracy of 1–4 K and relative accuracy of 0.3 K, and surface emissivity values an absolute accuracy of 0.05–0.1 and relative accuracy of 0.005 (TEWG, 1999). ASTER is an on-demand instrument, which means that data are only acquired over

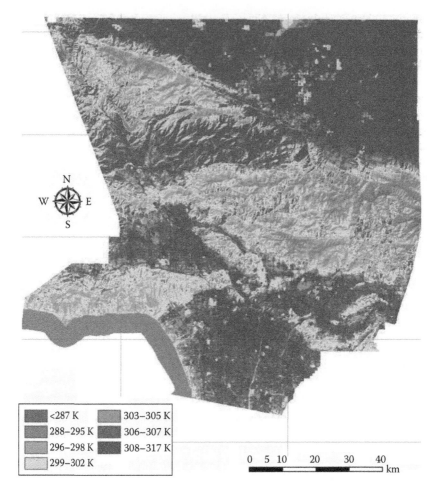

FIGURE 4.3 **(See color insert.)** Mean annual surface temperature in Los Angeles determined by an unconstrained nonlinear optimization with the Levenberg–Marquardt minimization scheme. LST measurements of all available 115 Landsat-5 TM scenes between 2000 and 2010 were used for modeling by a sine function. (From Weng, Q. and P. Fu, *Remote Sens. Environ.*, 2014, 140, 267.)

the requested locations. Terra satellite launched in December 1999 as part of NASA's Earth Observing System has a life expectancy of 6 years and is now also overdue.

Landsat TM and ETM+ TIR data have been extensively utilized to derive LSTs and to study UHIs (e.g., Nichol, 1994; Weng, 2001, 2003; Weng et al., 2004) for American and Asian cities. With ASTER imagery, Lu and Weng (2006) estimated hot-object and cold-object fractions and biophysical variables using linear spectral mixture analysis and analyzed their relationship across various spatial aggregations. Rajasekar and Weng (2009b) applied association rule mining for exploring the relationship between urban LST and biophysical/social parameters. Moreover, the landscape ecology approach was employed to assess the interplay between LST and LULC patterns in order to reach the optical scale for analysis (Liu and Weng, 2009). Because ASTER sensor collects both daytime and nighttime TIR images, analysis of LST spatial patterns has also been conducted for a diurnal contrast (Nichol, 2005).

Studies using satellite-derived LSTs have been termed surface temperature UHIs (Streutker, 2002). Moreover, satellite-derived LSTs are believed to correspond more closely with the canopy layer heat islands, although a precise transfer function between LST and the near-ground air temperature is not yet available (Nichol, 1994). Voogt and Oke (2003) criticized the slow progress in thermal remote sensing of urban areas, which has largely been limited to qualitative description of thermal patterns and simple correlations between LST and LULC types. Xiao et al. (2008) further noticed that little research has been done on the statistical relationship between LST and nonbiophysical factors. A key issue in the application of TIR data in urban climate studies is how to use LST measurements at the micro scale to characterize and quantify UHIs observed at the meso scale (Weng, 2009). Because medium-resolution sensors are typically associated with long-repeat-cycle satellites (16 days for both Landsat and Terra ASTER sensors), their TIR data are not readily useful for UHI monitoring. Bechtel (2012) found that it was feasible to extract mean annual surface temperature and yearly amplitude of surface temperature by modeling the ATC with Landsat data archive.

Looking into the near future, the LDCM may be the only option. It will have a TIR sensor acquiring data at 100 m resolution, but again with a low temporal resolution of 16 days (http://ldcm.nasa.gov/). The hyperspectral infrared imager (HyspIRI) has been defined as a mission with Tier 2 priority of the Decadal Survey (http://hyspiri.jpl.nasa.gov/). Its TIR imager is expected to provide seven bands between 7.5 and 12 μm and one band at 4 μm, all with 60 m resolution. The TIR sensor is intended for imaging global land and shallow water with a 5-day revisit at the equator (1-day and 1-night imaging). These improved capabilities would allow for a more accurate estimation of LST and emissivity and for deriving unprecedented information on biophysical characteristics, but HyspIRI has not yet set a definite time for launch due mainly to budget constraints.

4.4   SAR SENSORS

4.4.1   COARSE-RESOLUTION SAR SENSORS

SAR data in any ScanSAR mode are one of the most important sources of information for mapping purposes at the global level. The wide geographical coverage coupled with almost no blackout time mark SAR sensors as the best option for a number of land covers at the global level. For human settlements, the wide swath mode (WSM) data from the ASAR sensor on board ESA Envisat-1 are currently exploited in a semioperational way to globally map built-up arc extents (Gamba and Lisini, 2013). Table 4.2 provides ENVISAT ASAR sensor characteristics. Indeed, a global urban extraction of WSM data, with a spatial posting of 75 m per pixel, represents an excellent trade-off between detailed accuracy and computational load. They were collected as a sort of background mission whenever the satellite was not busy acquiring in a different mode. The number of acquisitions on the same area is thus variable

TABLE 4.2
ENVISAT ASAR Sensor Characteristics

Sensor	ASAR
Mission lifetime	2002–2012
Orbit	800 km altitude
	35 days orbit repeat cycle
	5–15 days revisit time (midlatitudes)
Range size	56–100 km (image and alternating polarization modes)
	400 km (wide swath and global monitoring modes)
Geometric resolution	30 m (image and alternating polarization modes)
	150 m (wide swath mode)
	1 km (global monitoring mode)
Spectral resolution	1 channel 5.331 GHz (C-band)
Polarizations	HH or VV (single pol)
	HH/VV or HH/HV or VV/VH (dual pol)

from one year to the other, but in general the yearly coverage is guaranteed for the whole globe, with a few exceptions. The same methodology will presumably be applied with minor adaptation to data from future missions, such as ESA Sentinel-1.

4.4.2 MEDIUM-RESOLUTION SAR SENSORS

The SAR sensors currently available have a spatial resolution in the range of 10–30 m, which has not been, until recently, considered useful for urban applications, or at least not very different from those (like urban extent extraction) equally achievable by coarse sensors. The main improvement that makes these systems useful for urban application is polarimetry, as it allows for the distinction among different scattering mechanisms. While this feature may be useful for mapping ULC, it is expected that very high-resolution SAR data can provide finer details of urban structures such as buildings and roads and thus further improve its application in urban analysis. In the following text, the characteristics of the RADARSAT-1 and -2 satellites are discussed.

Launched in November 1995 and December 2007, respectively, RADARSAT-1 and -2 are sophisticated Earth observation satellites developed by Canada to monitor environmental changes and the planet's natural resources. The C-band SAR sensors on board these satellites are operational radar systems capable of timely delivery of large amounts of data for many applications, including urban, marine surveillance, ice monitoring, disaster management, environmental monitoring, resource management, and mapping, in Canada and worldwide. The RADARSAT SAR system characteristics are listed in Table 4.3. Polarimetric SAR data have increasingly been used for urban analysis (Niu and Ban, 2012). Moreover, by exploiting the multitemporal capability of medium-resolution SAR with no limitation due to weather conditions, these SAR data have also been considered for urban change analysis (Niu and Ban, 2013). Finally, several studies have also been undertaken on the fusion of SAR and optical data, showing improved ULC mapping over SAR or optical data alone (Gamba and

TABLE 4.3

RADARSAT-1 and -2 Sensor Characteristics

Sensor	RADARSAT-1	RADARSAT-2
Mission lifetime	>15 years	7 years
Orbit	793–821 km	798 km
Range size	45 km (fine beam)	Selective polarization:
	100 km (standard beam)	50 km (fine beam)
	75 km (high incidence)	100 km (standard beam)
	170 km (low incidence)	75 km (high incidence)
	150 km (wide)	170 km (low incidence)
	300 km (ScanSAR narrow)	150 km (wide)
	500 km (ScanSAR wide)	300 km (ScanSAR narrow)
		500 km (ScanSAR wide)
		Polarimetric:
		25 km (fine quad pol)
		25 km (standard quad pol)
		Selective single polarization:
		20 km (ultrafine)
		18 km (SpotLight)
		50 km (multilook Fine)
Geometric resolution (m)	8 (fine beam)	Selective polarization:
	30 (standard beam)	10×9 (fine beam)
	18–27 (high incidence)	25×28 (standard beam)
	30 (low incidence)	40×28 (high incidence)
	30 (wide)	20×28 (low incidence)
	50 m (ScanSAR narrow)	25×28 (wide)
	100 m (ScanSAR wide)	50×50 (ScanSAR narrow)
		100×100 (ScanSAR wide)
		Polarimetric:
		11×9 (fine quad pol)
		25×28 (standard quad pol)
		Selective single polarization:
		3×3 (ultrafine)
		3×1 (SpotLight)
		11×9 (multilook fine)
Spectral resolution	1 channel	1 channel
	Center frequency 5.3 GHz	Center frequency 5.405 GHz
	(C-band), bandwidth: 30 MHz	(C-band), bandwidth: 100 MHz
Polarizations	HH	HH, VV, HV, VH

Houshmand, 2001; Ban et al., 2010; Ban and Jacob, 2013). The future is connected to the RADARSAT constellation. The three-satellite configuration will provide complete coverage of Canada's and most of the world's land and oceans, offering an average daily revisit as well as daily access to 95% of the world to Canadian and international users. The satellite launches are currently planned for 2018.

4.4.3 FINE-RESOLUTION SAR SENSORS

The last generation of SAR sensors has been developed to provide better spatial resolution characteristics, in the range of 1 m. This category of sensors has caused a boost in the use of SAR data in urban applications, especially connected to very detailed analysis (to the building level) of the urban environment. TerraSAR-X (TSX) and COSMO/SkyMed are examples of these systems. COSMO/SkyMed has been used for many applications, mainly related to risk mapping and interferometry (Ardizzone et al., 2012). In the following, a more detailed analysis of TSX is offered, because of its peculiar characteristics, connected to the twin TanDEM-X (TDX) mission.

The first German SAR satellite TSX was launched on June 15, 2007, in the context of a public–private partnership between the German Aerospace Center (DLR) and the EADS Astrium GmbH. Three years later, the TSX mission was amended by a second, almost identical X-band SAR satellite—TDX. For the TDX mission (TDM) (TSX Add-On for Digital Elevation Measurement), TSX and TDX are flying in an orbit at 514 km in a so-called helix formation with a typical distance of 250–500 m between the satellites. With this constellation, TDX is the first bistatic, spaceborne SAR mission. The primary mission is the generation of a consistent global digital elevation model with unprecedented accuracy. At the same time, TSX and TDX provide highly reconfigurable platforms for testing and demonstrating new SAR techniques and potential applications. TSX and TDX are scheduled for 5 years of operation, and they collect VHR data in three basic imaging modes—SpotLight (SL) and high-resolution SpotLight mode (HS), StripMap mode (SM), and ScanSAR mode (SC) (Roth et al., 2005). The characteristics of the imaging modes are listed in Table 4.4.

Due to the all-weather and day-and-night data acquisition capability of SAR sensors, the TDM allows to collect two global coverages of VHR images (SM mode) within a period of 3 years (2011–2013). With the unique spatial detail

TABLE 4.4
TerraSAR-X Sensor Characteristics

Sensor	TerraSAR-X
Mission lifetime	2007–2013 (at least)
Orbit	514 km altitude
	11 days orbit repeat cycle
	2–4 days revisit time (midlatitudes)
Range size	5–10 km (SpotLight mode)
	30 km (StripMap mode single-polarized)
	15 km (StripMap mode dual-polarized)
	100 km (ScanSAR mode)
Geometric resolution	1 m (SpotLight mode)
	3 m (StripMap mode)
	16 m (ScanSAR mode)
Spectral resolution	1 channel 9.65 GHz (X-band)
Polarizations	HH or VV (single-polarized)
	HH/VV or HH/HV or VV/VH (dual-polarized)

and temporal consistency of the data set in combination with the complementary characteristics of the VHR SAR data compared to medium- or high-resolution optical imagery used for urban analyses on a global scale so far, the TDM is predestined to substantially support the global mapping and future monitoring of human settlements. Hence, DLR's German Remote Sensing Data Center (DFD) has developed a fully operational processing chain—the urban footprint processor (UFP)—for the delineation of built-up areas from the TDM SAR database (Esch et al., 2012). The goal is to provide a public domain global coverage of binary settlement masks showing a spatial resolution of 3 arcsec (~50–75 m)—the global urban footprint (GUF). The accuracies of the binary GUF settlement masks usually range between 70% and 95% depending on the complexity of the landscape and the significance of the built-up environment (size, height, density, arrangement of houses, vegetation cover, etc.). Figure 4.4 shows an example of the GUF mosaic for the region of Accra, Ghana.

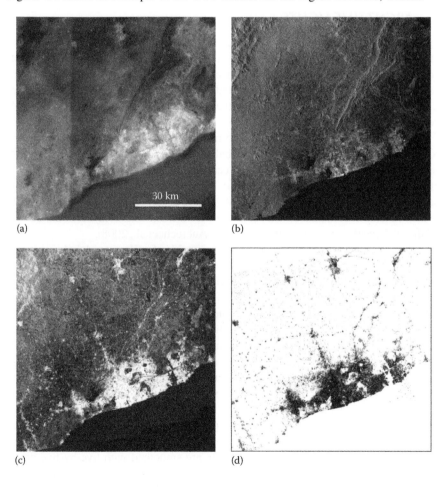

(a)

(b)

(c)

(d)

FIGURE 4.4 **(See color insert.)** Optical data from Google Earth (a), TerraSAR-X amplitude image (b), calculated texture image (c), and urban footprint mask derived from combined classification of amplitude and texture (d).

With the two global coverages of VHR SAR imagery collected in 2011/2012 and 2012/2013, the TDM data set represents a suitable baseline for future analyses of global urban sprawl. Apart from classic postclassification change detection approaches, the calculation of long-term coherences might serve as an effective method to improve the intended GUF product and to provide an alternative method for the mapping and monitoring of urban sprawl. Moreover, the extraction of building structures and the estimation of building densities based on texture measures and the modeling of building volume on the building block level using the VHR DEM generated in the context of the TDM hold further potential (Esch et al., 2012). From an applied perspective, the combined analysis global SAR and optical data sets or the combination of settlement masks derived from these complementary sources are highly interesting.

4.5   NIGHTTIME LIGHTS

Nighttime lights are a class of satellite observations and derived products based on the detection of anthropogenic lighting present at the Earth's surface. This style of product can only be produced using data from sensors that collect low-light imaging data in spectral bands covering emissions generated by electric lights. The standard "stable lights" product is a cloud-free composite that has been filtered to remove ephemeral fires and background noise. Nighttime lights are used to model the spatial distribution of variables that would be very difficult to measure in a globally consistent manner. Examples of nighttime lights–derived global grids include the spatial distribution of population (Doll, 2010; Sutton et al., 2010), economic activity (Ghosh et al., 2010), electrification rates (Elvidge et al., 2010), poverty mapping (Elvidge et al., 2009), density of constructed surfaces (Matsumura et al., 2009), food demand, stocks of steel and other metals (Takahashi et al., 2010), CO_2 emissions from fossil fuels (Rayner et al., 2009), and the ecological impact of artificial lighting (Aubrecht et al., 2008).

To date, there have been two systems flown capable of collecting global nighttime lights data. The original system is the Defense Meteorological Satellite Program (DMSP) operational linescan system (OLS). The more recently launched system is the Suomi National Polar Partnership (SNPP) VIIRS. In both cases, the low-light imaging was designed to serve the meteorological community, which has an interest in detecting moonlit clouds in the visible region to complement thermal observations.

The DMSP nighttime lights represent one of the most widely recognized global satellite data products and have proven valuable in a wide range of scientific applications. DMSP has flown low-light imaging sensors in polar orbits since the mid-1970s and has a digital archive that extends back to 1992. The OLS was designed to collect visible and TIR data, day and night, for use in observing weather systems and cloud cover. The "visible" band may be termed panchromatic, spanning the visible and near-infrared (NIR) from 0.5 to 0.9 μm. The DMSP low-light imaging is achieved using a photomultiplier tube. The global data are smoothed with five-by-five pixel averaging, which results in pixel footprints that are 5 km on a side at nadir and up to 8 km on a side at the edge of scan. The ground sample distance (GSD) is maintained at 2.7 km from nadir to edge of scan. Thus, there is substantial overlap between adjacent pixel footprints. The detection limit is estimated at 5E-10 Watts/cm²/sr.

The DMSP nighttime lights data have a set of well-known shortcomings (Elvidge et al., 2007): coarse spatial resolution, six-bit quantization, saturation on bright lights, lack of in-flight calibration, lack of spectral channels suitable for discrimination of thermal sources of lighting, and lack of low-light imaging spectral bands suitable for discriminating lighting types (Elvidge et al., 2010).

On October 28, 2011, NASA and NOAA launched the SNPP satellite carrying the first VIIRS. The VIIRS instrument includes a day/night band (DNB), which collects panchromatic (0.5–0.9 μm) low-light imaging data at night using a time delay and integration (TDI) charge-coupled device (CCD). The VIIRS instrument offers improvement in each of these shortcomings, except multispectral low-light imaging. The DNB pixel footprint is maintained at 742 m from nadir out to the edge of scan. Thus, the VIIRS DNB low-light imaging footprint is 45 times smaller than the DMSP-OLS footprint. The DNB data have a wide dynamic range, 14-bit quantization, and a detection limit estimated at 2E-10 Watts/cm^2/sr, which makes it possible for the VIIRS to detect clouds, snow, and bright playa lake beds with exceedingly dim airglow illumination when no moonlight is present (Miller et al., 2012). The DNB has an in-flight calibration capability. In addition, the VIIRS collects data at night in a SWIR band that detects combustion sources, but not nighttime lights (Zhizhin et al., 2013). This makes it possible to distinguish thermal sources of light from electric lighting. All of this results in a far superior nighttime lights product when compared to DMSP products (Figure 4.5).

In 2012 and 2013, nighttime lights were collected by both DMSP and VIIRS, making it possible to cross-calibrate the nighttime light products generated by the two systems. There are two more DMSP satellites to be launched; however, it is anticipated that these will fly in dawn–dusk orbits that are ill-suited for global mapping of nighttime lights. The second VIIRS is under construction, and NASA/NOAA is planning the third. There are good prospects for a continuing series of VIIRS instruments. While VIIRS nighttime lights are expected to yield substantial

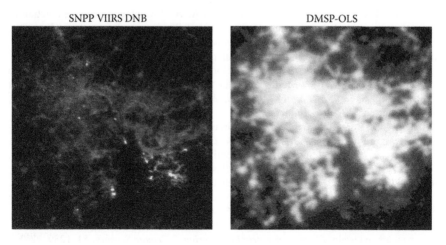

FIGURE 4.5 Comparison on DNB versus DMSP cloud-free composited nighttime lights of Guangzhou, China, in 2012.

advances in a range of science applications, there has yet to be a satellite mission dedicated to nighttime lights. Such a mission would likely have spatial resolution under 100 m and multispectral low-light imaging to enable discrimination of lighting types (Elvidge et al., 2007).

ACRONYMS

AATSR	Advanced along track scanning radiometer
ABI	Advanced baseline imager
AVHRR	Advanced very high resolution radiometer
ESA	European Space Agency
EUMETSAT	European Meteorological Satellite Organisation
GOES	Geostationary Operational Environmental Satellite
Metop	Meteorological Operational Satellite Programme
MODIS	Moderate-resolution imaging spectroradiometer
MSG	Meteosat second generation
MTSAT	Multifunction transport satellite
NASA	National Aeronautics and Space Administration
NOAA	National Oceanic and Atmospheric Administration
SEVIRI	Spinning-enhanced visible and infrared imager
SLSTR	Sea and land surface temperature radiometer

REFERENCES

Alam, K., Iqbal, M.J., Blaschke, T., Qureshi, S., and Khan, G. 2010. Monitoring the spatio-temporal variations in aerosols and aerosol-cloud interactions over Pakistan using MODIS data. *Advances in Space Research*, 46(9), 1162–1176.

Alam, K., Qureshi, S., and Blaschke, T. 2011. Monitoring spatio-temporal aerosol patterns over Pakistan based on MODIS, TOMS and MISR satellite data and a HYSPLIT model. *Atmospheric Environment*, 45(27), 4641–4651.

Ardila, J.P., Bijker, W., Tolpekin, V.A., and Stein, A. 2012. Context-sensitive extraction of tree crown objects in urban areas using VHR satellite images. *International Journal of Applied Earth Observation and Geoinformation*, 15(1), 57–69.

Ardizzone, F., Bonano, M., Giocoli, A., Lanari, R., Marsella, M., Pepe, A., Perrone, A. et al. 2012. Analysis of ground deformation using SBAS-DInSAR technique applied to COSMO-SkyMed images, the test case of Roma urban area. *Proceedings of SPIE 8536, SAR Image Analysis, Modeling, and Techniques XII*, 85360D, Edinburgh, U.K., November 21, 2012.

ASTER Temperature/Emissivity Working Group (TEWG). 1999. *Temperature/Emissivity Separation Algorithm Theoretical Basis Document, Version 2.4*. http://eospso.gsfc.nasa.gov/eos_homepage/for_scientists/atbd/docs/ASTER/atbd-ast-05-08.pdf, JPL, California Institute of Technology (last date accessed January 30, 2013).

Aubrecht, C., Elvidge, C.D., Longcore, T., Rich, C., Safran, J., Strong, A.E., Eakin, C.M. et al. 2008. A global inventory of coral reef stressors based on satellite observed nighttime lights. *Geocarto International*, 23(6), 467–479.

Balling, R.C. and Brazell, S.W. 1988. High resolution surface temperature patterns in a complex urban terrain. *Photogrammetric Engineering & Remote Sensing*, 54, 1289–1293.

Baltsavias, E., Mason, S., and Stallmann, D. 1995. Use of DTMs/DSMs and orthoimages to support building extraction. In *Automatic Extraction of Man-Made Objects from Aerial and Space Images*, eds. A. Gruen, O. Kuebler, and P. Agouris. Birkhäuser, Basel, Switzerland, pp. 199–210.

Ban, Y., Hu, H., and Rangel, I.M. 2010. Fusion of Quickbird MS and RADARSAT SAR for urban land-cover mapping: Object-based and knowledge-based approach. *International Journal of Remote Sensing*, 31(6), 1391–1410.

Ban, Y. and Jacob, A. 2013. Object-based fusion of multitemporal multi-angle ENVISAT ASAR and HJ-1 multispectral data for urban land-cover mapping. *IEEE Transaction on GeoScience and Remote Sensing*, 51(4), 1998–2006.

Bechtel, B. 2012. Robustness of annual cycle parameters to characterize the urban thermal landscapes. *IEEE Geoscience and Remote Sensing Letters*, 9(5), 876–880.

Bechtel, B., Zakšek, K., and Hoshyaripour, G. 2012. Downscaling land surface temperature in an urban area: A case study for Hamburg, German. *Remote Sensing*, 4(10), 3184–3200.

Benediktsson, J.A., Pesaresi, M., and Arnason, K. 2003. Classification and feature extraction for remote sensing images from urban areas based on morphological transformations. *IEEE Transactions on Geoscience and Remote Sensing*, 41(9 part I), 1940–1949.

Bengang, L. and Shu, T. 2000. Correlation between AVHRR NDVI and climate factors. *Acta Ecologica Sinica*, 20(5), 898–902.

Bovolo, F. and Bruzzone, L. 2005. A detail-preserving scale-driven approach to change detection in multitemporal SAR images. *IEEE Transactions on Geoscience and Remote Sensing*, 43(12), 2963–2972.

Bovolo, F., Marin, C., and Bruzzone, L. 2012. A novel approach to building change detection in very high resolution SAR images. In *Proceedings of SPIE 8537, Image and Signal Processing for Remote Sensing XVIII*, Bellingham, WA: SPIE, 2012. Atti di: SPIE 2012, Edinburgh, U.K., September 24–27, 2012.

Chaudhuri, D., Kushwaha, N.K., and Samal, A. 2012. Semi-automated road detection from high resolution satellite images by directional morphological enhancement and segmentation techniques. *IEEE Journal of Selected Topics in Applied Earth Observations and Remote Sensing*, 5(5), 1538–1544.

Chesnel, A.L., Binet, R., and Wald, L. 2008. Damage assessment on buildings using multisensor multimodal very high resolution images and ancillary data. In *Geoscience and Remote Sensing Symposium, 2008. IGARSS 2008*. IEEE International.

Clark, M.L., Aide, T.M., and Riner, G. 2012. Land change for all municipalities in Latin America and the Caribbean assessed from 250-m MODIS imagery (2001–2010). *Remote Sensing of Environment*, 126, 84–103.

Couloigner, I., Ranchin, T., Valtonen, V.P., and Wald, L. 1998. Benefit of the future SPOT-5 and of data fusion to urban roads mapping. *International Journal of Remote Sensing*, 19, 1519–1532.

Dash, P., Gottsche, F.-M., Olesen, F.-S., and Fischer, H. 2002. Land surface temperature and emissivity estimation from passive sensor data: Theory and practice current trends. *International Journal of Remote Sensing*, 23(13), 2563–2594.

DeFries, R.S., Rudel, T., Uriarte, M., and Hansen, M. 2010. Deforestation driven by urban population growth and agricultural trade in the twenty-first century. *Nature Geoscience*, 3, 178–181.

Dell'Acqua, F., Gamba, P., and Polli, D.A. 2011. Earthquake damage assessment from post-event VHR radar data: From Sichuan, 2008 to Haiti, 2010. In *Proceedings of 2011 Joint Urban Event (JURSE 2011)*, Munich, Germany, April 2011, pp. 201–204.

Deng, J., Wang, K., Li, J., and Deng, Y. 2009. Urban land use change detection using multisensor satellite images. *Pedosphere*, 19, 96–103.

Doll, C.N.H. 2010. Population detection profiles of DMSP-OLS night-time imagery by regions of the world. In *Proceedings of the 30th Asia-Pacific Advanced Network Meeting*, Hanoi, Vietnam, pp. 191–207.

Durieux, L., Lagabrielle, E., and Nelson, A. 2008. A method for monitoring building construction in urban sprawl areas using object-based analysis of Spot 5 images and existing GIS data. *ISPRS Journal of Photogrammetry & Remote Sensing*, 63, 399–408.

Elvidge, C.D., Baugh, K.E., Sutton, P.C., Bhaduri, B., Tuttle, B.T., Ghosh, T., Ziskin, D., and Erwin, E.H. 2010. Who's in the dark: Satellite based estimates of electrification rates. In *Urban Remote Sensing: Monitoring, Synthesis and Modeling in the Urban Environment*, ed. X. Yang. Wiley-Blackwell, Chichester, U.K,

Elvidge, C.D., Cinzano, P., Pettit, D.R., Arvesen, J., Sutton, P., Small, C., Nemani, R. et al. 2007. The Nightsat mission concept. *International Journal of Remote Sensing*, 28(12), 2645–2670.

Elvidge, C.D., Keith, D.M., Tuttle, B.T., and Baugh, K.E. 2010. Spectral identification of lighting type and character. *Sensors*, 10(4), 3961–3988.

Elvidge, C.D., Sutton, P.C., Ghosh, T., Tuttle, B.T., Baugh, K.E., Bhaduri, B., and Bright, E. 2009. A global poverty map derived from satellite data. *Computers and Geosciences*, 35, 1652–1660.

Engel-Cox, J.A., Holloman, C.H., Coutant, B.W., and Hoff, R.M. 2004. Qualitative and quantitative evaluation of MODIS satellite sensor data for regional and urban scale air quality. *Atmospheric Environment*, 38(16), 2495–2509.

Esch, T., Taubenböck, H., Roth, A., Heldens, W., Felbier, A., Thiel, M., Schmidt, M., Müller, A., and Dech, S. 2012. TanDEM-X mission-new perspectives for the inventory and monitoring of global settlement patterns. *Journal of Applied Remote Sensing*, 6(1), 061702, 21pp. doi: 10.1117/1.JRS.6.061702.

Ferretti, A., Prati, C., and Rocca, F. 2001. Permanent scatterers in SAR interferometry. *IEEE Transactions on Geoscience and Remote Sensing*, 39(1), 8–20.

Gallo, K.P., McNab, A.L., Karl, T.R., Brown, J.F., Hood, J.J., and Tarpley, J.D. 1993. The use of NOAA AVHRR data for assessment of the urban heat island effect. *Journal of Applied Meteorology*, 32(5), 899–908.

Gamba, P., Aldrighi, M., and Stasolla, M. 2011. Robust extraction of urban area extents in HR and VHR SAR images. *IEEE Journal on Selected Topics in Applied Earth Observation and Remote Sensing*, 4(1), 27–34.

Gamba, P. and Houshmand, B. 2001. An efficient neural classification chain of SAR and optical urban images. *International Journal of Remote Sensing*, 22(8), 1535–1553.

Gamba, P. and Lisini, G. 2013. Fast and efficient urban extent extraction using ASAR Wide Swath Mode data. *IEEE Journal of Selected Topics in Applied Earth Observation and Remote Sensing*, 6, 2184–2195.

Gerhardinger, A., Ehrlich, D., and Pesaresi, M. 2005. Vehicle detection from very high resolution satellite imagery. *International Archives of Photogrammetry, Remote Sensing and Spatial Information Sciences*, 36(Part 3/W24), 83–88.

Ghosh, T., Powell, R., Elvidge, C.D., Baugh, K.E., Sutton, P.C., and Anderson, S. 2010. Shedding light on the global distribution of economic activity. *The Open Geography Journal*, 3, 148–161.

Giada, S., De Groeve, T., Ehrlich, D., and Soille, P. 2003. Information extraction from very high resolution satellite imagery over Lukole refugee camp, Tanzania. *International Journal of Remote Sensing*, 24(22), 4251–4266.

Gillespie, A.R., Rokugawa, S., Matsunaga, T., Cothern, J.S., Hook, S.J., and Kahle, A.B. 1998. A temperature and emissivity separation algorithm for advanced spaceborne thermal emission and reflection radiometer (ASTER) images. *IEEE Transactions on Geoscience and Remote Sensing*, 36(4), 1113–1126.

Gueguen, L., Pesaresi, M., Soille, P., and Gerhardinger, A. 2009. Morphological descriptors and spatial aggregations for characterizing damaged buildings in very high resolution images. In *Proceedings of the ESA-EUSC-JRC 2009 Conference. Image Information Mining: Automation of Geospatial Intelligence from Earth Observation*, Madrid, Spain, 2009.

Guida, R., Iodice, A., Riccio, D., and Stilla, U. 2008. Model-based interpretation of high resolution SAR images of buildings. *IEEE Journal of Selected Topics in Applied Earth Observations and Remote Sensing*, 1(2), 107–119.

Haala, N. and Hahn, M. 1995. Data fusion for the detection and reconstruction of buildings. In *Automatic Extraction of Man-Made Objects from Aerial and Space Images*, eds. A. Gruen, O. Kuebler, and P. Agouris. Birkhäuser, Basel, Switzerland, pp. 211–220.

Hedman, K., Lisini, G., Gamba, P., and Stilla, U. 2010. Road network extraction in VHR SAR images of urban and suburban areas by means of class-aided feature-level fusion. *IEEE Transactions on Geoscience and Remote Sensing*, 48(3), 1294–1296.

Hu, H. and Ban, Y. 2012. Multitemporal RADARSAT-2 ultra-fine-beam SAR data for urban land cover classification. *Canadian Journal of Remote Sensing*, 38(01), 1–11, 10.5589/m12-008.

Hung, T., Uchihama, D., Ochi, S., and Yasuoka, Y. 2006. Assessment with satellite data of the urban heat island effects in Asian mega cities. *International Journal of Applied Earth Observation and Geoinformation*, 8(1), 34–48.

Hutchison, K.D., Smith, S., and Faruqui, S.J. 2005. Correlating MODIS aerosol optical thickness data with ground-based PM2.5 observations across Texas for use in a real-time air quality prediction system. *Atmospheric Environment*, 39(37), 7190–7203.

Imhoff, M.L., Zhang, P., Wolfe, R.E., and Bounoua, L. 2010. Remote sensing of the urban heat island effect across biomes in the continental USA. *Remote Sensing of Environment*, 114(3), 504–513.

Jenerowicz, M., Kemper, T., and Soille, P. 2011. An automated procedure for detection of IDP's dwellings using VHR satellite imagery. *Proceedings of SPIE—The International Society for Optical Engineering*, 8180, 818004.

Ji, L. and Peters, A.J. 2004. A spatial regression procedure for evaluating the relationship between AVHRR-NDVI and climate in the northern Great Plains. *International Journal of Remote Sensing*, 25(2), 297–311.

Jiménez-Muñoz, J.C. and Sobrino, J.A. 2008. Split-window coefficients for land surface temperature retrieval from low-resolution thermal infrared sensors. *IEEE Geoscience and Remote Sensing Letters*, 5(4), 806–809.

Jin, Z. and Xu, B. 2013. A novel compound smoother—RMMEH to reconstruct MODIS NDVI time series. *IEEE Geoscience and Remote Sensing Letter*, 10(4), 942–946.

Jonsson, P. 2004. Vegetation as an urban climate control in the subtropical city of Gaborone, Botswana. *International Journal of Climatology*, 24(10), 1307–1322.

Kasimu, A. and Tateishi, R. 2008. Global urban mapping using population density, MODIS and DMSP data with the reference of Landsat images. *The International Archives of the Photogrammetry, Remote Sensing and Spatial Information Sciences*, XXXVII(Part B7), 1523–1528.

Keramitsoglou, I., Daglis, I.A., Amiridis, V., Chrysoulakis, N., Ceriola, G., Manunta, P. et al. 2012. Evaluation of satellite-derived products for the characterization of the urban thermal environment. *Journal of Applied Remote Sensing*, 6, 061704.

Keramitsoglou, I., Kiranoudis, C.T., Ceriola, G., Weng, Q., and Rajasekar, U. 2011. Identification and analysis of urban surface temperature patterns in Greater Athens, Greece, using MODIS imagery. *Remote Sensing of Environment*, 115, 3080–3090.

Keramitsoglou, I., Kiranoudis, C.T., and Weng, Q. 2013. Downscaling geostationary land surface temperature imagery for urban analysis. *IEEE Geoscience and Remote Sensing Letters*, 10(5), 1253–1257.

Khoshelham, K., Nardinocchi, C., Frontoni, E., Mancini, A., and Zingaretti, P. 2010. Performance evaluation of automated approaches to building detection in multi-source aerial data. *ISPRS Journal of Photogrammetry and Remote Sensing*, 65(1), 123–133.

Kidder, S.Q. and Wu, H.T. 1987. A multispectral study of the St. Louis area under snow-covered conditions using NOAA-7 AVHRR data. *Remote Sensing of Environment*, 22, 159–172.

Kit, O., Lüdeke, M., and Reckien, D. 2012. Texture-based identification of urban slums in Hyderabad, India using remote sensing data. *Applied Geography*, 32(2), 660–667.

Kohli, D., Sliuzas, R., Kerle, N., and Stein, A. 2012. An ontology of slums for image-based classification. *Computers, Environment and Urban Systems*, 36(2), 154–163.

Langer, A., Miettinen, J., and Siegert, F. 2007. Land cover change 2002–2005 in Borneo and the role of fire derived from MODIS imagery. *Global Change Biology*, 13(11), 2329–2340.

Leinenkugel, P., Esch, T., and Kuenzer, C. 2011. Settlement detection and impervious surface estimation in the Mekong Delta using optical and SAR remote sensing data. *Remote Sensing of Environment*, 115, 3007–3019.

Leitloff, J., Hinz, S., and Stilla, U. 2010. Vehicle detection in very high resolution satellite images of city areas. *IEEE Transactions on Geoscience and Remote Sensing*, 48(7), 2795–2806.

Lin, C. and Nevatia, R. 1998. Building detection and description from a single intensity image. *Computer Vision and Image Understanding*, 72(2), 101–121.

Liu, H. and Weng, Q. 2009. Scaling-up effect on the relationship between landscape pattern and land surface temperature. *Photogrammetric Engineering & Remote Sensing*, 75(3), 291–304.

Lu, D. and Weng, Q. 2004. Spectral mixture analysis of the urban landscape in Indianapolis with Landsat ETM+ imagery. *Photogrammetric Engineering & Remote Sensing*, 70(9), 1053–1062.

Lu, D. and Weng, Q. 2006. Spectral mixture analysis of ASTER imagery for examining the relationship between thermal features and biophysical descriptors in Indianapolis, Indiana. *Remote Sensing of Environment*, 104(2), 157–167.

Lu, L., Guo, H., Corbane, C., Pesaresi, M., and Ehrlich, D. 2012. Rapid damage assessment of buildings with VHR optical airborne images in Yushu earthquake. In *Second International Conference on Remote Sensing, Environment and Transportation Engineering (RSETE)*.

Luo, L. and Mountrakis, G. 2010. Integrating intermediate inputs from partially classified images within a hybrid classification framework: An impervious surface estimation example. *Remote Sensing of Environment*, 114, 1220–1229.

Maktav, D., Erbek, F.S., and Jürgens, C. 2005. Remote sensing of urban areas. *International Journal of Remote Sensing*, 26, 655–659.

Matsumura, K., Hijmans, R.J., Chemin, Y., Elvidge, C.D., Sugimoto, K., Wu, W.B., Lee, Y.W., and Shibasaki, R. 2009. Mapping the global supply and demand structure of rice. *Sustainability Science*, 4(2), 301–313.

Michishita, R., Gong, P., and Xu, B. 2012b. Spectral mixture analysis for bi-sensor wetland mapping using Landsat TM and Terra MODIS data. *International Journal of Remote Sensing*, 33(11), 3373–3401.

Michishita, R., Jiang, Z., Gong, P., and Xu, B. 2012c. Bi-scale analysis of multi-temporal land cover fractions for wetland vegetation mapping. *ISPRS Journal of Photogrammetry and Remote Sensing*, 72, 1–15.

Michishita, R., Jiang, Z., and Xu, B. 2012a. Monitoring two decades of urbanization in the Poyang Lake area, China through spectral unmixing. *Remote Sensing of Environment*, 117(1), 3–18. doi: 10.1016/j.rse.2011.06.021.

Miller, S.D., Mills, S.P., Elvidge, C.D., Lindsey, D.T., Lee, T.F., and Hawkins, J.D. 2012. Suomi satellite brings to light a unique frontier of nighttime environmental sensing capabilities. *Proceedings of the National Academy of Science*, 109(39), 15706–15711.

Miyazaki, H., Shao, X., Iwao, K., and Shibasaki, R. An automated method for global urban area mapping by integrating ASTER satellite images and GIS data. *IEEE Journal of Selected Topics in Applied Earth Observations and Remote Sensing*, in press.

Mountrakis, G. and Luo, L. 2011. Enhancing and replacing spectral information with intermediate structural inputs: A case study on impervious surface detection. *Remote Sensing of Environment*, 115, 1162–1170.

NASA. 2013. Landsat 7 history. Available at: http://landsat.gsfc.nasa.gov/about/landsat7.html (last accessed January 19, 2014).

Netzband, M. and Stefanov, W.L. 2004. Urban land cover and spatial variation observation using ASTER and MODIS satellite image data. *The International Archives of the Photogrammetry, Remote Sensing, and Spatial Information Sciences*, 35, 1348–1353.

Nichol, J. and Lee, C.M. 2005. Urban vegetation monitoring in Hong Kong using high resolution multispectral images. *International Journal of Remote Sensing*, 26(5), 903–918.

Nichol, J. and Wong, M.S. 2007. Remote sensing of urban vegetation life form by spectral mixture analysis of high-resolution IKONOS satellite images. *International Journal of Remote Sensing*, 28(5), 985–1000.

Nichol, J.E. 1994. A GIS-based approach to microclimate monitoring in Singapore's high-rise housing estates. *Photogrammetric Engineering & Remote Sensing*, 60(10), 1225–1232.

Nichol, J.E. 2005. Remote sensing of urban heat islands by day and night. *Photogrammetric Engineering & Remote Sensing*, 71(5), 613–623.

Niu, X. and Ban, Y. 2012. An adaptive SEM algorithm for urban land cover mapping using multitemporal high-resolution polarimetric SAR data. *IEEE Journal on of Selected Topics in Applied Earth Observations and Remote Sensing*, 5(4), 1129–1139.

Niu, X. and Ban, Y. 2013. Multitemporal RADARSAT-2 polarimetric SAR data for urban land cover classification using object-based support vector machine and rule-based approach. *International Journal of Remote Sensing*, 34(1), 1–26.

NOAA. AVHRR. http://www.nsof.class.noaa.gov/release/data_available/avhrr/index.htm (last accessed February 2, 2013).

Orenstein, D., Bradley, B., Albert, J., Mustard, J., and Hamburg, S. 2011. How much is built? Quantifying and interpreting patterns of built space from different data sources. *International Journal of Remote Sensing*, 32, 2621–2644.

Ouyang, Z.T., Zhang, M.Q., Xie, X., Shen, Q., Guo, H.Q., and Zhao, B. 2011. A comparison of pixel-based and object-oriented approaches to VHR imagery for mapping saltmarsh plants. *Ecological Informatics*, 6(2), 136–146.

Ouzounis, G.K., Soille, P., and Pesaresi, M. 2011. Rubble detection from VHR aerial imagery data using differential morphological profiles. In *34th International Symposium on Remote Sensing of Environment*, Sydney, New South Wales, Australia, April 10–15, 2011.

Pagot, E. and Pesaresi, M. 2008. Systematic study of the urban postconflict change classification performance using spectral and structural features in a support vector machine. *IEEE Journal of Selected Topics in Applied Earth Observations and Remote Sensing*, 1(2), 120–128.

Parape, C.D.K., Premachandra, H.C.N., Tamura, M., and Sugiura, M. 2012. Damaged building identifying from VHR satellite imagery using morphological operators in 2011 Pacific coast of Tohoku earthquake and tsunami. In *Geoscience and Remote Sensing Symposium (IGARSS)*, 2012. IEEE International.

Peng, S.S., Piao, S.L., Ciais, P., Friedlingstein, P., Ottle, C., Breon, F.M., Nan, H.J., Zhou, L.M., and Myneni, R.B. 2012. Surface urban heat island across global big cities. *Environmental Science & Technology*, 46(2), 696–703.

Pesaresi, M. and Gerhardinger, A. 2011. Improved textural built-up presence index for automatic recognition of human settlements in arid regions with scattered vegetation. *IEEE Journal of Selected Topics in Applied Earth Observations and Remote Sensing*, 4(1), 16–26.

Pesaresi, M., Gerhardinger, A., and Haag, F. 2007. Rapid damage assessment of built-up structures using VHR satellite data in tsunami-affected areas. *International Journal of Remote Sensing*, 28, 13–14, 3013–3036.

Pesaresi, M., Gutjahr, K.H., and Pagot, E. 2008. Estimating the velocity and direction of moving targets using a single optical VHR satellite sensor image. *International Journal of Remote Sensing*, 29(4), 1221–1228.

Potere, D., Schneider, A., Angel, S., and Civco, D.L. 2009. Mapping urban areas on a global scale: Which of the eight maps now available is more accurate? *International Journal of Remote Sensing*, 30, 6531–6558.

Powell, R., Roberts, D., Dennison, P., and Hess, L. 2007. Sub-pixel mapping of urban land cover using multiple endmember spectral mixture analysis: Manaus, Brazil. *Remote Sensing of Environment*, 106(2), 253–267.

Quattrochi, D.A. and Luvall, J.C. 1999. Thermal infrared remote sensing for analysis of landscape ecological processes: Methods and applications. *Landscape Ecology*, 14(6), 577–598.

Rajasekar, U. and Weng, Q. 2009a. Urban heat island monitoring and analysis by data mining of MODIS imageries. *ISPRS Journal of Photogrammetry and Remote Sensing*, 64, 86–96.

Rajasekar, U. and Weng, Q. 2009b. Application of association rule mining for exploring the relationship between urban land surface temperature and biophysical/social parameters. *Photogrammetric Engineering & Remote Sensing*, 75(4), 385–396.

Rayner, P.J., Raupach, M.R., Paget, M., Peylin, P., and Koffi, E. 2009. A new global gridded dataset of CO_2 emissions from fossil fuel combustion: 1: Methodology and evaluation. *Journal of Geophysical Research*,

Rhinane, H., Hilali, A., Berrada, A., and Hakdaoui, M. 2011. Detecting slums from SPOT data in Casablanca Morocco using an object based approach. *Journal of Geographic Information System*, 3, 217–224.

Roth, A., Hoffmann, J., and Esch, T. 2005. TerraSAR-X: How can high resolution SAR data support the observation of urban areas? In *Proceedings of the ISPRS WG VII/1 "Human Settlements and Impact Analysis" Third International Symposium Remote Sensing and Data Fusion Over Urban Areas (URBAN 2005)* and *Fifth International Symposium Remote Sensing of Urban Areas (URS 2005)*, Tempe, AZ, March 14–16, 2005. CD-ROM.

Roth, M., Oke, T.R., and Emery, W.J. 1989. Satellite derived urban heat islands from three coastal cities and the utilisation of such data in urban climatology. *International Journal of Remote Sensing*, 10, 1699–1720.

Sandau, R., Brieß, K., and D'Errico, M. 2010. Small satellites for global coverage: Potential and limits. *ISPRS Journal of Photogrammetry and Remote Sensing*, 65(6), 492–504.

Schmidt, F. and Hinz, S. 2011. A scheme for the detection and tracking of people tuned for aerial image sequences. In *Lecture Notes in Computer Science (including subseries Lecture Notes in Artificial Intelligence and Lecture Notes in Bioinformatics)* 6952 LNCS, pp. 257–270.

Schmugge, T., Hook, S.J., and Coll, C. 1998. Recovering surface temperature and emissivity from thermal infrared multispectral data. *Remote Sensing of Environment*, 65(2), 121–131.

Schneider, A., Friedl, M.A., McIver, D.K., and Woodcock, C.E. 2003. Mapping urban areas by fusing multiple sources of coarse resolution remotely sensed data. *Photogrammetric Engineering & Remote Sensing*, 69(12), 1377–1386.

Schneider, M., Friedl, A., and Potere, D. 2009. A new map of global urban extent from MODIS satellite data. *Environmental Research Letters*, 4, 044003 (11pp). doi: 10.1088/1748-9326/4/4/044003.

Schwarz, N., Lautenbach, S., and Seppelt, R. 2011. Exploring indicators for quantifying surface urban heat islands of European cities with MODIS land surface temperatures. *Remote Sensing of Environment*, 115(12), 3175–3186.

Seto, K.C. and Shepherd, J.M. 2009. Global urban land-use trends and climate impacts. *Current Opinion in Environmental Sustainability*, 1, 89–95.

Shettigara, V., Kempinger, S., and Aitchison, R. 1995. Semi-automatic detection and extraction of man-made objects in multispectral aerial and satellite images. In *Automatic Extraction of Man-Made Objects from Aerial and Space Images*, eds. A. Gruen, O. Kuebler, and P. Agouris. Birkhäuser, Basel, Switzerland, pp. 63–72.

Sirmacek, B. and Reinartz, P. 2011. Automatic crowd analysis from very high resolution satellite images. *International Archives of Photogrammetry, Remote Sensing and Spatial Information Sciences*, 38(3/W22), 221–226.

Sirmacek, B. and Unsalan, C. 2011. A probabilistic framework to detect buildings in aerial and satellite images. *IEEE Transactions on Geoscience and Remote Sensing*, 49(1), 211–221.

Small, C. 2003. High spatial resolution spectral mixture analysis of urban reflectance. *Remote Sensing of Environment*, 88, 170–186.

Small, C. 2005. A global analysis of urban reflectance. *International Journal of Remote Sensing*, 26, 661–681.

Sobrino, J.A., Oltra-Carrió, R., Sòria, G., Bianchi, R., and Paganini, M. 2011. Impact of spatial resolution and satellite overpass time on evaluation of the surface urban heat island effects. *Remote Sensing of Environment*, 117, 50–56.

Soergel, U., Michaelsen, E., Thiele, A., Cadario, E., and Thoennessen, U. 2009. Stereo analysis of high-resolution SAR images for building height estimation in cases of orthogonal aspect directions. *ISPRS Journal of Photogrammetry and Remote Sensing*, 64(5), 490–500.

Stathopoulou, M. and Cartalis, C. 2009. Downscaling AVHRR land surface temperatures for improved surface urban heat island intensity estimation. *Remote Sensing of Environment*, 113(15), 2592–2605.

Stathopoulou, M., Cartalis, C., and Chrysoulakis, N. 2006. Using midday surface temperature to estimate cooling degree-days from NOAA-AVHRR thermal infrared data: An application for Athens, Greece. *Solar Energy*, 80, 414–422.

Stathopoulou, M., Cartalis, C., and Keramitsoglou, I. 2004. Mapping micro-urban heat islands using NOAA/AVHRR images and CORINE Land Cover: An application to coastal cities of Greece. *International Journal of Remote Sensing*, 25, 2301–2316.

Stefanov, W.L. and Netzband, M. 2005. Assessment of ASTER land cover and MODIS NDVI data at multiple scales for ecological characterization of an arid urban center. *Remote Sensing of Environment*, 99, 31–43.

Streutker, D.R. 2002. A remote sensing study of the urban heat island of Houston, Texas. *International Journal of Remote Sensing*, 23, 2595–2608.

Streutker, D.R. 2003. Satellite-measured growth of the urban heat island of Houston, Texas. *Remote Sensing of Environment*, 85, 282–289.

Sutton, P.C., Taylor, M.J., and Elvidge, C.D. 2010. Using DMSP OLS imagery to characterize urban populations in developed and developing countries. In *Remote Sensing of Urban and Suburban Areas*, eds. Rashed, T. and Jürgens, C. Springer, Dordrecht, the Netherlands, pp. 329–348.

Takahashi, K.I., Terakado, R., Nakamura, J., Adachi, Y., Elvidge, C.D., and Matsuno, Y. 2010. In-use stock analysis using satellite nighttime light observation data. *Resources, Conservation, and Recycling*, 55, 196–200.

Unsalan, C. and Boyer, K.L. 2004. A system to detect houses and residential street networks in multispectral satellite images. In *Proceedings of the 17th International Conference on Pattern Recognition, 2004 (ICPR 2004)*.

Voogt, J.A. and Oke, T.R. 2003. Thermal remote sensing of urban climate. *Remote Sensing of Environment*, 86, 370–384.

Weng, Q. 2001. A remote sensing-GIS evaluation of urban expansion and its impact on surface temperature in the Zhujiang Delta, China. *International Journal of Remote Sensing*, 22, 1999–2014.

Weng, Q. 2003. Fractal analysis of satellite-detected urban heat island effect. *Photogrammetric Engineering & Remote Sensing*, 69, 555–566.

Weng, Q. 2009. Thermal infrared remote sensing for urban climate and environmental studies: Methods, applications, and trends. *ISPRS Journal of Photogrammetry and Remote Sensing*, 64(4), 335–344.

Weng, Q. and Fu, P. 2014. Modeling annual parameters of land surface temperature variations and evaluating the impact of cloud cover using time series of Landsat TIR data. *Remote Sensing of Environment*, 140, 267–278.

Weng, Q. and Hu, X. 2008. Medium spatial resolution satellite imagery for estimating and mapping urban impervious surfaces using LSMA and ANN. *IEEE Transaction on Geosciences and Remote Sensing*, 46(8), 2397–2406.

Weng, Q. and Lu, D. 2009. Landscape as a continuum: An examination of the urban landscape structures and dynamics of Indianapolis city, 1991–2000. *International Journal of Remote Sensing*, 30(10), 2547–2577.

Weng, Q., Lu, D., and Schubring, J. 2004. Estimation of land surface temperature-vegetation abundance relationship for urban heat island studies. *Remote sensing of Environment*, 89, 467–483.

Wu, S.S., Xu, B., and Wang, L. 2006. Urban land use classification using variogram-based analysis with aerial photographs. *Photogrammetric Engineering & Remote Sensing*, 72(7), 813–822.

Xian, G., Crane, M., and McMahon, C. 2008. Quantifying multi-temporal urban development characteristics in Las Vegas from Landsat and ASTER data. *Photogrammetric Engineering & Remote Sensing*, 74, 473–481.

Xian, G. and Homer, C. 2010. Updating the 2001 National Land Cover Database impervious surface products to 2006 using Landsat imagery change detection methods. *Remote Sensing of Environment*, 114, 1676–1686.

Xian, G., Homer, C., Bunde, B., Danielson, P., Dewitz, J., Fry, J., and Pu, R. 2012. Quantifying urban land cover change between 2001 and 2006 in the Gulf of Mexico region. *Geocarto International*, 27, 479–497.

Xian, G., Homer, C., Dewitz, J., Fry, J., Hossain, N., and Wickham, J. 2011. Change of impervious surface area between 2001 and 2006 in the conterminous United States. *Photogrammetric Engineering & Remote Sensing*, 77, 758–762.

Xiao, R., Weng, Q., Ouyang, Z., Li, W., Schienke, E.W., and Zhang, W. 2008. Land surface temperature variation and major factors in Beijing, China. *Photogrammetric Engineering & Remote Sensing*, 74(4), 451–461.

Xu, B. and Gong, P. 2007. Land use/cover classification with multispectral and hyperspectral EO-1 data. *Photogrammetric Engineering & Remote Sensing*, 73(8), 955–965.

Xu, B. and Gong, P. 2008. Noise estimation in a noise-adjusted principal component transformation and hyperspectral image restoration. *Canadian Journal of Remote Sensing*, 34(3), 1–16.

Xu, B., Gong, P., Biging, G., Liang, S., Seto, E., and Spear, R. 2004. Snail density prediction for schistosomiasis control using IKONOS and ASTER images. *Photogrammetric Engineering & Remote Sensing*, 70(11), 1285–1294.

Xu, B., Gong, P., Seto, E., Liang, S., Yang, C.H., Wen, S., Qiu, D.C., Gu, X.G., and Spear, R. 2006. A spatial temporal model for assessing the effects of inter-village connectivity in schistosomiasis transmission. *Annals of the Association of American Geographers*, 96(1), 31–46.

Xu, B., Gong, P., Seto, E., and Spear, R. 2003. Comparison of gray level reduction schemes with a revised texture spectrum method for land-use classification using IKONOS imagery. *Photogrammetric Engineering & Remote Sensing*, 69(5), 529–536.

Yang, J. and Wang, Y. 2012. Classification of 10 m-resolution SPOT data using a combined Bayesian Network Classifier-shape adaptive neighborhood method. *ISPRS Journal of Photogrammetry and Remote Sensing*, 72, 36–45.

Zakšek, K. and Oštir, K. 2012. Downscaling land surface temperature for urban heat island diurnal cycle analysis. *Remote Sensing of Environment*, 117, 114–124.

Zhang, Q., Wang, J., Gong, P., and Shi, P. 2003. Study of urban spatial patterns from SPOT panchromatic imagery using textural analysis. *International Journal of Remote Sensing*, 24, 4137–4160.

Zhang, X., Friedl, M.A., Schaaf, C.B., Strahler, A.H., Hodges, J.C.F., Gao, F., Reed, B.C., and Huete, A. 2003. Monitoring vegetation phenology using MODIS. *Remote Sensing of Environment*, 84(3), 471–475.

Zhizhin, M., Elvidge, C.D., Hsu, F.-C., and Baugh, K. 2013. Using the short-wave infrared for nocturnal detection of combustion sources in VIIRS data. *Proceedings of the Asia-Pacific Advanced Network*, 35, xx–zz. http://dx.doi.org/10.7125/APAN.35.x.

Zhu, C., Shi, W., Pesaresi, M., Liu, L., Chen, X., and King, B. 2005. The recognition of road network from high-resolution satellite remotely sensed data using image morphological characteristics. *International Journal of Remote Sensing*, 26(24), 5493–5508.

Zhu, G. and Blumberg, D.G. 2002. Classification using ASTER data and SVM algorithms: The case study of Beer Sheva, Israel. *Remote Sensing of Environment*, 80, 233–240.

Section II

Global Urban Footprint

Data Sets and Products

Section II

5 Mapping Global Human Settlements Pattern Using SAR Data Acquired by the TanDEM-X Mission

Thomas Esch, Mattia Marconcini, Andreas Felbier, Achim Roth, and Hannes Taubenböck

CONTENTS

5.1 TanDEM-X MISSION

DLR's first Earth observation radar satellite, TerraSAR-X, was launched in June 2007, followed by the almost identical TanDEM-X satellite 3 years later. Both systems are the basis for the first bistatic spaceborne SAR mission TanDEM-X (TerraSAR-X add-on for Digital Elevation Measurement), with TerraSAR-X and TanDEM-X orbiting in a unique helix formation with a typical distance of 250–500 m between the two satellites (Krieger et al., 2007). The primary goal of the TanDEM-X mission, which started in 2011, is the generation of a consistent global digital elevation model (DEM) with unprecedented spatial detail and accuracy. TerraSAR-X and TanDEM-X acquire HR/VHR SAR data in three basic imaging modes, namely, SpotLight (SL) and High-Resolution SpotLight mode (HS), StripMap mode (SM), and ScanSAR mode (SC) (Roth et al., 2005).

The targeted accuracy of the global DEM generated in the context of the TanDEM-X mission follows the High-Resolution Terrain Information 3 (HRTI-3) specifications. The elevation product will be provided in geographic coordinates. The grid spacing is 0.4 arcsec in latitude and longitude corresponding to a ground resolution of ~12 × 12 m. The longitude spacing is varied every 10° beginning at 50° and every 5° from 80° to 90° in order to compensate for the convergence of the meridians at the poles. Hence, the coarsest spacing is 4 arcsec between 85° and 90° latitude. In the TanDEM-X mission, the TerraSAR-X and TanDEM-X satellites record data in the single-polarized StripMap mode with a resolution of 3.3 m in azimuth and 1.2 m in range. Depending on the incidence angle, the range resolution converts into 3.0–3.5 m ground resolution (Eineder et al., 2010). The position of the strip acquired during the second acquisition phase (2012/2013) is shifted by approximately 50% compared to the first year (2011/2012).

The TanDEM-X mission will provide a global DEM (to be finished by mid-2015) that will boost research requiring detailed relief information across the globe. Moreover, it provides a highly reconfigurable platform for testing and demonstrating new SAR techniques and potential thematic applications. This includes, among others, along-track interferometry or experiments exploring the so-called dual receive antenna mode required for polarimetric interferometry. A comprehensive description of the potential applications, products, and data access procedures are provided in the TanDEM-X Science Plan (Hajnsek et al., 2010).

5.2   WORLDWIDE MAPPING OF HUMAN SETTLEMENT PATTERNS USING TanDEM-X DATA

The beginning of the twenty-first century represents a historic moment in the development of humankind since the number of urban residents has exceeded the rural population for the first time in history, marking the start of an "urban century" (UN, 2011). On the basis of this development, it is expected that the near future will be characterized by rapid and continuous global urbanization with almost two-thirds of the world's population living in cities by 2030. Hence, the monitoring of urban and peri-urban development is a key issue to analyze and understand the complexity, cross-linking, and increasing dynamics of sprawling and transforming settlements and settlement patterns in order to ensure their sustainable future development.

5.2.1   GLOBAL URBAN FOOTPRINT INITIATIVE

Spaceborne Earth observation has successfully been established as a technology to provide the required global and up-to-date geoinformation on the position, distribution, and development of human settlements (Potere and Schneider, 2009). However, the geometric resolution of the current global human settlement data sets (GHSDs)—for example, GLC2000 (Bartholomé and Belward, 2005), GLOBCOVER 2009 (Bontemps et al., 2011), or MODIS 500 (Schneider et al., 2009)—is limited to 300–1000 m. Therefore, it is not possible to detect villages or even smaller cities—a limitation that clearly hampers the potential to provide data that can be used to analyze important processes such as peri-urbanization. Moreover, most of the GHSDs are derived from optical data; therefore, especially in tropical regions, the data are collected over a comparably long period of time due to the almost permanent cloud coverage. However, recent activities aim at the generation of spatially and thematically enhanced GHSDs—for example, based on new Night Lights 2012 data (NASA, 2012), a built-up index applied to high-resolution (HR) optical and/or SAR data (Pesaresi et al., 2011, 2012), by integrating ASTER satellite images and GIS data (Miyazaki et al., 2013) or by properly analyzing global Envisat-ASAR imagery (Gamba and Lisini, 2012).

With two global coverages of VHR SAR data acquired at 3 m spatial resolution and collected within a period of about 1 year, the German TanDEM-X mission is predestined to be included among the new initiatives aiming at the provision of innovative GHSDs. Accordingly, the German Remote Sensing Data Center (DFD) of the German Aerospace Center (DLR) has developed and implemented a fully automated, operational image processing and analysis procedure—the Urban Footprint Processor (UFP)—that detects and delineates built-up areas from the global TanDEM-X mission data. The outputs of the UFP are binary settlement masks—the Urban Footprint (UF) masks—indicating built-up and non-built-up areas at a spatial resolution of 0.4 arcsec (~12 m). The global coverage of UF data sets will then be used to generate a worldwide inventory of human settlements—the Global Urban Footprint (GUF) layer—that is also intended to be made publicly available at a spatial resolution of 3.0 arcsec (~75 m). Therewith, the GUF will be of paramount importance in supporting worldwide analyses of settlement patterns in urban and rural areas, and will complement existing GHSDs and global land-cover maps that are mostly based on medium- (MR) or high-resolution (HR) optical imagery.

5.2.2 Urban Footprint Processor

A number of analyses have demonstrated the general suitability of VHR and HR SAR imagery to serve as a basis for the identification and mapping of built-up areas (Esch et al., 2006; Stasolla and Gamba, 2008). In this framework, the basic methodological components for the identification and delineation of built-up areas from TerraSAR-X/TanDEM-X StripMap data have been introduced by the authors in Esch et al. (2010, 2011, 2012). A texture measure is performed as a first step, followed by a classification procedure that focuses on the combined analysis of local backscattering characteristics (intensity) and local image heterogeneity (texture). Since this basic approach is primarily focused on the analysis of strong, local backscattering centers (i.e., corner reflections) appearing in a highly heterogeneous image region (i.e., urban areas), the algorithm is mainly sensitive toward the detection of man-made structures showing a vertical dimension, whereas flat, smooth urban entities such as runways or paved squares will not be classified as built-up areas.

In order to effectively process the TanDEM-X mission mass data set of about 300 TB—one coverage comprises ~180,000 complex SAR images with each image having an average size of ~50,000 × 40,000 pixels—the previously cited approach has systematically been enhanced and transformed into a fully automatic processing chain with several additional modules and functionalities (Esch et al., 2013).

This processing chain—the so-called Urban Footprint Processor—takes single look slant range complex (SSC) StripMap data of one TanDEM-X mission coverage (2011/2012) as input. The image analysis and classification module consists of three main components: (1) feature extraction, (2) classification, and (3) mosaicking and postediting. The input data as well as the intermediate layers and the final products are stored in a specific database W42 (Habermeyer et al., 2009), facilitating the effective handling and querying of the data. A schematic view on the UFP concept is given in Figure 5.1.

5.2.2.1 Feature Extraction

The first module of the UFP is dedicated to the extraction of the so-called speckle divergence (Esch et al., 2010), which allows one to derive texture information suitable for highlighting regions characterized by heterogeneous and highly structured built-up area. In particular, due to the strong scattering from double-bounce effects in urban areas typical of SAR data, the focus is on the analysis of the local speckle and its development is estimated accounting for the local image heterogeneity $\mathcal{H}$ (Esch et al., 2012) defined as:

$$\mathcal{H} = \frac{\sigma_{\mathcal{A}}}{\mu_{\mathcal{A}}} \tag{5.1}$$

where $\sigma_{\mathcal{A}}$ and $\mu_{\mathcal{A}}$ represent the standard deviation and mean, respectively, of the original backscattering amplitude image $\mathcal{A}$ (stored inside a single look slant range complex image product, SSC) computed in a local neighborhood. According to Lee et al. (1992), the image heterogeneity $\mathcal{H}$, the fading texture $\mathcal{F}$ (which represents the heterogeneity caused by speckle), and the true image texture $\mathcal{T}$ are related as follows:

$$\mathcal{H}^2 = \mathcal{T}^2\mathcal{F}^2 + \mathcal{T}^2 + \mathcal{F}^2 \tag{5.2}$$

On the one side, homogeneous surfaces without true structuring (e.g., noncultivated bare soil, grasslands) exhibit almost no true texture $\mathcal{T}$; hence, $\mathcal{H}$ approximates the fading texture $\mathcal{F}$, which results in very low values with the backscattering almost randomly distributed. On the other side, textured surfaces (e.g., urban environments, woodlands) show a significant amount of directional, non-Gaussian backscattering and are characterized by distinct structures, which result in high values for $\mathcal{T}$, and hence in high $\mathcal{H}$. This behavior is taken into consideration for defining a robust measure that properly describes the true image texture $\mathcal{T}$. In particular, we account for the difference between the scene-specific heterogeneity $\mathcal{F}$ and the measured local image heterogeneity $\mathcal{H}$ and, reasonably, we assume that $\mathcal{T}$ grows with the increasing amount of real structures within the resolution cell. According to Equation 5.2, the square of the local true image texture $\mathcal{T}^2$ (i.e., the speckle divergence $\mathcal{S}$) is obtained as

$$\mathcal{T}^2 = \mathcal{S} = \frac{\left(\mathcal{H}^2 - \mathcal{F}^2\right)}{\left(1 - \mathcal{F}^2\right)} \tag{5.3}$$

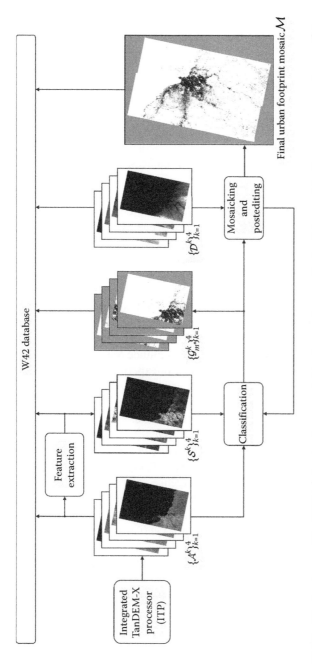

FIGURE 5.1 Block scheme of the UFP ($\mathcal{A}$ denotes the geocoded backscattering amplitude, $\mathcal{S}$ the speckle divergence texture feature, $\mathcal{G}$ the produced binary UF map, and $\mathcal{D}$ the digital elevation model). The example refers to the four scenes available for Dar es Salaam (Tanzania).

It is worth noting that $\mathcal{F}$ can be computed as a function of the number of looks N of a given SAR image. In particular, as done in Huang and Genderen (1996), we approximated it as $\hat{\mathcal{F}} = 0.5233 \cdot N^{-0.5}$. Since we considered radiometrically unenhanced SSC products, it holds that $N = N_A \cdot N_R$, where N_A and N_R represent the number of looks in azimuth and range, respectively. In the case of the complex SAR data used in this study, both N_A and N_R are equal to 1, finally leading to a final constant value $\mathcal{F} = 0.5233$ (Auquiere, 2001).

In order to reduce the amount of data (due to technical restrictions), a multilooking is finally performed for rescaling both $\mathcal{A}$ and $\mathcal{S}$ to a spatial resolution of ~0.4 arcsec (~12 m), that is, concurrently the highest resolution in which the global DEM produced in the context of the TanDEM-X mission will be made available. For this purpose, we used the method described in Eineder et al. (2004), which has already been implemented in the TerraSAR-X multimode SAR processor.

5.2.2.2   Classification

The aim of the second module of the UFP is to derive a binary settlement layer (built-up, non-built-up) for the investigated scene provided as input with the backscattering image $\mathcal{A}$ and the corresponding speckle divergence $\mathcal{S}$. This is carried out by means of a fully automatic and unsupervised technique whose main features are as follows.

Pixels showing high values of $\mathcal{S}$ correspond to urbanized areas, while those associated with lower values correspond to non-built-up structures. Therefore, for each investigated scene, the objective is to determine a specific optimal threshold for $\mathcal{S}$ capable of effectively discriminating between built-up and non-built-up areas.

All those pixels with a backscattering amplitude lower than the prefixed threshold $Th^A = 100$ (i.e., derived from a number of experiments on hundreds of different images and corresponding to ~−10 dB) are initially marked as nonurban and are associated with information classes not belonging to built-up areas (e.g., surfaces with a smooth mesoscale roughness, water bodies).

A set of M candidate thresholds for $\mathcal{S}$, $\{Th^S_m\}^M_{m=1}$, $Th^S_1 > \cdots > Th^S_M$, is then determined based on the specific image dynamics. For each of them, samples are categorized into urban ($\mathcal{U}_m$) or nonurban ($\mathcal{L}_m = \mathcal{L}_m(x_A, x_S)$) candidates depending on whether the corresponding speckle divergence value is greater or lower than Th^S_m, respectively. Afterward, we compute the Jensen–Shannon divergence $D_{JS}[U_m \| L_m]$ (Lin, 1991) accounting for both $\mathcal{A}$ and $\mathcal{S}$, which allows us to estimate the "distance" between $U_m = U_m(x_A, x_S)$ and $L_m = L_m(x_A, x_S)$ (i.e., the probability distributions of $\mathcal{U}_m$ and $\mathcal{L}_m$, respectively), where $x_A \in \mathcal{A}$ and $x_S \in S$.

$D_{JS}[U_m \| L_m]$ increases with increasing Th^S_m, while it decreases as the threshold gets lower. As soon as the two distributions U_m and L_m start to significantly overlap, there is a consistent fall in $D_{JS}[U_m \| L_m]$. When this happens, it means that the corresponding threshold $Th^S_{m^*}$ allows optimal separation between the urban and nonurban distributions based on the specific statistics of the image under analysis. Then the subset $\mathcal{U}_{m^*}$ is employed for training a one-class classifier based on support vector data description—SVDD (Tax and Duin, 2004). SVDD is based on the principles of support vector machines (SVM) and aims at determining the hypersphere with minimum radius enclosing all the training samples available for the unique class of interest. In our case, samples falling inside the boundary are finally associated with built-up areas, whereas all the others are labeled as nonurban. This approach permits increasing generalization and obtaining a more consistent and reliable final UF map $\mathcal{G}_{m*}$.

5.2.2.3 Mosaicking and Postediting

The last module of the UFP is dedicated to specific mosaicking and masking operations that are applied in an automated postediting phase to further improve the quality of the generated UF product. Let us assume that K different SAR images $\{\mathcal{I}^k\}_{k=1}^K$ are available and that the corresponding geocoded UF maps $\{\mathcal{G}_{m*}^k\}_{k=1}^K$ (obtained as output of the second module of the UFP) are employed for creating the final UF mosaic $\mathcal{M}$ for the investigated study region. The criterion adopted for selecting the optimal threshold for $\mathcal{S}$ generally results in a robust performance. Nevertheless, in some cases it could happen that one or a few UFs exhibit slight under- or overestimation of urban areas with respect to corresponding neighboring UFs in mosaic $\mathcal{M}$ (which might then show a sort of striping effect). This phenomenon mostly occurs for scenes located at the coastline and generally includes mainly water with just a few land areas; indeed, this leads to extreme distributions that hinder the proper definition of accurate classification settings. In order to solve this problem, we implemented a simple but effective technique, which accounts for the fact that each TanDEM-X image partly overlaps with at least four of its neighbors. In particular, by comparing the number of samples categorized as urban falling in the intersections between neighboring scenes, we can identify UFs showing under- or overestimation. We assume under- or overestimation as soon as the analyzed UF shows a significant systematic trend in the deviation of the estimated urban area compared to all its neighboring UF scenes. In case of underestimation, we generate three additional versions $\{\mathcal{G}_{m*+i}^{\bar{k}}\}_{i=1}^3$ (where $\bar{k}$ denotes the UF outlier tile to refine) by choosing as many relaxed thresholds for $\mathcal{S}$, whereas in case of overestimation, we generate three additional maps $\{\mathcal{G}_{m*-i}^{\bar{k}}\}_{i=1}^3$ by choosing as many stricter thresholds for $\mathcal{S}$. In both circumstances, we then select the version that fits best with its neighbors (i.e., the one exhibiting the lowest difference with respect to the neighboring UFs in terms of the number of urban samples).

It is worth noting that sometimes highly mountainous areas are misclassified as built-up regions since they show high values for both $\mathcal{A}$ and $\mathcal{S}$ as an effect of the high backscattering and texture in areas showing layover or foreshortening. To solve this issue, a dedicated mask has been implemented by taking into consideration the ASTER Global DEM (NASA, 2013) and marking all those pixels showing a slope (i.e., the maximum rate of height change between each pixel and its closest eight neighbors computed as in Horn [1981]) higher than $20°$ in the neighborhood of a local peak as nonurban. With this approach, the rate of false detections in areas with layover or foreshortening could be decreased significantly. As soon as the high-resolution TanDEM-X mission DEM becomes available in 2015, this new data set will be used to further optimize the presented procedure for the correction of topographic effects.

5.2.3   URBAN FOOTPRINT SETTLEMENT MASKS

So far, the UFP has produced both $\mathcal{A}$ and $\mathcal{S}$ for a total of 140,000 scenes acquired in the context of the first TanDEM-X mission coverage (2011/2012) with each scene covering an area of ~50 × 30 km. Moreover, a number of globally distributed test runs for the final UF/GUF generation have been performed investigating either single scenes or extensive mosaics consisting of several hundred images. An accuracy assessment of the corresponding results showed that the overall accuracies mostly lie in the range of 70%–90%. The validation was performed by visually comparing 1500 randomly distributed reference points for the study regions to VHR satellite data, aerial imagery, or other suitable reference data available. These results are in line with the outcomes of earlier studies investigating the methodological precursors of the UFP technique (Esch et al., 2010, 2011, 2012; Taubenböck et al., 2011, 2012).

Figure 5.2 shows three mosaics for the region of Japan based on TanDEM-X backscattering amplitude images ($\mathcal{A}$), the derived texture data ($\mathcal{S}$), and the UF/GUF

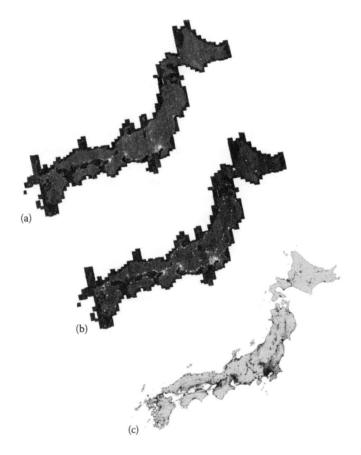

FIGURE 5.2   Japan mosaics (608 scenes) generated from (a) the original backscattering amplitude images, (b) the extracted speckle divergence texture features, and (c) the final settlement masks obtained with the UFP.

classification, respectively. Each mosaic is composed of 608 single data sets covering a total area of ~378.000 km². The UF/GUF for entire Japan clearly visualizes the concentration of urban centers along the coastline. The large urban hubs such as Osaka or Tokyo stick out, with Tokyo, the largest metropolitan area in the world with approximately 36 million inhabitants, showing an immense urban spread into the hinterland. To further pinpoint the characteristics of the UF/GUF product, Figure 5.3 illustrates several exemplary spatial zooms to specific regions of the

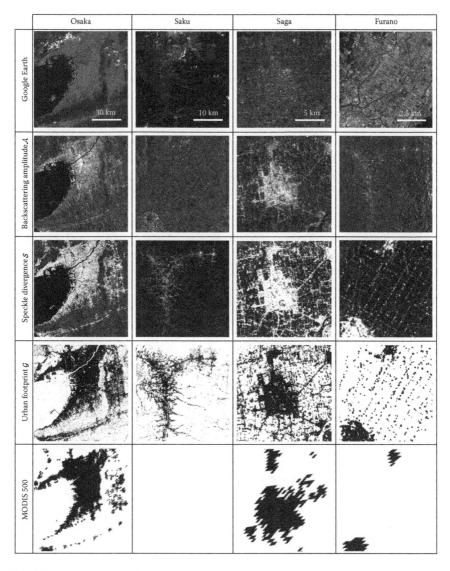

FIGURE 5.3 (See color insert.) Optical data (from Google Earth), TanDEM-X mission backscattering amplitude $\mathcal{A}$, speckle divergence $\mathcal{S}$, urban footprint $\mathcal{G}$, and the corresponding MODIS 500 urban class map for the Japanese cities of Osaka, Saku, Saga, and Furano.

mosaic of Japan. The UFP products $\mathcal{A}$, $\mathcal{S}$, and UF/GUF are opposed to a VHR optical image and the urban area classification provided by the global MODIS 500 layer. The UF/GUF masks highlight the spatial diversity among the settlement patterns. Osaka, a large coastal city, clearly features an urban center with complex patterns in the outskirts due to the hilly terrain in the hinterlands. In Saku, a comparatively small city, the orographic situation of a valley determines the longitudinal urban outline with axial development directions. Saga, also a small urbanized area, shows the most typical structure, with a clear urban center and decreasing densities in suburban areas. In the case of the rural Furano region, the UF/GUF clearly indicates the two urban settlements in that region along with several small villages distributed in between them. It is interesting to note that the MODIS 500 results of the Urban Footprints generally captured the same basic urban extents—at least for the large city of Osaka. However, the capability to delineate the settlement patterns is limited to the urban core, while the details in the hinterlands with lower densities are not adequately recognized. This becomes obvious for the Saku case, as MODIS 500 did not detect any urbanized areas in this complex terrain or in the Furano region where not all of the small villages were identified. The examples highlight that the UF/GUF products hold certain potential to provide valuable geoinformation for a spatially more detailed delineation of urban, peri-urban, and rural settlement patterns compared to currently available GHSDs.

Settlement patterns across the globe are highly diverse. This becomes obvious by looking at the globally distributed UF/GUF masks generated for the regions of Rome (Italy), Izmir (Turkey), Oklahoma City (United States), and Zanzibar (Tanzania) shown in Figure 5.4. The different scales of the given examples allow us to assess the capabilities of the UF/GUF to see large area settlement patterns in their spatial configurations with, for example, Rome or Izmir as dominant urban centers and a hierarchical system of smaller urban centers or low-density rural settlements. At the same time, structural details of cities are captured, as the examples of Zanzibar and Oklahoma display. The capability of the algorithm to even ignore open spaces without any vertical structures such as buildings or green belts within the urban centers becomes obvious. The examples stress that the settlement patterns can be extracted and characterized for diverse geographical regions and landscape types using the TanDEM-X mission imagery and the UFP technique, respectively.

However, the previous validation studies of the UF/GUF products have shown that errors mainly comprise errors of commission originating from extreme local topographies typically occurring in mountainous or alpine regions as well as areas showing strong erosion patterns or a high local mesoscale roughness (e.g., particular forms of stone deserts or glacier ice fields). Nevertheless, if spatially not too detailed, the corresponding errors can be eliminated at the DEM-based postprocessing stage. For the remaining cases, we work on specific pattern recognition techniques to identify the nonurban high-texture regions. In addition to the effects mentioned earlier, man-made objects with a vertical dimension—for example, poles of power lines, railway tracks, fortified watersides—also lead to false detections of assumed settlements. Here, the specific pattern recognition techniques are also supposed to help reduce the number of false alarms.

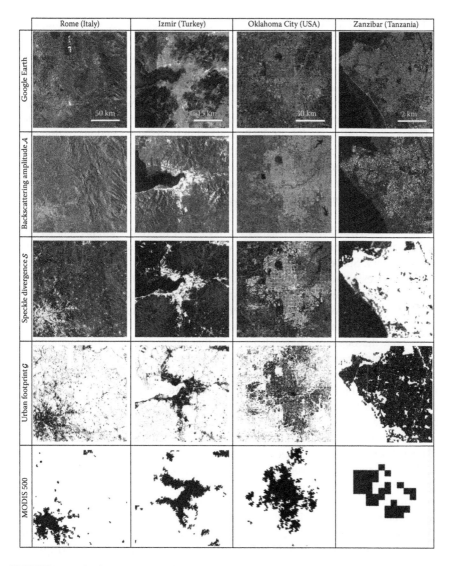

FIGURE 5.4 Optical data (from Google Earth), TanDEM-X mission backscattering amplitude $\mathcal{A}$, speckle divergence $\mathcal{S}$, urban footprint $\mathcal{G}$, and the corresponding MODIS 500 urban class map for the cities of Rome (Italy), Izmir (Turkey), Oklahoma City (United States), and Zanzibar (Tanzania).

5.3 CONCLUSIONS AND OUTLOOK

In this chapter, we have highlighted the new perspectives for the generation of a worldwide inventory of human settlements arising from the analysis of the unique set of VHR SAR imagery collected by the TanDEM-X mission. In this context, we presented DLR's GUF idea along with its technical implementation, that is, the fully automatic processing and image analysis system of the UFP. Using the

TanDEM-X mission data of the first coverage (~140.000 images), we could test the performance of the processing chain and assess the resulting UF/GUF mask on the basis of several thousand globally distributed images. The results of these tests demonstrate the high potential of the GUF approach to provide a spatially detailed map of global settlement patterns for urbanized areas as well as for rural regions. Moreover, the acquisition of the global data set within the comparably short period of ~12 months qualifies this layer as an interesting baseline product for future comparative studies of the urbanization in different regions of the world from a local up to a global level. With these properties, the GUF data set holds a certain potential to boost applications and analyses in the context of urbanization patterns, peri-urbanization, spatiotemporal dynamics of settlement development as well as population estimation, vulnerability assessment, or the modeling of global change.

With the described characteristics, the GUF will provide a unique data set that is to some extent complementary to the existing GHSDs derived from MR or HR optical imagery. Spaceborne SAR systems, as opposed to optical sensors, are active imaging devices operating in a single wavelength or frequency. Hence, the appearance of surfaces or objects in the SAR imagery is determined by geometrical and dielectric properties of the illuminated objects than by the biophysical or chemical characteristics as in the case of optical data.

However, the outcomes have also shown that the accuracy of the methodology can still be improved, in particular in the context of reducing the false alarms rate in highly textured areas showing backscattering characteristics that are locally quite similar to that of built-up areas (e.g., regions featuring many man-made infrastructures not representing settlements, extreme topographies, or peculiar forms of rice fields and forest areas). Hence, we are currently enhancing and extending the automated postediting module in order to specifically improve the performance in critical areas. Moreover, a systematic, worldwide validation campaign based on globally distributed in situ ground-truth information is scheduled for the near future. In the overall perspective, the completion of a first global GUF data set is planned for 2014. In addition to the full-resolution 12 m product, a public domain version will be made available at a spatial resolution of 3.0 arcsec (~75 m).

Regarding follow-on research and development, it is planned to adapt the UFP to all TerraSAR-X/TanDEM-X imaging modes (ScanSAR, SpotLight) as well as to other SAR satellites such as Sentinel-1 or Radarsat-2. Moreover, the calculation and consideration of long-term coherences will be investigated—between the first and second TanDEM-X mission coverage, but also based on data from satellites like Sentinel-1. Initial studies have also shown the potential to characterize building structures and estimate building densities based on texture measures or the modeling of building volume on building block level using the VHR DEM data generated on the basis of TanDEM-X mission imagery. Considering ongoing preparations of various HR GHSDs by different teams, a systematic comparison of the methodologies and results might help in identifying synergies as well as potentials for the enhancement of each GHSD by improving the underlying techniques or by combining or including intermediate or final products of the different approaches.

ACKNOWLEDGMENTS

The authors wish to thank the TerraSAR-X and TanDEM-X Science Teams for providing the SAR data for this study. We also thank the German Federal Ministry of Economics and Technology (BMWi) and the European Commission (EC) for their financial support in the research projects TerraSAR-X urban, Geoland2—Euroland, and SENSUM.

REFERENCES

Auquiere, E. (2001). SAR temporal series interpretation and backscattering modeling for maize growth monitoring, PhD thesis, Presses universitaires de Louvain, Louvain, Belgium.

Bartholomé, E. and Belward, A. S. (2005). GLC2000: A new approach to global land cover mapping from Earth observation data. *International Journal of Remote Sensing*, 26(9), 1959–1977.

Bontemps, S., Defourny, P., Van Bogaert, E., Arino, O., Kalogirou, V., and Ramos Perez, J. J. (2011). GLOBCOVER 2009 products description and validation report, Université catholique de Louvain (UCL) & European Space Agency (ESA), issue 2.2.

Eineder, M., Boerner, E., Breit, H., Holzner, J., Freitz, T., Palubinskas, G., and Balss, U. (2004). TerraSAR-X—Payload ground segment—TMSP design. TX-PGS-DD-300, issue 1.0. http://ophrtsxgss.intra.dlr.de/ (accessed January 31, 2013).

Eineder, M., Fritz, T., Mittermayer, J., Roth, A., Börner, E., Breit, H., and Bräutigam, B. (2010). TerraSAR-X ground segment basic product specification document. TX-GS-DD-3302, issue 1.7. http://sss.terrasar-x.dlr.de (accessed January 31, 2013).

Esch, T., Marconcini, M., Felbier, A., Roth, A., Heldens, W., Huber, M., Schwinger, M., and Müller, A. (2013). Urban footprint processor—Fully automated processing chain generating settlement masks from global data of the TanDEM-X mission. *Geoscience and Remote Sensing Letters*, Special Stream EORSA2012, 10(6), 1617–1621.

Esch, T., Roth, A., and Dech, S. (2006). Analysis of urban land use pattern based on high resolution radar imagery. *Proceedings of the 2006 IEEE Geoscience and Remote Sensing Symposium (IGARSS 2006)*, Denver, CO, 31 July–4 August, pp. 3615–3618.

Esch, T., Schenk, A., Ullmann, T., Thiel, M., Roth, A., and Dech, S. (2011). Characterization of land cover types in TerraSAR-X images by combined analysis of speckle statistics and intensity information. *IEEE Transactions on Geoscience and Remote Sensing*, 49(6), 1911–1925.

Esch, T., Taubenböck, H., Roth, A., Heldens, W., Felbier, A., Thiel, M., Schmidt, M., Müller, A., and Dech, S. (2012). TanDEM-X mission—New perspectives for the inventory and monitoring of global settlement patterns. *Journal of Applied Remote Sensing*, 6(1), 1–21.

Esch, T., Thiel, M., Schenk, A., Roth, A., Müller, A., and Dech, S. (2010). Delineation of urban footprints from TerraSAR-X data by analyzing speckle characteristics and intensity information. *IEEE Transactions on Geoscience and Remote Sensing*, 48(2), 905–916.

Gamba, P. and Lisini, G. (2012). A robust approach to global urban area extent extraction using ASAR Wide Swath Mode data. *Proceedings of Tyrrhenian Workshop on Advances in Radar and Remote Sensing (TyWRRS)*, Naples, Italy, 12–14 September, pp. 1–5.

Habermeyer, M., Marschalk, U., and Roth, A. (2009). W42—A scalable spatial database system for holding digital elevation models. *Proceedings of the 17th International Conference of Geomatics*, Fairfax, VA, 12–14 August.

Hajnsek, I., Busche, T., Fiedler, H., Krieger, G., Buckreuss, S., Zink, M., Moreira, A., Wessel, B., Roth, A., and Fritz, T. (2010). TanDEM-X ground TanDEM-X science plan. TD-PD-PL-0069, issue 1.0. http://tandemx-science.dlr.de (accessed January 31, 2013).

Horn, B. (1991). Hill shading and the reflectance map. *Proceedings of the IEEE*, 69, 14–47.

Huang, Y. and Genderen, J. L. (1996). Evaluation of several speckle filtering techniques for ERS-1 & 2 Imagery. *International Archives of Photogrammetry and Remote Sensing*, vol. XXXI, Part B2, Vienna, Austria.

Krieger, G., Moreira, A., Fiedler, H., Hajnsek, I., Werner, M., Younis, M., and Zink, M. (2007). TanDEM-X: A satellite formation for high resolution SAR interferometry. *IEEE Transactions on Geoscience and Remote Sensing*, 45(11), 3317–3341.

Lee, J.-S., Hoppel, K., and Mango, S. A. (1992). Unsupervised estimation of speckle noise in radar images. *International Journal of Imaging Systems and Technology*, 4(4), 298–305.

Lin, J. (1991). Divergence measures based on the Shannon entropy. *IEEE Transactions on Information Theory*, 37(1), 145–151.

Miyazaki, H., Shao, X., Iwao, K., and Shibasaki, R. (2013). An automated method for global urban area mapping by integrating ASTER satellite images and GIS data. *IEEE Journal of Selected Topics in Applied Earth Observations and Remote Sensing*, 6(2), 1–27.

National Aeronautics and Space Administration (NASA). (2012). Night Lights 2012. http://earthobservatory.nasa.gov/IOTD/view.php?id = 79803 (accessed January 31, 2013).

National Aeronautics and Space Administration (NASA). (2013). ASTER Global Digital Elevation Model Version 2. http://asterweb.jpl.nasa.gov/gdem.asp (accessed January 31, 2013).

Pesaresi, M., Blaes, X., Ehrlich, D., Ferri, S., Gueguen, L., Haag, F., Halkia, M. et al. (2012). A global human settlement layer from optical high resolution imagery—Concepts and first results. Scientific and Technical Research Reports, Publications Office of the European Union, DOI: 10.2788/74059 (print) 10.2788/73897 (online).

Pesaresi, M., Ehrlich, D., Caravaggi, I., Kauffmann, M., and Louvrier, C. (2011). Towards global automatic built-up area recognition using optical VHR imagery. *IEEE Journal of Selected Topics in Applied Earth Observations and Remote Sensing*, 4(4), 923–934.

Potere, D. and Schneider, A. (2009). Comparison of global urban maps. In: Gamba, P. and Herold, M. (Eds.), *Global Mapping of Human Settlements: Experiences, Data Sets, and Prospects*, Taylor & Francis Group, Boca Raton, FL, pp. 269–308.

Roth, A., Hoffmann, J., and Esch, T. (2005). TerraSAR-X: How can high resolution SAR data support the observation of urban areas? *Proceedings of the ISPRS WG VII/1 Human Settlements and Impact Analysis—Third International Symposium on Remote Sensing and Data Fusion Over Urban Areas (URBAN 2005) and Fifth International Symposium on Remote Sensing of Urban Areas (URS 2005)*, Tempe, AZ, 14–16 March.

Schneider, A., Friedl, M. A., and Potere, D. (2009). A new map of global urban extent from MODIS data. *Environmental Research Letters*, 4(4), 044003.

Stasolla, M. and Gamba, P. (2008). Spatial indexes for the extraction of formal and informal human settlements from high-resolution SAR images. *IEEE Journal of Selected Topics in Applied Earth Observation and Remote Sensing*, 1(2), 98–106.

Taubenböck, H., Esch, T., Felbier, A., Roth, A., and Dech, S. (2011). Pattern-based accuracy assessment of an urban footprint classification using TerraSAR-X data. *IEEE Geoscience and Remote Sensing Letters*, 8(2), 278–282.

Taubenböck, H., Esch, T., Felbier, A., Wiesner, M., Roth, A., and Dech, S. (2012). Monitoring of mega cities from space. *Remote Sensing of Environment*, 117, 162–176.

Tax, D. M. J. and Duin, R. P. W. (2004). Support vector data description. *Machine Learning*, 54(1), 45–66.

United Nations (UN) (2011). *World Urbanization Prospects—The 2009 Revision*, United Nations, New York.

6 National Trends in Satellite-Observed Lighting
1992–2012

Christopher D. Elvidge, Feng-Chi Hsu,
Kimberly E. Baugh, and Tilottama Ghosh

CONTENTS

6.1 INTRODUCTION

Nighttime lights are a class of urban remote sensing products derived from satellite sensors with specialized low light imaging capabilities. To date, two sensors have collected global nighttime lights data. The original instrument is the operational linescan system (OLS) flown by the U.S. Air Force Defense Meteorological Satellite Program (DMSP). The earliest version of the OLS began collecting data in the early 1970s. A digital OLS data archive was established in 1992 at the National Oceanic and Atmospheric Administration (NOAA) National Geophysical Data Center (NGDC). The second instrument flown with a global collection capability for low light imaging data is the visible infrared imaging radiometer suite (VIIRS) flown on the NASA/NOAA SNPP satellite, launched in 2011. The VIIRS offers substantial improvements in spatial resolution, radiometric calibration, and usable dynamic range when compared to the DMSP low light imaging data.

The DMSP nighttime lights provide the longest continuous time series of global urban remote sensing products, now spanning 21 years. The flagship product is the stable lights, an annual cloud-free composite of average digital brightness value for the detected lights, filtered to remove ephemeral lights and background noise. The stable lights present a panoramic view of humanity from space (Figure 6.1). At a glance, one gets a sense of how population, commerce, and resource consumption are distributed. There are thousands of points of light forming clusters of various shapes, surrounded by the darkness of rural and ocean areas (Elvidge et al., 2001).

Other global urban remote sensing products focus on mapping areas of dense infrastructure, producing binary grids, assigning each grid cell to urban or nonurban. Small towns and development in rural areas are neglected due to a lack of diagnostic signals. In contrast, the DMSP and VIIRS measure a human activity—artificial lighting—which is commonly present wherever there is built infrastructure. DMSP and VIIRS are able to detect faint light sources from small towns and exurban development that are substantially smaller than the ground footprint of the observations. One of the advantages of nighttime lights over the binary urban maps is that they offer the user the option for additional processing to meet the objectives of their specific application. For instance, to prepare the nighttime lights for a study of dense urban cores, dim lighting from the sparse edges of urban areas can be removed by applying a threshold.

Because of their global extent, standardized production, and the relative ease with which DMSP nighttime lights can be accessed, they have been widely used as a proxy for other more difficult to measure variables. The logic is that urban processes are highly correlated with each other. If one process or activity can be measured well, it can be used to make reasonable estimates of others. As examples, nighttime lights have been used to map economic activity (Ghosh et al., 2010), fossil fuel carbon emissions (Rayner et al., 2010), spatial distribution of population (Doll, 2008; Sutton, 1997), poverty mapping (Elvidge et al., 2009a), density of constructed surfaces (Elvidge et al., 2007), food demand (Matsumura et al., 2009), water use (Zhao et al., 2011), and stocks of steel and other metals (Hsu et al., 2013).

The NGDC recently reprocessed the DMSP time series producing 33 annual products from six satellites spanning 21 years. This is referred to as the v.4 DMSP stable lights time series. One of the key questions for the science community is whether it is possible to reliably analyze changes in lighting across the time series. This capability is not assured since the visible band on the OLS has no in-flight calibration. From preflight calibrations, it is known that the sensors differ from each other in terms of radiometric performance. There are also minor differences in spectral bandpasses of the six instruments. Even at launch, the sensors had different detection limits and saturation radiances. The saturation radiance issue is serious since bright urban cores saturate in this time series. In addition, the individual sensors invariably degrade in optical throughput over time. To address all of these issues, we have developed an intercalibration that is designed to convert data values from individual satellite products into a common range defined by reference year.

This chapter has two primary objectives. The first objective is to evaluate whether changes in nighttime lights can be quantitatively analyzed with the intercalibrated v.4 DMSP stable lights time series. The second objective is to search for underlying

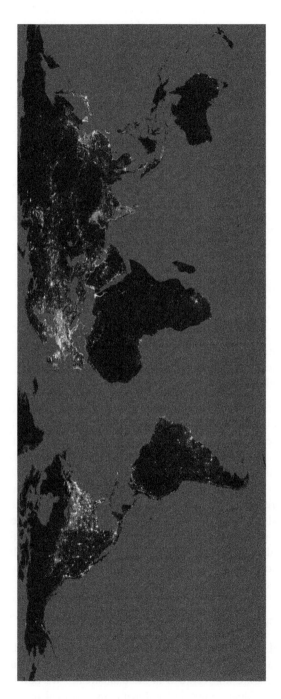

FIGURE 6.1 (See color insert.) DMSP stable lights for the year 2010.

variables that cause changes in satellite-observed lighting. Since the lighting observed by the satellite is from illuminated infrastructure, processes that result in the construction of new infrastructure are the causes of increase in satellite-observed lighting. We decided to explore the population and gross domestic product (GDP) as candidate variables since population growth and economic expansion are logical drivers for the expansion of infrastructure. The question on the behavior of lighting is whether there is a standard relationship between population and economic activity—or do countries differ in the manner in which lighting behaves in response to changes in population and economic activity levels?

6.2   REPROCESSING OF THE DMSP ARCHIVE

The DMSP archive was processed on annual increments using the data from each available satellite to produce a time series of global cloud-free stable lights products using the methods described in Baugh et al. (2010). Each orbit is analyzed in a stepwise fashion to identify the pixel set meeting the following criteria:

1. Center half of orbital swath (best geolocation, reduced noise, and sharpest features)
2. No sunlight present
3. No moonlight present
4. No solar glare contamination
5. Cloud-free (based on thermal detection of clouds)
6. No contamination from auroral emissions
7. Normal gain settings (no reduced gain data)
8. Masking of lighting in gas flaring areas

Nighttime image data from individual orbits that meet the aforementioned criteria are added into a global latitude–longitude grid (Platte Carree projection) having a resolution of 30 arcsec. This grid cell size is approximately a square kilometer at the equator. The total number of coverages and number of cloud-free coverages are also tallied. In the typical annual cloud-free composite, most areas have 20–100 cloud-free observations, providing a temporal sampling of the locations and brightness variation of lights present on the earth's surface. This is then filtered to remove fires based on their high digital number (DN) values and lack of persistence. Background noise is removed by setting thresholds based on visible band values found in areas known to be free of detectable lighting. In many years, data were available from two satellites, and two global composites were included in the analysis. The result is a set of 30 global products spanning 18 years (Table 6.1).

6.3   INTERCALIBRATION AND DATA EXTRACTION

Unfortunately, the OLS has no on-board calibration for the visible band. The preflight calibration for the individual instruments is of little value because the gain commands made to the instrument are not recorded in the data stream. We developed an intercalibration for the individual composites following an empirical procedure (Elvidge et al., 2009b). This is a regression-based procedure, relying on the assumption that

TABLE 6.1

Stable Lights Time Series Includes Data from Six Satellites and Spans 21 Years

Year	F-10	F-12	F-14	F-15	F-16	F-18
1992	F101992					
1993	F101993					
1994	F101994	F121994				
1995		F121995				
1996		F121996				
1997		F121997	F141997			
1998		F121998	F141998			
1999		F121999	F141999			
2000			F142000	F152000		
2001			F142001	F152001		
2002			F142002	F152002		
2003			F142003	F152003		
2004				F152004	F162004	
2005				F152005	F162005	
2006				F152006	F162006	
2007				F152007	F162007	
2008					F162008	
2009					F162009	
2010						F182010
2011						F182011
2012						F182012

the brightness of lighting in a reference area has changed little over time. Samples of lighting from human settlements (cities and towns) were extracted from numerous candidate reference areas and examined. In reviewing the data, it was found that the data from satellite year F121999 had the highest digital values. Because there is saturation (DN = 63) in the bright cores of urban centers and large gas flares, F121999 was used as the reference, and the data from all other satellite years were adjusted to match the F121999 data range. In examining the candidate intercalibration areas, it was found that many had a cluster of very high values (including saturated data with DN = 63) and a second cluster of very low values. We concluded that having a wide spread of DN values would be a valuable characteristic since it would permit a more accurate definition of the intercalibration equation. By examining the scattergrams of the DN values for each year versus F121999, we were able to observe evidence of changes in lighting based on the width of the primary data axis and quantity of outliers away from the primary axis. Our interpretation was that areas having very little change in lighting over time would have a clearly defined diagonal axis with minimal width. Of all the areas examined, Sicily had the most favorable characteristics—an even spread of data across the full dynamic range and a clearly defined diagonal cluster of points. A second-order regression model was developed for each satellite year, as shown in Table 6.2.

TABLE 6.2

Coefficients for the Intercalibration Applied the Digital Values in the Time Series

Satellite	Year	C_0	C_1	C_2	R^2	Number
F10	1992	−2.0570	1.5903	−0.0090	0.9075	35,720
F10	1993	−1.0582	1.5983	−0.0093	0.9360	38,893
F10	1994	−0.3458	1.4864	−0.0079	0.9243	36,494
F12	1994	−0.6890	1.1770	−0.0025	0.9071	34,485
F12	1995	−0.0515	1.2293	−0.0038	0.9178	37,571
F12	1996	−0.0959	1.2727	−0.0040	0.9319	35,762
F12	1997	−0.3321	1.1782	−0.0026	0.9245	37,413
F12	1998	−0.0608	1.0648	−0.0013	0.9536	37,791
F12	1999	0.0000	1.0000	0.0000	1.0000	39,157
F14	1997	−1.1323	1.7696	−0.0122	0.9101	36,811
F14	1998	−0.1917	1.6321	−0.0101	0.9723	36,701
F14	1999	−0.1557	1.5055	−0.0078	0.9717	38,894
F14	2000	1.0988	1.3155	−0.0053	0.9278	37,888
F14	2001	0.1943	1.3219	−0.0051	0.9448	38,558
F14	2002	1.0517	1.1905	−0.0036	0.9203	36,964
F14	2003	0.7390	1.2416	−0.0040	0.9432	38,701
F15	2000	0.1254	1.0452	−0.0010	0.9320	38,831
F15	2001	−0.7024	1.1081	−0.0012	0.9593	38,632
F15	2002	0.0491	0.9568	0.0010	0.9658	38,035
F15	2003	0.2217	1.5122	−0.0080	0.9314	38,788
F15	2004	0.5751	1.3335	−0.0051	0.9479	36,998
F15	2005	0.6367	1.2838	−0.0041	0.9335	38,903
F15	2006	0.8261	1.2790	−0.0041	0.9387	38,684
F15	2007	1.3606	1.2974	−0.0045	0.9013	37,036
F16	2004	0.2853	1.1955	−0.0034	0.9039	36,856
F16	2005	−0.0001	1.4159	−0.0063	0.9390	38,984
F16	2006	0.1065	1.1371	−0.0016	0.9199	37,204
F16	2007	0.6394	0.9114	0.0014	0.9511	37,759
F16	2008	0.5564	0.9931	0.0000	0.9450	37,469
F16	2009	0.9492	1.0683	−0.0016	0.8918	33,895
F18	2010	2.3430	0.5102	0.0065	0.8462	36,445
F18	2011	1.8956	0.7345	0.0030	0.9095	36,432
F18	2012	1.8750	0.6203	0.0052	0.9392	37,576

An extraction, which summed up the DN values for the lighting found in each country for each of the satellite years, was performed. Lighting from gas flares was masked out based on methods described by Elvidge et al. (2009). The extraction performs two important adjustments to the values for each grid cell. First, the inter-calibration is applied based on the offsets and coefficients listed in Table 6.2. The form of the calculation is $Y = C_0 + C_1X + C_2X^2$. Calculated values that run over 63 are truncated at 63. Thus, the application of the intercalibration increases the number of saturated pixels (DN = 63). The resulting values are then adjusted to compensate for the change in surface area in the 30 arc second grid. The digital values are then concatenated to derive the "sum-of-lights" index value (or SOL) from each satellite year for each country. To exclude the dim lighting detected in rural areas and to compensate for difference in the detection limits of the different products, only DNs of six or larger are added to the SOL. The output of the SOL extraction is a CSV file, which is then converted to a spreadsheet with charts showing the time series results for individual countries.

The objective of the intercalibration is to make it possible to detect changes in the brightness of lights across the time series. One indication of a successful intercalibration is the convergence of SOL values in years where two satellite products are available. Another indication of a successful intercalibration is the emergence of clear trajectories such as continuous growth in lighting across the time series. In most countries, the intercalibration yielded substantial convergence. The raw versus intercalibrated SOL for Egypt, showing both convergence of SOL values in individual years and the emergence of a clear upward trajectory in the steady growth in lighting from year to year, is shown in Figure 6.2. From reviewing the results for many other countries, it is clear that the intercalibration brought about substantial convergence in many countries. The lack of convergence within single year pairs in some countries may have been caused by changes in lighting activity between the overpass times of the satellites, which can differ by as much as 2 h.

6.4 DEFINING NIGHTTIME LIGHTS BEHAVIOR

Regressions were run between the annual SOL data versus annual population and annual GDP. The population data were drawn from U.S. Census Bureau, International Data Base (http://www.census.gov/ipc/www/idb/). As an indicator of economic activity, we use GDP data, normalized through a purchasing power parity (PPP) analysis, drawn from the World Development Indicators database (http://data.worldbank.org/data-catalog/world-development-indicators). In the data analysis, we make use of the correlation coefficient, or Pearson's R, which is a measure of the correlation (linear dependence) between two variables.

By examining the correlation coefficients for SOL with population and SOL and GDP (Figure 6.3), it is clear that most countries (90%) fall on a primary diagonal axis. This axis is defined by points where the R values with population and GDP are approximately the same, generally within 0.3 of each other. There are two sets of outlier countries, where the sign of the R values for population and GDP do not match. We divided the countries into seven groups based on the behavior of their lighting relative to population and GDP: (1) rapid growth, (2) moderate growth, (3) population centric, (4) GDP centric, (5) stable, (6) erratic, and (7) antipole. Each of the seven will be discussed next. The decision tree used to sort the countries into the seven groups is shown in Figure 6.4.

6.4.1 COUNTRIES WITH RAPID GROWTH IN LIGHTING

In the upper tip of the primary axis shown in Figure 6.3 is a tightly packed cluster of countries with very high correlation coefficients for both GDP and population. Examination of the SOL versus time charts for these countries indicates that they have had rapid growth in lighting over a substantial portion of the 20-year temporal record. Also, there

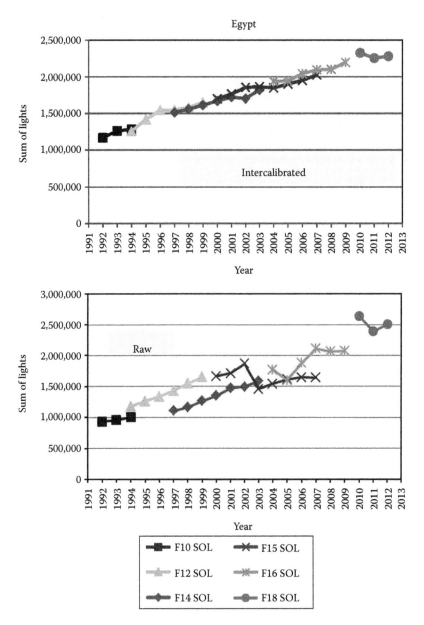

FIGURE 6.2   Raw versus intercalibrated SOL versus time results for Egypt. Note the convergence of the SOL values from different satellites for the same year. The growth trend is clear in the intercalibrated data.

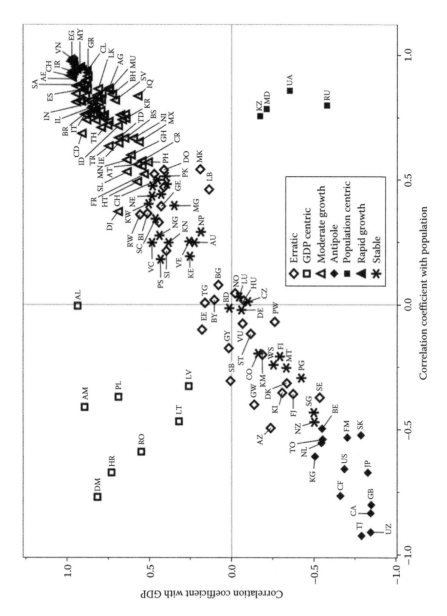

FIGURE 6.3 Scattergram of R values for SOL with population and GDP.

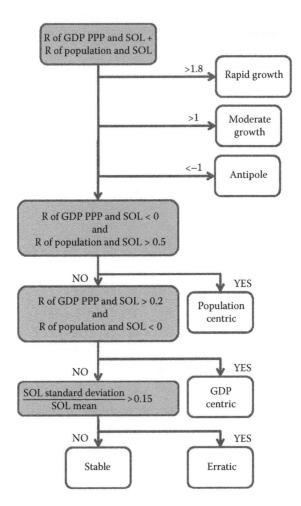

FIGURE 6.4 Decision tree developed for categorizing countries based on the behavior of their satellite-observed lighting over time in relation to GDP and population.

is very little dispersion in SOL values between satellite products from the same year. We define this group as the "rapid growth countries." They are identified as the set of countries where the sum of the GDP and population correlation coefficients exceeds 1.8. This category includes 31 countries, starting from the country with the highest sum of R values to the lowest: Mali, Portugal, Egypt, Vietnam, Qatar, Libya, Botswana, Oman, Iran, Morocco, China, Cyprus, Burkina Faso, Mozambique, Ethiopia, Bhutan, Yemen, Senegal, Jordan, Afghanistan, Grenada, Malaysia, Cape Verde, Liberia, St. Lucia, United Arab Emirates, Tunisia, Laos, Greece, Chile, and Bolivia. These countries have expansions in both population and GDP. The concomitant expansion in lighting is an indication that infrastructure has been built to enable higher standards of living. China can be considered an exemplar of this group of countries (Figure 6.5). Note that the SOL in China grew steadily from 1992 to 2011. There is an indication of minor declines in lighting in 2012 relative to 2011.

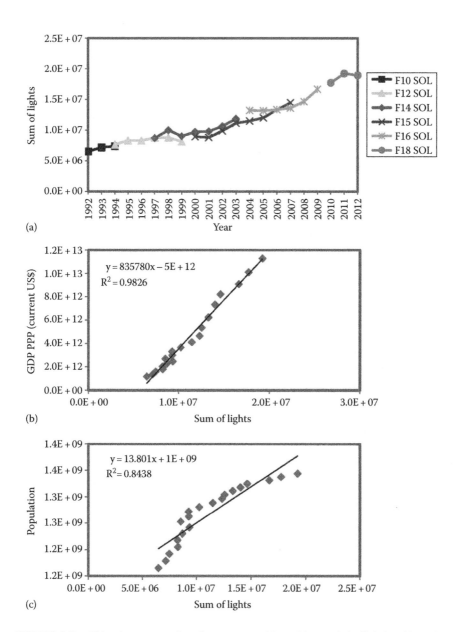

FIGURE 6.5 China is an exemplar of a country with rapid growth in lighting. Note that SOL tripled in 20 years. The correlation coefficients with GDP and population are very high: (a) Sum of lights versus time, (b) sum of lights versus GDP, and (c) sum of lights versus population.

6.4.2   COUNTRIES WITH MODERATE GROWTH IN LIGHTING

Moderate-growth countries are defined as those where the sum of the GDP and population correlation coefficients was larger than 1 and less than 1.8. This group includes 65 countries (in descending order): Saudi Arabia, Honduras, Angola, Belize, Argentina, Guatemala, Malawi, Spain, Eritrea, Benin, Sudan, Trinidad & Tobago, Equatorial Guinea, Sri Lanka, the Gambia, Panama, India, Italy, Suriname, Brazil, Peru, Syria, Israel, Congo, Turkmenistan, Ecuador, Antigua & Barbuda, Barbados, Algeria, Swaziland, Mauritius, Zambia, Congo DRC, Lesotho, Bahrain, Cambodia, Paraguay, Mauritania, Bosnia & Herzegovina, El Salvador, Gabon, Thailand, Cote d'Ivoire, Namibia, South Africa, Iraq, Indonesia, Maldives, South Korea, Turkey, Chad, Ireland, the Bahamas, Nicaragua, Mongolia, Mexico, Ghana, Sierra Leone, Austria, Costa Rica, Uganda, Philippines, Haiti, Djibouti, and France. These countries exhibit some growth in lighting over time, but the percentage growth is lower, and often the duration of the growth is confined to a specific set of years. There are typically one or more clear breaks in slope in the SOL versus time records. Spain is an exemplar for countries having moderate growth in lighting (Figure 6.6). Note that lighting grew by about a third from 1992 to 2008. Lighting has declined slightly from 2008 to 2012, a period of economic downturn and government austerity in Spain.

6.4.3   ANTIPOLE LIGHTING COUNTRIES

The rapid and moderate growth countries occupy the quadrant of Figure 6.3 where both the GDP and population correlation coefficients are positive. In the opposite quadrant of Figure 6.3, there is a loose cluster of 12 countries where both correlation coefficients are negative. The negative sign indicates that lighting either declines or remains stable despite growth in population and GDP. We define this group as the antipole countries and identify them as the set where the sum of the GDP and population correlation coefficients is –1 or greater. There are 13 countries in this category, a small number compared to the total for the moderate and rapid growth categories. The antipole group includes several countries from the former Soviet Union—Tajikistan, Uzbekistan, and Kyrgyzstan. These were among the poorest republics of the Soviet Union and remain among the poorest countries in Central Asia today. It appears that over time, lighting has been decommissioned, probably due to government inability to provide services to the population. Also in the antipole group are some of the richest and most well-developed countries in the world—Canada, the United Kingdom, Japan, the United States, the Netherlands, and Belgium. In these countries, aggregate lighting has declined despite expansions in population numbers and GDP. Our interpretation of this decline in lighting is that ongoing improvements in lighting efficiency are offsetting the growth in illuminated infrastructure. Over the past two decades, there has been a proliferation of state and local regulations designed to conserve energy and protect the night sky from light pollution. This includes banning incandescent lights and improving shielding to limit the quantity of light that shines directly into the sky. Uzbekistan is an exemplar of the antipole group (Figure 6.7).

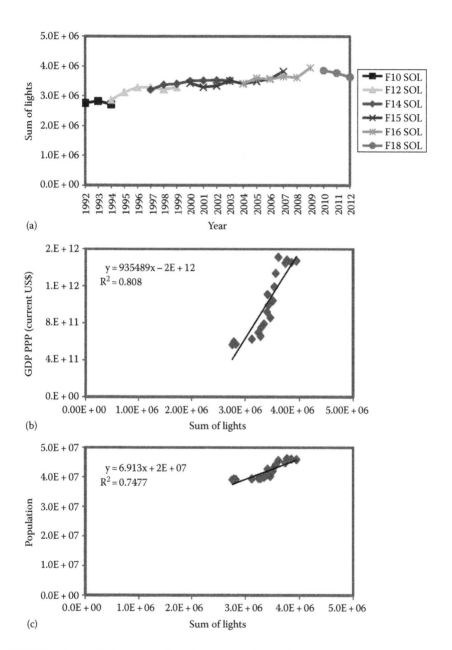

FIGURE 6.6 Spain is an exemplar of a country with moderate growth in nighttime lights: (a) Sum of lights versus time, (b) sum of lights versus GDP, and (c) sum of lights versus population.

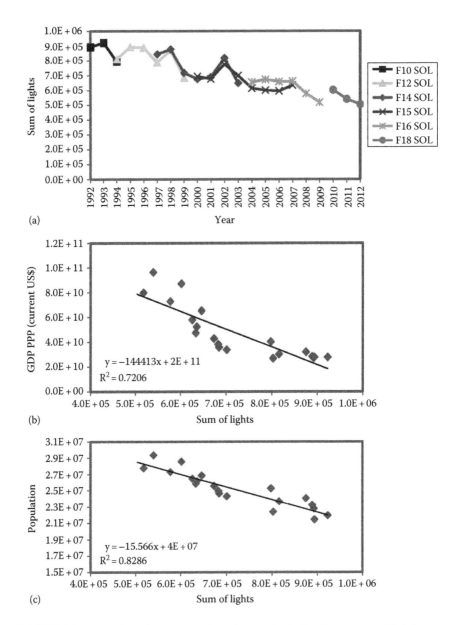

(a)

(b)

(c)

FIGURE 6.7 Uzbekistan is an exemplar of an antipole lighting country. Lighting has declined over time, and the correlation coefficients with GDP and population are negative: (a) Sum of lights versus time, (b) sum of lights versus GDP, and (c) sum of lights versus population.

6.4.4   Countries with Stable or Erratic Lighting

There is a large group of countries falling in the middle of the primary axis in Figure 6.3. Their SOL values lack strong correlation to either GDP or population. We divide this cluster into two groups. In the stable lighting group, we place the countries with SOL values that are highly consistent between satellites observing them in the same year. This includes 27 countries: Cameroon, Switzerland, Niger, Pakistan, Kuwait, Uruguay, Tanzania, Seychelles, Madagascar, St. Vincent & the Grenadines, Nigeria, St. Kitts & Nevis, Gaza Strip, Venezuela, Nepal, Australia, Kenya, Bangladesh, Luxembourg, Germany, Colombia, Samoa, Finland, Malta, Papua New Guinea, Singapore, and New Zealand. Australia is an exemplar for stable lighting (Figure 6.8).

Co-mingled with the stable lighting countries is a group of 29 countries with erratic lighting, where there is a zigzag pattern in the SOL values over time. The discrimination between the stable lighting and erratic lighting countries is based on an analysis of dispersion in SOL values in the years where two satellites collected data. For each country, a calculation is made of the percent dispersion around the mean, using the pairs of observations made within single years. Countries with less than 20% dispersion in this metric are labeled as stable. Countries with more than 20% dispersion are labeled as erratic. The erratic country set includes some of the poorest countries in the world such as Rwanda, Burundi, and Timor Liste. Also included are several European countries such as Sweden, Norway, and Denmark. It is suspected that satellite-observed nighttime lights are unstable at high latitudes due to annual variations in the extent of snow cover. The Czech Republic is an exemplar of the countries with erratic lighting (Figure 6.9).

6.4.5   Countries with GDP-Centric Lighting

There is a cluster of eight countries having a positive correlation coefficient with GDP and a negative (or zero) correlation coefficient with population: Albania, Armenia, Poland, Croatia, Dominica, Romania, Latvia, and Lithuania. These are in the upper left quadrant of Figure 6.3. Because of the positive correlation coefficient with GDP, these are referred to as "GDP-centric countries." In these countries, GDP and lighting have been growing, whereas population has declined or has lagged relative to GDP. Except for Dominica, all of these countries were under the influence of the Soviet Union until the early 1990s. Dominica is an exemplar for GDP-centric lighting (Figure 6.10).

6.4.6   Countries with Population-Centric Lighting

There is a set of four countries (Kazakhstan, Moldova, Ukraine, and Russia) where lighting has a positive correlation coefficient with population and negative correlation coefficient with GDP. These are in the lower right quadrant of Figure 6.3. Each of the countries was part of the former Soviet Union. In these countries, lighting has declined over time even though GDP has grown, resulting in the negative correlation coefficients between SOL and GDP. Moldova is an exemplar for this group of countries (Figure 6.11).

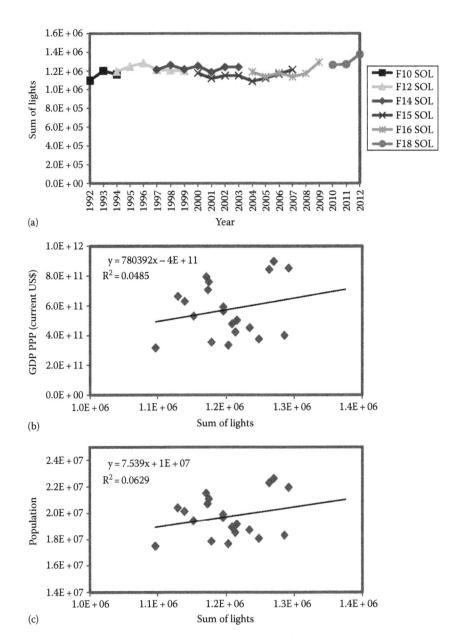

(a)

(b)

(c)

FIGURE 6.8 Australia is an exemplar of a country with stable lighting: (a) The SOL pattern from each available satellite from 1992 through 2009, (b) scattergram of the SOL versus GDP levels with a trend line, and (c) scattergram of the SOL versus population levels with a trend line.

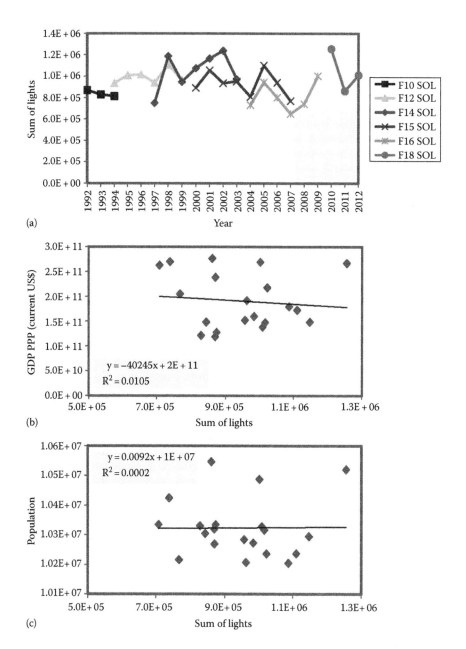

FIGURE 6.9 The Czech Republic is an exemplar of a country with erratic lighting: (a) The SOL pattern from each available satellite from 1992 through 2009, (b) scattergram of the SOL versus GDP levels with a trend line, and (c) scattergram of the SOL versus population levels with a trend line.

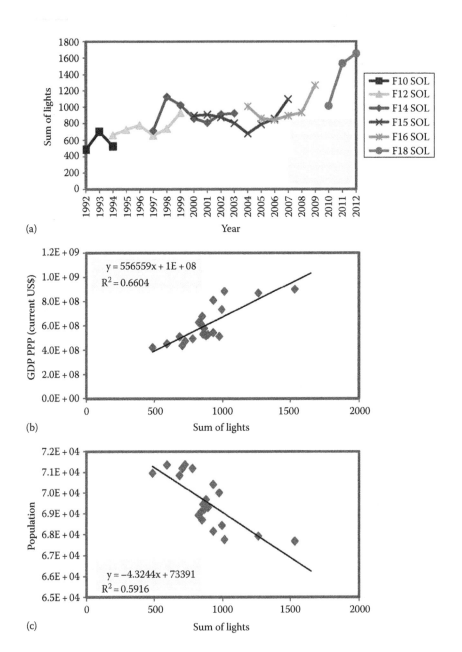

(a)

(b)

(c)

FIGURE 6.10 Dominica is an exemplar of a country with GDP-centric lighting. There is an increase in lighting over time, a positive correlation coefficient with GDP, and a negative correlation coefficient with population: (a) The SOL pattern from each available satellite from 1992 through 2009, (b) scattergram of the SOL versus GDP levels with a trend line, and (c) scattergram of the SOL versus population levels with a trend line.

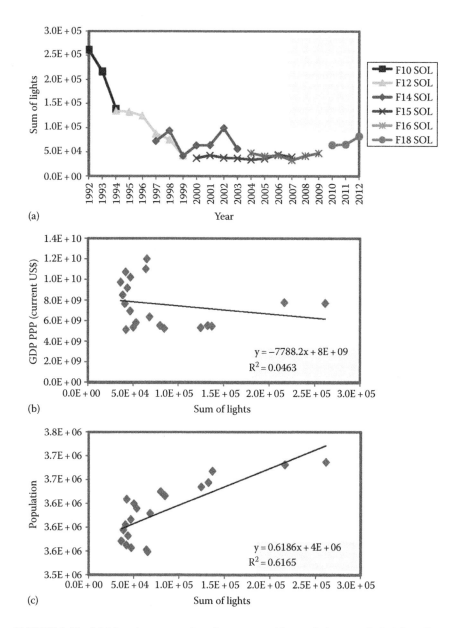

FIGURE 6.11 Moldova is an exemplar of a country with population-centric lighting. There is a decline in lighting over time, a negative correlation coefficient with GDP, and a positive correlation coefficient with population: (a) The SOL pattern from each available satellite from 1992 through 2009, (b) scattergram of the SOL versus GDP levels with a trend line, and (c) scattergram of the SOL versus population levels with a trend line.

6.5 DISCUSSION

Figure 6.12 summarizes the relationships between satellite-observed nighttime lights, GDP, and population. Lighting in the majority of countries has relatively equal affinity for GDP and population, forming the primary axis in Figure 6.12. The countries along this axis have a positive correlation between GDP and population. The densest concentration of countries is in the upper right-hand corner, at the growth tip of the primary axis. In these countries, lighting is in tight synchronization with growth in GDP and population. This synchronization begins to break down as correlation coefficient values decline, moving toward the lower left corner of the chart. In the lower left corner, the SOL is negatively correlated with population and GDP. These are countries where the SOL has either declined or has been stable despite growth in population and GDP. A plausible explanation for this behavior is that improvements in lighting technology, such as improved shielding, have constrained SOL growth.

There is a second axis defined by countries that are dispersed away from the primary axis. For countries along the secondary axis, population and GDP are out of synchronization. With one exception, all the countries that are not part of the primary axis were subjected to economic collapse following the breakup of the Soviet Union. After initially declining in the early 1990s, GDP in these countries began to rise and eventually surpassed the Soviet era levels. However, population has either declined or shifted to a stable level. Because GDP and population are not in synchronization, a secondary axis is formed. If lighting increases, the points plot in the upper left quadrant. If lighting decreases, the points plot in the lower right quadrant.

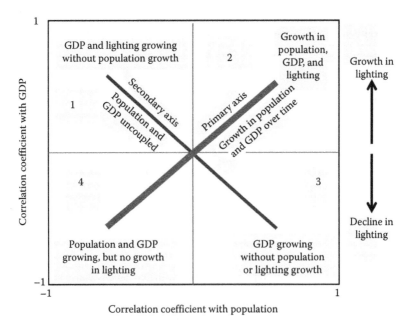

FIGURE 6.12 Summary chart describing the four quadrants and the primary and secondary axes.

6.6 CONCLUSION

An examination of the raw stable lights time series finds that there is a lack of radiometric consistency that confounds the analysis of changes in the brightness of surface lighting. These issues can be reasonably resolved by applying an intercalibration. Evidence that the intercalibration works as intended include the convergence of SOL values in years where two satellites have products and the emergence of steady, consistent upward trends in SOL values for countries undergoing rapid economic growth such as China and Vietnam. Our conclusion is that the intercalibrated v.4 DMSP nighttime lights time series is suitable for the analysis of changes in urban nighttime lights.

We found that population growth and economic growth are the primary variables contributing to the expansion of nighttime lights. In examining the two-decade record, we found that total lighting grew for more than half of the countries examined. This expansion in lighting is an indication of the expansion of built infrastructure, driven both by population growth and economic expansion. This suggests that the DMSP time series could be used to map expansion in infrastructure. Population and economic activity levels are typically reported as national or subnational statistics. Our results indicate that nighttime lights are good spatial proxies for use in mapping both population and economic activity.

At the aggregate level, total lighting is not growing in the world's most developed nations, such as the United States, Japan, and some European countries. This is surprising, especially in the United States where the built infrastructure has grown substantially over the past two decades. The best explanation for this lack of growth in total lighting is that technological advances in lighting efficiency have kept pace with infrastructure expansion. To further test this hypothesis, the area of lighting could be examined, a different variable from the SOL studied here.

Satellite-observed lighting can contract in response to catastrophic events such as war, economic collapse, and de-population. Many of the countries associated with the former Soviet Union exhibited contraction in lighting during the 1990s. Satellite observations of lighting may be useful in identifying the urban areas that have been most heavily impacted by catastrophic events and in tracking their recovery.

Despite the numerous deficiencies of the DMSP-OLS, it is possible to extract coherent trends in satellite-observed lighting from the time series. The DMSP plans to fly the remaining two satellites into dawn–dusk orbits that have low value for nighttime lights. The DMSP F18 satellite is expected to collect usable nighttime data for several more years. In the long term, VIIRS will be the primary instrument collecting low light imaging data suitable for monitoring urban areas worldwide. The overlap in VIIRS data collection with DMSP suggests that the current two-decade record of nighttime lights can be extended to three or more decades, providing urban scientists an extended record for analysis.

The temporal patterns of satellite-observed nighttime lights can be viewed as a signature tracking the metabolic processes of a nation. In underdeveloped countries, the lighting may go up and down from year to year in an erratic pattern or it may be more stable showing neither an upward nor a downward trend. Lighting can be lost following catastrophic events (e.g., political dissolution, war, or economic collapse), and such information can be used for tracking either economic or population losses.

Countries that have shifted from being underdeveloped to rapid economic growth typically have rapid growth in lighting, in synchronization with population and economic growth. Developed countries tend to have stable and in some cases declining lighting, despite their continued economic and population growth.

We attribute the stability in lighting in developed countries to improvements in lighting efficiency. While there is continuing growth in urban and suburban infrastructure, the lighting being installed is more efficient. At the same time, older inefficient lighting is being replaced by higher efficiency lighting. For example, older lighting may have had either no shielding or limited shielding to prevent light from escaping directly into the sky. New outdoor lighting in developed countries tends to be "full-cutoff," with shielding extending below the position of the light source, blocking the direct escape of light into the sky. The improvements in lighting technology are driven by the drive toward energy efficiency and environmental concerns over outdoor lighting (Rich and Longcore, 2006). This is an example of a Kuznets curve, where a society's environmental impacts decline as prosperity spreads, and there is increased environmental awareness and actions taken to protect the environment (Stern, 2004). One can imagine that in the future, the drive toward energy efficiency and minimizing human impacts on the environment will lead to decline in satellite-observed lighting in certain developed countries.

The findings of the national trends in satellite-observed lighting have implications for the approaches to be used in modeling either population or economic activity levels in individual countries. For instance, if a country has an erratic pattern of satellite-observed lighting, there may be wide error bars placed on estimates of GDP based on nighttime lights. In contrast, for a country undergoing rapid growth, the evidence indicates that nighttime lights will be a very good predictor of economic activity levels. For developed countries with stable lighting, the quantity of lighting at an aggregate level has very little linkage to the annual changes in population and GDP. One final consideration regarding the findings is that the presented results are at a national level, and there may be subnational variation that should be considered to fully utilize the information content of the nighttime lights.

ACKNOWLEDGMENT

The DMSP orbital data used in this study were provided to the NGDC by the U.S. Air Force Weather Agency (AFWA).

REFERENCES

Baugh, K., Elvidge, C., Ghosh, T., and Ziskin, D. (2010) Development of a 2009 Stable Lights Product using DMSP-OLS data. *Proceedings of the Asia-Pacific Advanced Network*, 30, 114–130.

Doll, C. N. H. (2008) *CIESIN Thematic Guide to Night-Time Light Remote Sensing and Its Applications*, Center for International Earth Science Information Network of Columbia University, Palisades, NY.

Elvidge, C. D., Imhoff, M. L., Baugh, K. E., Hobson, V. R., Nelson, I., Safran, J., Dietz, J. B., and Tuttle, B. T. (2001) Nighttime lights of the world: 1994–95. *ISPRS Journal of Photogrammetry and Remote Sensing*, 56, 81–99.

Elvidge, C. D., Sutton, P. C., Ghosh, T., Tuttle, B. T., Baugh, K. E., Bhaduri, B., and Bright, E. A. (2009a) Global poverty map derived from satellite data. *Computers and Geosciences*, 35, 1652–1660.

Elvidge, C. D., Tuttle, B. T., Sutton, P. C., Baugh, K. E., Howard, A. T., Milesi, C., Bhaduri, B., and Nemani, R. (2007) Global distribution and density of constructed impervious surfaces. *Sensors*, 7, 1962–1979.

Elvidge, C. D., Ziskin, D., Baugh, K. E., Tuttle, B. T., Ghosh, T., Pack, D. W., Erwin, E. H., and Zhizhin, M. (2009b) A fifteen year record of global natural gas flaring derived from satellite data. *Energies*, 2(3), 595–622.

Ghosh, T., Powell, R., Elvidge, C. D., Baugh, K. E., Sutton, P. C., and Anderson, S. (2010) Shedding light on the global distribution of economic activity. *The Open Geography Journal*, 3, 148–161.

Hsu, F.-C., Elvidge, C. D., and Matsuno, Y. (2013) Exploring and estimating in-use steel stocks in civil engineering and buildings from night-time lights. *International Journal of Remote Sensing*, 34(2), 490–504. DOI:10.1080/01431161.2012.712232.

Matsumura, K., Hijmans, R. J., Chemin, Y., Elvidge, C. D., Sugimoto, K., Wu, W. B., Lee, Y. W., and Shibasaki, R. (2009) Mapping the global supply and demand structure of rice. *Sustainability Science*, 4(2), 301–313.

Rayner, P. J., Raupach, M. R., Paget, M., Peylin, P., and Koffi, E. (2010) A new global gridded data set of CO_2 emissions from fossil fuel combustion: Methodology and evaluation. *Journal of Geophysical Research*, 115, D19306. DOI:10.1029/2009JD013439.

Rich, C. and Longcore, T. (2006) *Ecological Consequences of Artificial Night Lighting*, Island Press, Washington, DC.

Stern, D. I. (2004) The rise and fall of the environmental Kuznets curve. *World Development*, 32(8), 1419–1439.

Sutton, P. (1997) Modeling population density with night-time satellite imagery and GIS. *Computers Environment and Urban Systems*, 21, 227–244.

Zhao, N., Ghosh, T., Currit, N. A., and Elvidge, C. D. (2011) Relationships between satellite observed lit area and water footprints. *Water Resources Management*, 25(9), 2241–2250.

7 Development of a Global Built-Up Area Map Using ASTER Satellite Images and Existing GIS Data

Hiroyuki Miyazaki, Xiaowei Shao, Koki Iwao, and Ryosuke Shibasaki

CONTENTS

7.1 INTRODUCTION

7.1.1 BACKGROUND

Urbanization is a major issue in regional and global environmental changes [1] and socioeconomic problems [2]. Global urban area maps are used in various types of studies to assess the impacts of urbanization on the natural and human environments and to evaluate the critical aspects of urbanization such as the size, scale, and form of cities [3]. Remote sensing plays an important role in monitoring such geographic aspects of urbanization. Several global urban area maps and global land cover maps have been developed at coarse resolutions ranging from 300 to 1000 m using coarse-resolution satellite images (e.g., Advanced Very High Resolution Radiometer [AVHRR] [4,5], VEGETATION [6], Moderate Resolution Imaging Spectroradiometer [MODIS] [7–9], Defense Meteorological Satellite Program Operational Linescan System [DMSP-OLS] [10,11], and Medium Resolution Imaging Spectrometer [MERIS] [12]). These global land cover maps provide valuable information on urbanization for grid-based population estimates [13,14], studying food problems [15], predicting epidemics [16,17], estimating ecological footprints [18], estimating tsunami mortality [19], and assessing damage from rising sea levels [20], especially for less documented regions.

As studies on urbanization have progressed, however, the spatial resolution of global urban area maps has been found to be insufficient for measuring the spatial structure of urban areas [2], for modeling land use conversion resulting from socioeconomic impacts [21], and for measuring disaster risks in coastal regions [22]. For such purposes, which require finer spatial data for urban areas, medium-resolution satellite data, such as Landsat and Advanced Spaceborne Thermal Emission and Reflection radiometer (ASTER) data, have commonly been used for urban studies at city scale [2,23]. Those data are valuable sources of information for identifying geographic features of urbanization because of their fine spatial resolution. In addition, a global coverage of the archive is advantageous in comprehensive and comparative studies on urbanization. Several studies have developed urban area maps from medium-resolution satellite images (e.g., [2,23–25]); however, these maps were made for specific purposes and regions, and the map-making procedures used are not easily applicable to other purposes and regions. Development of a global urban area map at medium resolution is greatly needed for further progress in studies on urbanization.

To meet the demand for a medium-resolution global urban area map, we have demonstrated the development of a global built-up area map from ASTER data. In this chapter, we present the development of the global built-up area map according to the following steps: (1) development of a ground truth database, (2) development of an automated algorithm to generate mosaic image data for cities around the world, (3) development of an automated algorithm for extracting built-up areas, (4) system development with grid computing, and (5) the resulting built-up area map on a global scale. This chapter concludes with perspectives on further development of the global built-up area map.

Because we focused on the development of a comprehensive system, some component-specific problems remain, such as the baseline year used for scene selection and the accuracy of the built-up area extraction. These problems will be addressed in ongoing development efforts discussed in Section 7.7.

7.1.2 Definition of Urban

To develop the global urban area map, we had to define "urban." The socioeconomic literature defines an urban area by demographic and economic attributes [26]. However, identifying an urban area using such a definition depends greatly on the administrative units used because socioeconomic statistics are often available by administrative unit. When this type of definition is applied, urban development cannot be monitored at a smaller scale than that of the administrative unit, which produces imprecision and time inconsistencies among regions and countries [26].

For the definition to be consistent throughout the world, we have to define urban areas by their physical aspects. Urban areas are commonly defined in the remote-sensing literature as places characterized by a built-up environment, consisting of nonvegetative, human-constructed elements (e.g., roads, buildings, runways, and industrial facilities) [2,8,24]. Nonurban areas are defined as places without any built-up environment (e.g., open spaces, forests, and agricultural fields). This definition has the advantage of being comparable across or within nations.

7.1.3 ASTER/VNIR Images

We employed surface reflectance image data derived from the visible and near-infrared (VNIR) subsystem of ASTER (ASTER/VNIR), which have often been used for monitoring the urban environment [25,27,28]. The 15 m spatial resolution of ASTER/VNIR is much finer than that of existing global built-up area maps; thus, built-up area maps derived from ASTER/VNIR images permit the measurement of complex spatial structures at a finer scale. Moreover, ASTER/VNIR has been in operation since December 1999 with the goal of generating complete global cloud-free coverage [29]. We therefore chose ASTER/VNIR as the most suitable source of images to use in generating medium-resolution global built-up area maps.

We used ASTER/VNIR images archived and processed on the Global Earth Observation Grid at the National Institute of Advanced Industrial Science and Technology in Japan [30].

7.2 DEVELOPMENT OF GROUND TRUTH DATABASE OF URBAN SITES

Built-up area mapping using remote sensing data is performed with satellite imagery and ground truth data. However, for global built-up area mapping, the development of ground truth data would involve enormous costs for field surveys and visual interpretation of aerial photos or high-resolution satellite data. Publicly available ground truth databases are helpful for global land cover mapping; however, those databases do not have enough urban sites for global human settlement mapping. For example, the MOD12Q1 V003 Land Cover Product [31], the Global Land Cover Ground Truth database [32], and the Degree Confluence Project ground truth validation databases [33] include 0, 3, and 11 ground truth data points, respectively, for urban areas. These ground truth data are obviously insufficient for validating global built-up area maps, which estimate the global urban area to be from 276×10^3 to 3524×10^3 km^2 [34].

To develop such a database, we employed as a primary data source the Global Rural–Urban Mapping Project (GRUMP) Alpha Version Settlement Points, which is a gazetteer of populated places with latitude and longitude coordinates derived from various types of maps. We assumed that built-up areas existed at the point coordinates of the populated places within that database because those points were previously used as primary input data for an urban area map [10]. In addition, the gazetteer has a large number of place names of populated places covering the entire world. Thus, we regard this gazetteer as a suitable source of data for a ground truth database for urban sites.

7.2.1 SAMPLING SCHEME FOR URBAN SITES

An unbiased sampling scheme is a primary requirement of ground truth data. Spatially balanced systematic sampling at a $1° × 1°$ grid of latitude and longitude has been proposed for global land cover classification [35]. However, this systematic sampling method does not result in a sufficient number of ground truth data points in urban areas, which tend to be strongly concentrated in a very small area of the earth's surface.

To solve the problems posed by the concentrated geographic distribution of urban areas, we employed the GRUMP Settlement Points (GSP) gazetteer [36], which is a database of place names with point coordinates and an estimated population of 55,412 places with populations greater than 1,000. Because the point coordinates in GSP are provided to allow location of a city [37], the point coordinates corresponding to a place name can be regarded as a point chosen randomly within the geographic extent represented by the place name. In addition, the place names, attribute data, and geographic coordinates were manually linked by human decision. This direct human input is indispensable for accurate association of place names with geographic data because insufficient information from the source prevents automatic matching [38]. The precision of the point coordinates for place names in the GSP is approximately 1 km. Such precision is enough to represent the urban area of a populated place because the urban area of a city is typically more than 1 km².

7.2.2 RESULT OF DEVELOPING GROUND TRUTH DATA

We retrieved the GSP data for approximately 3734 populated places, each inhabited by more than 0.1 million people. We visually interpreted the point coordinates of these populated places using false color composite images of ASTER/VNIR. For these images, the near-infrared band (0.52–0.60 μm) was assigned to the red channel, the red band (0.63–0.69 μm) was assigned to the green channel, and the green band (0.76–0.86 μm) was assigned to the blue channel. Three trained operators visually interpreted the presence of urbanization at each point on the false color composite images based on color tone and texture. For a point to be interpreted as built-up, two of the three operators had to interpret it as being built-up.

As a result of this analysis, 2144 of the 3734 points were interpreted as built-up, 1388 were interpreted as non-built-up, 10 were interpreted as being in between built-up and non-built-up, and 192 could not be interpreted because of clouds or shadows on the image or discrepancies in interpretations among the operators. Thus, the number of ground truth data points identified as built-up was much larger than the number of urban points in existing databases. We used 3,532 of the data points with ground truth data derived from visual interpretation of false color composite images of ASTER/VNIR for 249 cities, including 9,228 built-up and 64,593 non-built-up areas, for the development of the global built-up area map.

7.3   DEVELOPMENT OF AN AUTOMATED ALGORITHM FOR SELECTING ASTER DATA FOR MOSAIC

For cities with urban areas broader than the coverage of a single scene of ASTER data, multiple scenes of ASTER data needed to be selected and mosaicked. Because the global built-up area mapping involved thousands of cities, we had to automate image selection from the ASTER data for all of the cities. We defined the process as consisting of three steps: defining the search extent, searching for ASTER data that overlap the search extent, and selecting ASTER data that would maximize the quality of the mosaic output. Here, we describe the automated algorithms used in these steps.

7.3.1   Defining the Search Extent

ASTER data for a city can be found by searching the data's coverage, that is, the extent to which the data overlap the search extent of cities. Because the search extents of cities were not given, we had to define them with existing Geographic Information System (GIS) data. Existing global built-up area maps would be a good reference for search extents of cities; however, those maps could omit cities with small urban extents [39].

We used GSPs [36] as reference data because of their completeness and defined the search extent with a buffer zone of point coordinates. We set the buffer distance at 30 km (half of the swath length of ASTER) to ensure complete coverage of the search extent by ASTER data.

7.3.2   Determining Combination of ASTER Data

Image data for the mosaic are often selected manually because cloud contamination must be checked to ensure the quality of the mosaic output. In addition, ASTER data are not necessarily aligned to path-row grids because of the wide range of the view angle. To automate the image selection process, we clarified the minimum requirements of the ASTER data to be chosen for the mosaic: overlapping with any segment of search extent and having the least cloud contamination among the ASTER data overlapping the segment.

Based on these criteria, we employed the following algorithm to determine the least cloud-contaminated combination of ASTER data for the mosaic (Figure 7.1):

1. Search scenes of ASTER data overlapping with search extent from archives of ASTER.
2. Sort the scenes in descending order by percentage of cloud contamination.
3. Check whether the first scene is necessary to maintain complete coverage of the search extent.
   a. If the mosaic completely covers the extent even when the scene is removed, the scene is considered to be unnecessary and removed from the set of scenes.
   b. Otherwise, keep the scene with the set of scenes.

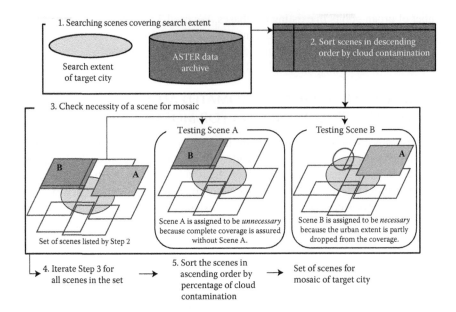

FIGURE 7.1   Steps to determine the least cloud-contaminated combination of ASTER data for the mosaic.

4. Iterate Step 3 for all of the scenes in order.
5. Sort the scenes kept within the set in ascending order by percentage of cloud contamination so that the scene with the least cloud contamination is laid on the top.

7.3.3   Results of Automated Selection of ASTER Data for the Mosaic

We attempted to develop mosaic ASTER data by city because the output had to be associated with city profiles. We used the 3734 data points of cities with more than 0.1 million people from the GSP data. With search extents overlapping each other, these were merged to create a single search extent to eliminate duplicated data processing of overlapping scenes. Merging the search extents in this way resulted in the assembly of 2,214 extents. We searched ASTER data from the archives with cloud contamination of less than 20%. We also constrained the search period to be between March 2000 and March 2008 to exclude scenes with degraded accuracy in cloud detection due to the malfunctioning of shortwave infrared (SWIR) beginning in April 2008.

As a result, for 1,951 of the 2,214 extents, 11,802 scenes of ASTER data were successfully arranged for mosaic processing. For the other 263 extents, no scene was assigned due to cloud contamination. For 1340 of the 1951 extents, the mosaic of ASTER images was well organized; however, for the other 611 extents, the coverage was incomplete or considerably contaminated with cloud cover. We manually selected scenes of ASTER data for the 611 search extents.

Because the purpose of this study is to demonstrate a comprehensive system development, we overlooked inconsistencies within an observation year among neighboring scenes although such inconsistencies should be considered, especially for rapidly growing megacities in developing countries. Further investigation of the availability of cloud-free ASTER data for each city would be required for the search period to be precise (e.g., 2000–2002, 2003–2005).

7.4 DEVELOPMENT OF AN AUTOMATED ALGORITHM FOR EXTRACTING BUILT-UP AREAS FROM ASTER IMAGES

7.4.1 EXTRACTING BUILT-UP AREAS FROM ASTER/VNIR SATELLITE IMAGES

Classification of satellite image pixels as built-up or non-built-up requires two basic steps: clustering and labeling. For automated clustering, an unsupervised clustering method, such as Iterative Self-Organizing Data Analysis Technique (ISODATA), has been employed in the past for land cover classification [2,40]. To label clusters correctly, the classifier requires external training data. In the conventional method, training data for labeling clusters have been acquired by human visual interpretation [2,40]. However, it is not feasible to conduct human visual interpretations of all of the cities of the world because the labor costs would be prohibitive.

To automate cluster labeling, we employed existing global built-up area maps as training data. Well-classified built-up area maps should be good training data for clustered satellite images of medium resolution because of their overall classification accuracy rates of 0.83–0.98 [39].

For the roughly labeled training data, an iterative machine learning algorithm should work effectively. We employed the Learning with Local and Global Consistency (LLGC) algorithm [41], which determines classifiers with an infinite number of iterations of spectral clustering. Zhou et al. [41] presented an analytical solution to the problem of incorporating infinite iterations into the operations of linear algebra, thus reducing computation costs through a few operations of linear algebra.

We arranged the LLGC algorithm to fit with our built-up area mapping using ASTER satellite image and coarse-resolution land cover maps (left column of Figure 7.2) by the following process.

- Perform ISODATA clustering on ASTER/VNIR surface reflectance image data, including the near-infrared band, the red band, and the green band and calculate the mean surface reflectance for each band by cluster. The feature vectors (mean surface reflectance) were normalized using the following equation:

$$x_{ij} = \frac{\rho_{ij} - \mu_j}{\sigma_j} \tag{7.1}$$

where
  ρ_{ij} is the mean surface reflectance of band j for cluster i
  μ_j is the mean value of band j among the clusters
  σ_j is the standard deviation of band j among the clusters

- Partition the clusters with the boundary of the coarse-resolution map, called the initial built-up area map (IBAM). We termed the partitioned cluster layer the ASTER-IBAM cluster layer.

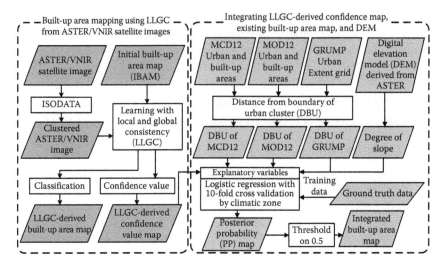

FIGURE 7.2 Overview of the automated algorithm of built-up area mapping. (Reprinted with permission from Miyazaki, H., Shao, X., Iwao, K., and Shibasaki, R., An automated method for global urban area mapping by integrating ASTER satellite images and GIS data, *IEEE J. Select. Top. Appl. Earth Observ. Remote Sens.*, 6(2), 1004–1019, 2012. Copyright 2012 IEEE.)

- Form an affinity matrix and a Laplacian normalized by numbers of pixels, defined as follows:

$$W_{ij} = \exp\left(\frac{-\|x_i - x_j\|^2}{2\sigma^2}\right) \quad \text{if } i \neq j \text{ and } W_{ij} = 0 \tag{7.2}$$

$$L = D^{-1/2}N^{-1/2}WN^{-1/2}D^{-1/2} \tag{7.3}$$

where

    W is an $n \times n$ affinity matrix

    σ^2 is the distance weight of the feature vector among the data (smaller values reduce the distance effect)

    D is a diagonal matrix, called the degree matrix, with its (i, i) element equal to the sum of the ith row of W

- In Equation 7.2, i and j indicate the index of a cluster of the ASTER-IBAM cluster layer, and x_i is the feature vector or mean surface reflectance for cluster i.
- Calculate F, defined as follows:

$$F = (I - \alpha L)^{-1} NY \tag{7.4}$$

where

    I is an $n \times n$ unit matrix

    Y is an $n \times 2$ matrix, in which, at the initial step, if cluster i is built-up, $Y_{i1} = 0$ and $Y_{i2} = 1$ and otherwise, $Y_{i1} = 1$ and $Y_{i2} = 0$

The *i*th row of F and Y corresponds to the *i*th cluster of the ASTER-IBAM cluster layer.

- Classify clusters as built-up or non-built-up based on F. For the *i*th cluster to be classified as built-up, F_{i2} must be greater than F_{i1}, and vice versa.

Basically, the clusters were classified as built-up or non-built-up by comparing F_{i1} and F_{i2}; however, that classification discards information on the compositions of built-up and non-built-up classifications in the clusters. Retaining that information is useful in identifying built-up areas in regions where relatively few urbanized clusters are dominant. We proposed a confidence value that ranged from 0 to 1.

$$\text{LLGC Confidence}_i = \frac{F_{i2}}{\left(F_{i2} + F_{i1} \right)} \tag{7.5}$$

The confidence value was calculated for each cluster to introduce it into the map integration discussed in Section 7.4.2. We call this map the LLGC confidence map.

We applied a built-up area map extracted from MODIS Terra+Aqua Land Cover Type Yearly L3 Global 500 m SIN Grid product for 2001 (MCD12) [42] as the IBAM because of its two main advantages: it has a resolution of 500 m, which is somewhat finer than that of other existing maps, and it was found to be the most accurate map in an accuracy assessment of 140 cities [39]. Previous research has shown that MCD12 is a good reference data source for LLGC with ASTER/VNIR, yielding good results at a resolution of 15 m although the results depend considerably on the accuracy of MCD12 [43].

7.4.2 INTEGRATING MAPS USING LOGISTIC REGRESSION

Although the clusters were successfully classified by LLGC, the results included misclassifications resulting from similarities in surface reflectance among different land covers. Cloud contamination in satellite images also leads to misclassifications. These disturbances stem from heterogeneities of landscape and image quality among ASTER/VNIR images. To reduce the number of misclassifications, other map resources with lower levels of uncertainty should be used to complement the ASTER/VNIR images.

However, even though a map has less uncertainty, the existing uncertainties would still result in discrepancies with other maps. As a solution to this problem, the use of posterior probability (PP) has been shown to be an effective tool to represent the likelihood that a disputed pixel is actually built up for each combination of conditions given by maps.

Logistic regression methods are the most basic and classical methods for estimating PP. We defined the following logistic model to estimate the PP of the presence of a built-up area in a pixel:

$$P_i(\text{urban}) = \frac{\exp(U_i)}{1 + \exp(U_i)} \tag{7.6}$$

where
    $P_i(\text{urban})$ is the probability of the presence of a built-up area at a pixel
    U_i is defined in the form of a polynomial expression of the explanatory variables

As an explanatory variable, we introduced the LLGC confidence map. In contrast to the binary classification result, the LLGC confidence map has more information on the likelihood of built-up or non-built-up categorization, especially on the gradient transition between an urban center and suburban areas. We believed that this feature would contribute to a better estimate of PP values.

We also considered the geographic heterogeneity of accuracy in existing built-up area maps. The accuracy of satellite-based estimates of built-up areas at urban centers is thought to be higher than the accuracy of estimates in rural areas [9]. This indicates that pixels close to the urban center are more likely to be urban than those close to the urban area boundary.

To reflect geographic heterogeneity within and between land cover classes, we calculated the distance from the boundary of an urban cluster (DBU). DBU has high negative values at the centers of urban clusters and high positive values in rural regions far from any urban center. Calculating DBU with the same resolution as that of ASTER/VNIR images would be useful in determining the likelihood that a pixel should be built up. We calculated DBU from urban clusters of the MCD12, MOD12Q1 V004 Land Cover Product (MOD12) [44] and GRUMP Urban Extent Grid [10].

Terrain is also a significant factor in the presence of built-up areas [45]. We therefore introduced the degree of slope as an explanatory variable in the logistic regression. The degree of slope was calculated from a digital elevation model (DEM) derived from the orthorectification of the ASTER/VNIR data.

In summary, the polynomial in Equation 7.6 was defined as follows (see also right column of Figure 7.2):

$$U_i = \beta_1 LLGC_i + \beta_2 MCD12_i + \beta_3 MOD12_i + \beta_4 GRUMP_i + \beta_5 SLP_i \qquad (7.7)$$

where
   β_k is the coefficient for each variable
   $LLGC_i$ is the confidence value of LLGC at pixel i
   $MCD12_i$, $MOD12_i$, and $GRUMP_i$ are the DBU at pixel i in MCD12, MOD12, and
      GRUMP, respectively (the values are positive outside of urban areas and nega-
      tive inside urban areas)
   SLP_i is the slope (degrees) at pixel i

In the logistic regression, ground truth data representing the presence of built-up areas in a pixel are needed for the response variable, which is defined as 1 for definitely urban pixels and 0 for definitely non-built-up pixels.

Because regionally tuned models yield better accuracy [6], we defined regions for estimation of logistic models. The regions were defined by an iterative merge of 10-degree grid cells including the least amount of ground truth data with neighboring grid cells so that each region had enough ground truth data.

As a result, we defined 30 regions to maintain the minimum required amount of ground truth data, more than 100 urban sites for each.

7.4.3   GRID COMPUTING FOR THE DEVELOPMENT OF GLOBAL BUILT-UP AREA MAP

The global built-up area mapping used over 11,802 scenes of ASTER data and requires huge computer resources. We employed grid computing to perform the large number computations required for the global built-up area map. Grid computing is the federation of distributed computer resources to achieve a common goal [46]. We used GEO Grid, which is an infrastructure operated by the National Institute of Advanced Industrial Science and Technology of Japan [30], for the development of the global built-up area map. GEO Grid stores 200 TB of archives of ASTER data, which have been archived since the launch of ASTER in 1999. GEO Grid is also operated with a PC cluster, which contributes to the generation of orthorectified ASTER data and other high-level data products. The storage system and the PC cluster are connected by a fast network so that data processing is performed seamlessly between the storage system and computation with PC Cluster.

We implemented a system of global built-up area mapping using free and open-source software for geospatial (FOSS4G; Table 7.1) and general open-source software on the GEO grid. Command-line operation of the FOSS4G and other open-source software improved the interoperability of data handling among the components of the global built-up area mapping.

TABLE 7.1

Free and Open-Source Software for Geospatial (FOSS4G) Used for Global Urban Area Mapping

Software	Website	Role in This Study
GDAL	http://www.gdal.org/	File format conversion of raster data; image mosaic
GRASS	http://grass.osgeo.org/	Clustering of satellite image; raster data processing
R	http://www.r-project.org/	Logistic regression; accuracy assessment
Octave	http://www.gnu.org/software/octave/	Linear algebra operation of LLGC
PostgreSQL	http://www.postgresql.org/	Relational database management
PostGIS	http://postgis.refractions.net/	Geospatial data extension of PostgreSQL
SQLite	http://www.sqlite.org/	Database operation
SpatiaLite	http://www.gaia-gis.it/spatialite	Geospatial data extension of SQLite
MapServer	http://mapserver.org/	Map rendering engine of processed output

7.5   RESULTS AND DISCUSSION

7.5.1   Built-Up Area Map

We applied the algorithm for extracting built-up area to 1,951 mosaic ASTER images generated with 11,802 scenes for 3,372 cities around the world. LLGC and logistic regression were successfully applied to the mosaic ASTER images. Figure 7.3 shows examples of the results, illustrating the spatial structures of built-up areas in detail at a 15 m spatial resolution. For example, the built-up area map shows the distribution of minor non-built-up areas, such as open areas, parks, urban forests, and water bodies, which were not shown in previously developed global built-up area maps. In addition, some major roads were also extracted as built-up areas.

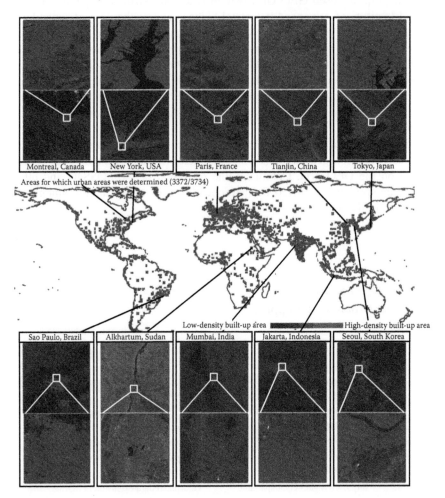

FIGURE 7.3   (See color insert.) Examples of the results of the global urban area maps. Red pixels in the close-up figures indicate urban areas.

Goodness of fit was evaluated with Nagelkerke R^2 for each region. Minimum, median, and maximum of the R^2 were 0.22, 0.52, and 0.77, respectively, indicating that fitness of the models varies by regions and the accuracy for regions poor in model fitness could be improved by finer partition of the regions.

7.5.2 ACCURACY ASSESSMENT

An 85% accuracy rate is widely accepted as a reasonable target. However, setting a target for a specific application may be more appropriate [47]. For built-up area mapping, because of the lack of a specific target, we deemed that comparing the accuracy with that of existing built-up area maps would be appropriate to determine the level of improvement. We therefore conducted accuracy assessments on the LLGC-derived maps, the integrated maps, the GRUMP, and the MOD12. Although the spatial resolution of the accuracy assessment should be the same as that of classified maps [48], we conducted accuracy assessments at a resolution of 15 m for all of the built-up area maps because we had to make the assessment protocols equivalent among the maps.

Figure 7.4 shows the result of the accuracy assessment in terms of the producer's accuracy, the user's accuracy, the overall accuracy, and the kappa coefficient of accuracy, which are commonly used for accuracy assessment [49]. The integrated map had the highest overall accuracy and kappa coefficient among the maps.

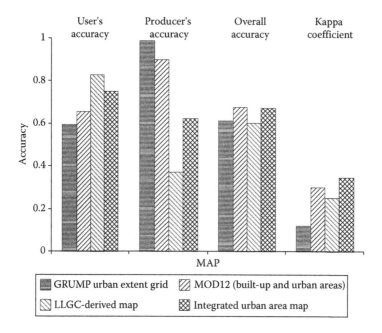

FIGURE 7.4 Accuracy assessment of the existing urban area maps (GRUMP and MOD12), LLGC-derived map and integrated map.

The relatively high user's accuracy and low producer's accuracy of the integrated map and LLGC-derived map indicate that these maps are likely to underestimate the global built-up area.

The user's accuracy of the integrated map was lower than that of the LLGC-derived map, while the producer's accuracy of the integrated map was higher than that of the LLGC-derived map. This indicates that the integration with existing built-up area maps captured built-up areas omitted by LLGC. The improvement, in terms of overall accuracy and the kappa coefficient, of the integrated map over GRUMP and MOD12 is notable.

Figure 7.5 shows an accuracy assessment by continent. For all of the continents except Asia, the kappa coefficient of the integrated map was higher than that of MOD12. For Asia, however, the kappa coefficient of the integrated map was lower than that of MOD12. These results indicate the significant impact of cloud contamination on the built-up area extraction for Asia. This might be due to the limited amount of cloud-free ASTER data for Asia [50].

Figure 7.6 shows the accuracy assessment by climatic zone. The kappa coefficient of the integrated map for the tropical zone was lower than that of MOD12, indicating less availability of cloud-free ASTER data [50], as in the case of the integrated map for Asia.

For dry zones, the integrated map had a lower kappa coefficient than MOD12. This might be due to misclassifications resulting from similarities in surface reflectance among built-up areas, sand, and bare land.

7.5.3 Agreement with Existing Built-Up Area Maps

Accuracy assessments using ground truth data can only capture agreement at sampled sites. Therefore, such assessments are unsuitable for assessing geographic trends in quality on a global scale. To illustrate a geographic trend in quality, we assessed the degree of agreement between the integrated map and existing built-up area maps, assuming that the existing built-up area maps were sufficiently accurate to serve as references.

We assessed the degree of agreement by cross tabulation between the classification results (the integrated map) and the reference data (MCD12). From the output of the cross tabulation, we calculated the producer's agreement, the user's agreement, the overall agreement, and the kappa coefficient of agreement, which corresponded to the producer's accuracy, the user's accuracy, the overall accuracy, and the kappa coefficient of accuracy assessment using point-based ground truth data. The indexes were calculated by 10-degree grid.

Figure 7.7 shows the distribution of the agreement between the integrated map and MCD12. The results show poor user's agreement in the southern part of Africa, Southeast Asia, the northeastern part of Europe, and the southern part of North America, indicating high rates of commission errors in those regions. This might be due to overestimation of built-up areas by LLGC. The results also suggest poor user's accuracy for the LLGC-derived map in Africa (Figure 7.5c) and poor producer's agreement for South America, Europe, and South Asia, indicating high rates of omission errors in those regions.

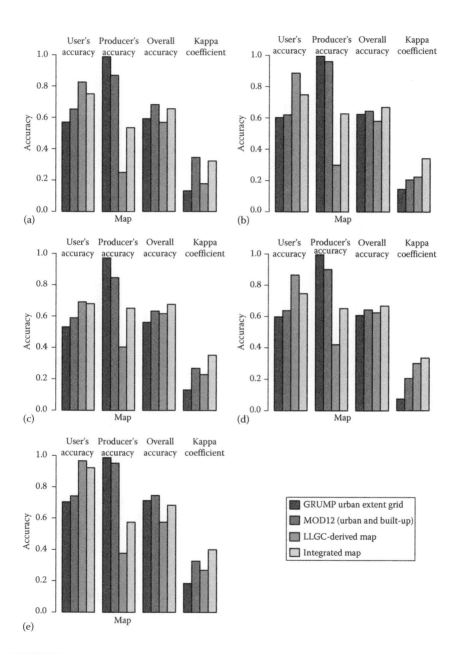

FIGURE 7.5 Accuracy assessment of the built-up area maps by continent: (a) Asia, (b) Europe, (c) Africa, (d) North America, and (e) South America.

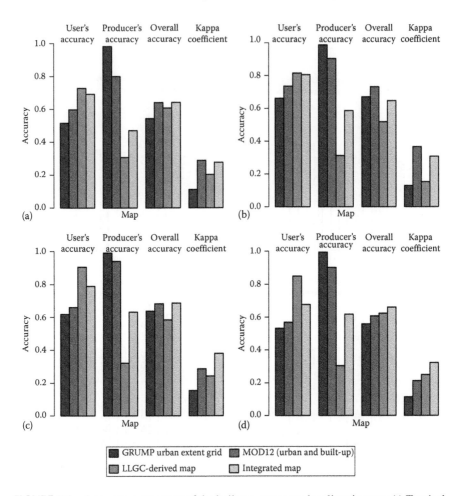

FIGURE 7.6 Accuracy assessment of the built-up area maps by climatic zone: (a) Tropical, (b) dry, (c) temperate, and (d) cold.

The overall agreement was low in the southeastern part of Africa, the northeastern part of Europe, and Southeast Asia, indicating high rates of both commission and omission errors in those regions. The kappa coefficient of agreement was high for North America, the western part of Europe, and eastern Asia, indicating accurate results for both the integrated map and MCD12.

7.6 EXPERIMENTAL RELEASE OF THE GLOBAL BUILT-UP AREA MAP WITH WEB-MAPPING SYSTEM

As a form of publication of the global built-up area map, we developed a publicly available web-mapping system for the global built-up area map (http://maps.geogrid.org/examples/basemap/). The web mapping system provides a function for downloading a subset GeoTiff of the global built-up area map for an extent selected by the end user. This function makes it easy for end users to use the global built-up area map with GIS software.

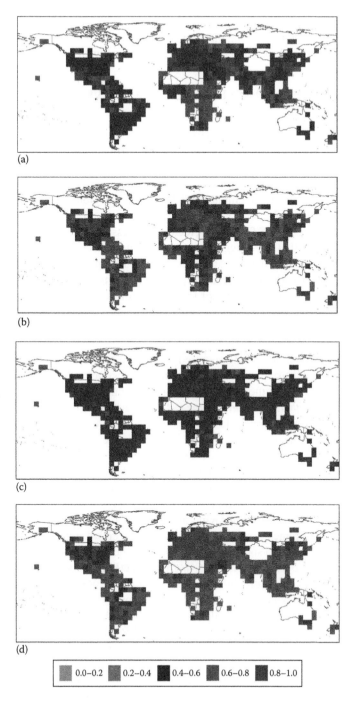

FIGURE 7.7   (See color insert.) Distribution of the agreements between the integrated map and the MCD12: (a) User's agreement, (b) producer's agreement, (c) overall agreement, and (d) kappa coefficient of agreement.

In addition, the service also supports Web Map Service (WMS), which is a standardized protocol for transferring map images over the Internet. Therefore, the global built-up area map can be used in the development of new web map services.

7.7   CONCLUSIONS AND FUTURE PERSPECTIVES

In this chapter, we presented the development of a global built-up area map using ASTER satellite data and existing built-up area maps. The development consisted of four steps: development of ground truth data, development of an automated algorithm to generate mosaic ASTER images, development of an automated algorithm for extracting built-up areas from ASTER images, and system implementation with grid technologies. The resulting global built-up area map was developed for 3,374 cities with populations of more than 0.1 million people using 11,802 scenes of ASTER data. Accuracy assessments of the global built-up area map showed that the integrated map was more accurate than the LLGC-derived map, indicating that the accuracy of the built-up area map was improved by the integration with existing built-up area maps. In addition to assessing the accuracy, we assessed the quality of the global built-up area map by comparing it with the MCD12. This assessment helped to provide an overview of the quality with respect to geographic trend, which would not be well illustrated with point-based ground truth data alone.

Although the data processing required to develop the global built-up area map was conducted well using a consistently automated system, there is considerable room for improvement in the map's quality, including the following:

Introducing synthetic aperture radar (SAR) data into the algorithm: cloud contamination is the greatest obstacle to developing a high-quality global built-up area map. For cities with heavy cloud contamination, the application of SAR data for urban monitoring has been proposed and examined in several studies [51–53]. Integration of the ASTER-based built-up area map and a SAR-based built-up area map would yield much better accuracy for cloud-contaminated regions, such as Southeast Asia.

Development of ground truth data by crowd sourcing: The amount of ground truth data available is important to improving the accuracy of built-up area mapping. With regard to this requirement for a large amount of data, crowd sourcing is expected to be a good solution. Crowd sourcing is a method of collecting data through the efforts of many operators over the Internet [54]. OpenStreetMap is the most active crowd sourcing project for geospatial information on roads [55]. Volunteers can contribute to developing a worldwide road map by digitizing roads using high-resolution satellite images. This approach would also be applicable to built-up area mapping, with operators performing visual interpretation of ASTER images.

We also suggest the development of population grid data at high resolution using the global built-up area map as source data. Currently, the population grid dataset at 1 km resolution (e.g., the GRUMP Population Grid [36] and LandScan [14]) is the most detailed such data available. However, population grid data with higher resolution is urgently needed for use in disaster management, especially in coastal regions where the sea level rising by a few meters constitutes a disaster with considerable impact [56]. The global built-up area map could contribute to the improvement of the spatial resolution of population grid data.

ACKNOWLEDGMENTS

This research used ASTER Data beta processed by the AIST GEO Grid from ASTER Data owned by the Ministry of Economy, Trade and Industry of Japan. This work was supported by Grant-in-Aid for JSPS Fellows (22-2598).

REFERENCES

1. J. A. Foley, R. DeFries, G. P. Asner, C. Barford, G. Bonan, S. R. Carpenter, F. S. Chapin et al., Global consequences of land use, *Science*, 309(5734), 570–574, 2005.
2. S. Angel, S. C. Sheppard, and D. L. Civco, The dynamics of global urban expansion, 2005, The World Bank, Washington D.C. Available at http://go.worldbank.org/58A0YZVOV0 (accessed October 6, 2013).
3. G. Laumann, Science plan: Urbanization and global environmental change, 15, 2005, IHDP Report No. 15, Bonn, Germany, pp. 20–21. Available at http://www.ihdp.unu.edu/article/read/ugec-science-plan (accessed October 6, 2013).
4. M. C. Hansen, R. S. Defries, J. R. G. Townshend, and R. Sohlberg, Global land cover classification at 1 km spatial resolution using a classification tree approach, *International Journal of Remote Sensing*, 21, 1331–1364, 2000.
5. T. R. Loveland, B. C. Reed, J. F. Brown, D. O. Ohlen, Z. Zhu, L. Yang, and J. W. Merchant, Development of a global land cover characteristics database and IGBP DISCover from 1 km AVHRR data, *International Journal of Remote Sensing*, 21(6–7), 1303–1330, 2000.
6. E. Bartholome and A. S. Belward, GLC2000: A new approach to global land cover mapping from Earth observation data, *International Journal of Remote Sensing*, 26(9), 1959–1977, 2005.
7. A. Schneider, M. A. Friedl, and D. Potere, Mapping global urban areas using MODIS 500-m data: New methods and datasets based on 'urban ecoregions,' *Remote Sensing of Environment*, 114(8), 1733–1746, 2010.
8. A. Schneider, M. A. Friedl, and D. Potere, A new map of global urban extent from MODIS satellite data, *Environmental Research Letters*, 4(4), 44003, 2009.
9. A. Schneider, M. A. Friedl, D. K. McIver, and C. E. Woodcock, Mapping urban areas by fusing multiple sources of coarse resolution remotely sensed data, *Photogrammetric Engineering & Remote Sensing*, 69(12), 1377–1386, 2003.
10. Center for International Earth Science Information Network, Columbia University, International Food Policy Research Institute, The World Bank, and Centro Internacional de Agricultura Tropical, *Global Rural-Urban Mapping Project (GRUMP), Alpha Version: Urban Extents*, Socioeconomic Data and Applications Center (SEDAC), Columbia University, NY. Available at http://sedac.ciesin.columbia.edu/gpw (accessed March 10, 2011).
11. C. D. Elvidge, B. T. Tuttle, P. C. Sutton, K. E. Baugh, A. T. Howard, C. Milesi, B. Bhaduri, and R. Nemani, Global distribution and density of constructed impervious surfaces, *Sensors*, 7, 1962–1979, 2007.
12. S. Bontemps, P. Defourny, E. Van Bogaert, O. Arino, V. Kalogirou, and J. R. Perez, 2011, GLOBCOVER 2009 products description and validation report, European Space Agency, Paris, France. Available at http://due.esrin.esa.int/globcover/LandCover2009/GLOBCOVER2009_Validation_Report_2.2.pdf (accessed October 4, 2013).

13. Center for International Earth Science Information Network, Columbia University, and Centro Internacional de Agricultura Tropical, *Gridded Population of the World Version 3 (GPWv3)*. Palisades, NY: Socioeconomic Data and Applications Center (SEDAC), Columbia University, 2005.

14. J. E. Dobson, E. A. Bright, P. R. Coleman, R. C. Durfee, and B. A. Worley, LandScan: A global population database for estimating populations at risk, *Photogrammetric Engineering & Remote Sensing*, 66(7), 849–857, 2000.

15. D. Balk, A. Storeygard, M. Levy, J. Gaskell, M. Sharma, and R. Flor, Child hunger in the developing world: An analysis of environmental and social correlates, *Food Policy*, 30(5), 584–611, 2005.

16. S. Brooker, A. C. A. Clements, P. J. Hotez, S. I. Hay, A. J. Tatem, D. A. P. Bundy, and R. W. Snow, The co-distribution of *Plasmodium falciparum* and hookworm among African schoolchildren, *Malaria Journal*, 5(1), 99, 2006.

17. J. A. Omumbo, C. A. Guerra, S. I. Hay, and R. W. Snow, The influence of urbanisation on measures of *Plasmodium falciparum* infection prevalence in East Africa, *Acta Tropica*, 93(1), 11–21, 2005.

18. P. C. Sutton, S. J. Anderson, C. D. Elvidge, B. T. Tuttle, and T. Ghosh, Paving the planet: Impervious surface as proxy measure of the human ecological footprint, *Progress in Physical Geography*, 33(4), 510–527, 2009.

19. S. Doocy, Y. Gorokhovich, G. M. D. Burnham, D. Balk, and C. Robinson, Tsunami mortality estimates and vulnerability mapping in Aceh, Indonesia, *American Journal of Public Health*, 97(S1), S146, 2007.

20. S. Dasgupta, B. Laplante, C. Meisner, D. Wheeler, and J. Yan, The impact of sea level rise on developing countries: A comparative analysis, *Climatic Change*, 93(3), 379–388, 2009.

21. G. C. Nelson and R. D. Robertson, Comparing the GLC2000 and GeoCover LC land cover datasets for use in economic modelling of land use, *International Journal of Remote Sensing*, 28(19), 4243–4262, 2007.

22. C. Small, V. Gornitz, and J. E. Cohen, Coastal hazards and the global distribution of human population, *Environmental Geosciences*, 7(1), 3–12, 2000.

23. A. Schneider and C. E. Woodcock, Compact, dispersed, fragmented, extensive? A comparison of urban growth in twenty-five global cities using remotely sensed data, pattern metrics and census information, *Urban Studies*, 45, 659–692, 2008.

24. D. Orenstein, B. Bradley, J. Albert, J. Mustard, and S. Hamburg, How much is built? Quantifying and interpreting patterns of built space from different data sources, *International Journal of Remote Sensing*, 32(9), 2621–2644, 2011.

25. W. L. Stefanov and M. Netzband, Assessment of ASTER land cover and MODIS NDVI data at multiple scales for ecological characterization of an arid urban center, *Remote Sensing of Environment*, 99(1), 31–43, 2005.

26. M. R. Montgomery, The urban transformation of the developing world, *Science*, 319(5864), 761–764, 2008.

27. D. Lu and Q. Weng, Spectral mixture analysis of ASTER images for examining the relationship between urban thermal features and biophysical descriptors in Indianapolis, Indiana, USA, *Remote Sensing of Environment*, 104, 157–167, 2006.

28. G. Zhu and D. G. Blumberg, Classification using ASTER data and SVM algorithms: The case study of Beer Sheva, Israel, *Remote Sensing of Environment*, 80(2), 233–240, 2002.

29. Y. Yamaguchi, A. B. Kahle, H. Tsu, T. Kawakami, and M. Pniel, Overview of Advanced Spaceborne Thermal Emission and Reflection Radiometer (ASTER), *IEEE Transactions on Geoscience and Remote Sensing*, 36(4), 1062–1071, 1998.

30. N. Yamamoto, R. Nakamura, H. Yamamoto, S. Tsuchida, I. Kojima, Y. Tanaka, and S. Sekiguchi, GEO grid: Grid infrastructure for integration of huge satellite imagery and geoscience data sets, in *Proceedings of the Sixth IEEE International Conference on Computer and Information Technology*, Los Alamitos, CA, 2006, p. 75.

31. J. Hodges, Validation of the consistent-year V003 MODIS land cover product, Boston University, Boston, MA. Available at http://www-modis.bu.edu/landcover/userguidelc/consistent.htm (accessed March 10, 2011).

32. R. Tateishi, Global land cover ground truth database (GLCGT database) Version 1.2, Center for Environmental Remote Sensing (CEReS), Chiba University, Chiba, Japan. Available at ftp://geoinfo.cr.chiba-u.jp/pub/geoinfo/globalproducts/GLCGT/ (accessed October 3, 2013).

33. K. Iwao, K. Nishida, T. Kinoshita, and Y. Yamagata, Validating land cover maps with degree confluence project information, *Geophysical Research Letters*, 33, L23404, 2006.

34. D. Potere and A. Schneider, A critical look at representations of urban areas in global maps, *GeoJournal*, 69(1), 55–80, 2007.

35. H. Miyazaki, K. Iwao, and R. Shibasaki, Development of a new ground truth database for global urban area mapping from a gazetteer, *Remote Sensing*, 3(6), 1177–1187, 2011.

36. Center for International Earth Science Information Network, Columbia University, International Food Policy Research Institute, The World Bank, and Centro Internacional de Agricultura Tropical, *Global Rural-Urban Mapping Project (GRUMP), Alpha Version: Population Density Grid*, Socioeconomic Data and Applications Center (SEDAC), Columbia University, NY. Available at http://sedac.ciesin.columbia.edu/gpw (accessed March 10, 2011).

37. GEOnet Names Server, *NGA GEOnet Names Server (GNS)*, vol. 2010. Bethesda, MD: NGA, 2008.

38. M. Doerr and M. Papagelis, A method for estimating the precision of placename matching, *IEEE Transactions on Knowledge and Data Engineering*, 19(8), 1089–1101, 2007.

39. D. Potere, A. Schneider, S. Angel, and D. L. Civco, Mapping urban areas on a global scale: Which of the eight maps now available is more accurate? *International Journal of Remote Sensing*, 30(24), 6531–6558, 2009.

40. G. T. Koeln, T. B. Jones, and J. E. Melican, GeoCover LC: Generating global land cover from 7600 frames of Landsat TM data, *Proceedings of ASPRS 2000 Annual Conference*, Washington, DC, 2000.

41. D. Zhou, O. Bousquet, T. N. Lal, J. Weston, and B. Schölkopf, Learning with local and global consistency, *Advances in Neural Information Processing Systems*, 16, 321–328, 2003.

42. D. LP, Land cover type yearly L3 global 500 m SIN grid, 2009. [Online]. Available at https://lpdaac.usgs.gov/lpdaac/products/modis_products_table/land_cover/yearly_l3_global_500_m/mcd12q1 (accessed on August 2, 2011).

43. H. Miyazaki, X. Shao, K. Iwao, and R. Shibasaki, An automated method for global urban area mapping by integrating ASTER satellite images and GIS data, *IEEE Journal of Selected Topics in Applied Earth Observations and Remote Sensing*, 6(2), 1004–1019, 2012.

44. J. Hodges, MOD12Q1 land cover product, 2002. [Online]. Available at http://www-modis.bu.edu/landcover/userguidelc/lc.html (accessed on August 2, 2011).

45. K. C. Clarke, S. Hoppen, and L. Gaydos, A self-modifying cellular automaton model of historical urbanization in the San Francisco Bay area, *Environment and Planning B: Planning and Design*, 24(2), 247–261, 1997.

46. I. Foster, The grid: A new infrastructure for 21st century science, *Physics Today*, 55(2), 42–47, 2002.

47. G. M. Foody, Harshness in image classification accuracy assessment, *International Journal of Remote Sensing*, 29(11), 3137–3158, 2008.

48. A. H. Strahler, L. Boschetti, G. M. Foody, M. A. Friedl, M. C. Hansen, M. Herold, P. Mayaux, J. T. Morisette, S. V. Stehman, and C. E. Woodcock, *Global Land Cover Validation: Recommendations for Evaluation and Accuracy Assessment of Global Land Cover Maps*. Luxembourg: Office for Official Publication of the European Communities, 2006.

49. G. M. Foody, Status of land cover classification accuracy assessment, *Remote Sensing of Environment*, 80(1), 185–201, 2002.

50. H. Tonooka, K. Omagari, H. Yamamoto, T. Tachikawa, M. Fujita, and Z. Paitaer, ASTER cloud coverage reassessment using MODIS cloud mask products, in *Proceedings of the SPIE 7862, Earth Observing Missions and Sensors: Development, Implementation, and Characterization*, Incheon, Republic of Korea, 78620S, 2010.

51. H. Taubenbock, T. Esch, A. Felbier, A. Roth, and S. Dech, Pattern-based accuracy assessment of an urban footprint classification using TerraSAR-X data, *IEEE Geoscience and Remote Sensing Letters*, 8(2), 278–282, 2011.

52. T. Esch, H. Taubenböck, A. Roth, W. Heldens, A. Felbier, M. Thiel, M. Schmidt, A. Müller, and S. Dech, TanDEM-X mission—New perspectives for the inventory and monitoring of global settlement patterns, *Journal of Applied Remote Sensing*, 6(1), 61701–61702, 2012.

53. M. Kajimoto and J. Susaki, Urban-area extraction from polarimetric SAR images using polarization orientation angle, *IEEE Geoscience and Remote Sensing Letters*, 10(2), 337–341, 2013.

54. E. Hand, Citizen science: People power, *Nature*, 466(7307), 685–687, August 2010.

55. M. Haklay and P. Weber, OpenStreetMap: User-generated street maps, *Pervasive Computing, IEEE*, 7(4), 12–18, 2008.

56. K. Chen and J. McAneney, High-resolution estimates of Australia's coastal population, *Geophysical Research Letters*, 33, L16601, 2006.

8 Building of a Global Human Settlement Layer from Fine-Scale Remotely Sensed Data

*Martino Pesaresi, Vasileios Syrris, Daniele Ehrlich,
Matina Halkia, Thomas Kemper, and Pierre Soille*

CONTENTS

8.1 INTRODUCTION

Is the production of national, continental, or global landmass fine-scale mapping using high-resolution remotely sensed imageries feasible with today's remote sensing technology? In particular, is fine-scale large-area mapping of built-up areas feasible? Even if these questions may be considered trivial and already demonstrated for data scenarios including low- and moderate-resolution images, they are still far from being solved if applied to high- and very-high-resolution (HR/VHR) optical images. The reasons behind this are linked to the specific characteristics of the input data scenarios including HR/VHR data and to the specific characteristics of the physical targets associated with the human settlement information to be recognized and analyzed. Quadratic increase of input data volume, exponential increase of computational complexity due to the necessity to process multiscale structural (shape/size, morphological, and textural) image descriptors and the necessity of spatial and thematic uncertainty management, and increase of thematic complexity and automatic learning strategies are some of the challenges that must be addressed in order to solve the two questions regarding the class of HR/VHR input image data. These two main questions were explored during the first operational test made by the Joint Research Center (JRC) during 2012 within the framework of the Global Human Settlement Layer (GHSL) production. The extent of the test area was 24.3 millions of square kilometers and covered parts of four continents. The imagery was collected by a variety of optical satellite and airborne sensors with spatial resolution ranging from 0.5 to 10 m (Pesaresi et al. 2013). It is the largest known automatic image classification involving this kind of image input (Figure 8.1).

This chapter discusses the methodological choices made during the design of the GHSL experiment and summarizes the main results and conclusions. The chapter is organized as follows: Section 8.2 discusses the rationale; Section 8.3 then highlights the main technological and methodological challenges addressed by the GHSL experiment. Thereafter, Section 8.4 discusses the adopted methodological solutions and Section 8.5 introduces the GHSL production workflow. In Section 8.6, the principal results are presented from different perspectives. Finally, Section 8.7 discusses the lesson learned and concludes the chapter.

8.2   RATIONALE

Satellite imagery today could potentially provide information about the built environment worldwide, due to advances in computational and storage capacity, as well as data availability and cost-effectiveness. Despite this potentiality of remote sensing technologies, there are few global data sets that can be used to map human settlement. Examples of available thematic proxies to the global human settlement information include world nighttime lights based on the DMSP-OLS sensor (Elvidge et al. 2001), MODIS 500 based on land use/land cover (LU/LC) classifications (Bartholome and Belward 2005; Schneider et al. 2010), and global population data sets like LandScan (Dobson et al. 2000) or the gridded population of the Columbia University Center for International Earth Science Information Network in collaboration with Centro Internacional de Agricultura Tropical (2012). Since 2011, the Suomi National Polar-Orbiting Partnership (SUOMI NPP) satellite has been producing night-lights at 750 m spatial resolution.* An overview, a comparison, and an analysis of eight global data sets are provided by Schneider et al. (2010). These available information layers can be categorized under two families: (1) those derived from low-resolution (LR) satellite sensors (ranging from 3 to 0.5 km) and (2) those derived from a mixture of census administrative, cartographic, and GIS sources merged with ad hoc models.

---

* National Aeronautics and Space Administration, Goddard Space Flight Center, National Polar-orbiting Partnership, http://npp.gsfc.nasa.gov/index.html.

FIGURE 8.1 **(See color insert.)** An overview of HR/VHR image data processed during the GHSL experiment. (From Pesaresi, M. et al., *IEEE J. Select. Top. Appl. Earth Observ. Remote Sens.*, 6(5), 2102, 2013.)

Because of the nature of the sources and the methodologies used for the production of these data, they show an implicit and sometimes explicit generalization scale in the range of 1:500 K until 1:2 M. This practically means that BU patches with sizes smaller than 100 m or even 1 km have relatively less probability to be represented than compact BU patches having a size greater than 1 km. This fact may introduce bias in the estimation of total areas occupied by the human settlement, and the bias will depend on the spatial settlements patterns existing in different regions. This bias will influence the output of the spatial modeling (risk, impact, population) using settlements information in input. In general, it can be observed (Tenerelli and Ehrlich 2011) that these sources introduce an underestimation bias for the total settlement footprint that is proportional to the degree of sprawl or dispersion of the settlement under analysis. From this point of view, the use of these sources for quantification of settlement sprawl or dispersion as required in several models is arguable.

Figure 8.2 shows the results of the analysis of the influence of spatial resolution in the representation of BU areas in the region of Toscana, Italy. This region has a total surface area of about 23,000 km^2 and a total population of 3,667,780 (census 2012). The regional settlement pattern comprises small/medium towns with a few hundreds of thousands of people and scattered settlements in rural areas and in some productive areas (Prato). The input data used in the analysis are derived from the official regional cartography made from an aerial photogrammetric survey with a production scale of 1:2000. The vector data related to the building footprints of the whole region has been initially rasterized with a grid cell of 2.5 m. At this spatial resolution, the total BU surface is estimated at 234.32 km^2, which corresponds to ~1% of the total surface. The curve in Figure 8.2 was calculated by decreasing the spatial resolution (equivalently increasing the grid cell size) of the representation in a stepwise manner and by the subsequent application of a classification rule selecting the BU/NBU class of the cell according to the majority of share covered by, respectively, the BU and NBU classes in the same cell. Surfaces belonging to BU omission and commission errors were calculated thus. In this example, the bias introduced by the spatial resolution used to represent the settlement information is clearly noticeable as dominated by increasing omission error generated by increasing the cell size (decreasing the spatial resolution)

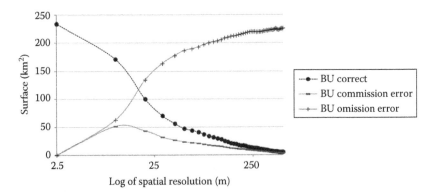

FIGURE 8.2   Analysis of the effect of the spatial resolution parameter in the representation of the BU areas of the region of Toscana, Italy.

in the raster representation of the information. Omission and commission errors are almost comparable only until 10 m of spatial resolution, and then they tend to diverge for greater cell sizes. The effect is a dramatic underestimation of the total BU surface in the region: if we sum the BU correctly detected and the BU accounted in the commission error as total BU surface accounted by the model in the specific region, we obtain, for example, 102.79 and 21.59 km^2 for, respectively, 30 and 250 m of cell size. They are only 43.87% and 9.22% of total BU surface calculated with 2.5 m of spatial resolution in the same region.

Due to their increased spatial resolution, HR/VHR input image data can potentially contribute toward mitigating the issues described earlier, allowing much finer scale analysis than obtained with moderate-resolution imageries. The HR/VHR input image data can allow fine-scale recognition and characterization of all basic components of the human settlement, namely, BU structures or buildings (Shettigara et al. 1995; Lin and Nevatia 1998; Benediktsson et al. 2003; Unsalan and Boyer 2004; Zhu et al. 2005; Khoshelham et al. 2010; Sirmacek and Unsalan 2011; Chaudhuri et al. 2012) and open spaces such as city squares, public and private gardens and parks, walking areas, and parking lots. Although HR and even VHR data with an almost global coverage are available with different sensors, so far no consistent global information has ever been extracted from these data. Mapping and monitoring of urban areas at HR and VHR scales are mostly limited in terms of temporal and spatial coverage and remain at the stage of case studies for individual or a few cities, often providing only a single time step (Niebergall et al. 2008; Baud et al. 2010; Ehrlich and Bielski 2011). The lack of a consistent global layer with HR/VHR spatial resolution can be attributed mainly to two reasons. First, the data availability of HR/VHR satellite data: most, if not all, HR/VHR satellite missions are operated on a commercial basis and consequently global coverage is costly. Second, the information extraction capacity from HR/VHR data: to date, no system has demonstrated the capacity to automatically extract global information layers about human settlements from HR/VHR satellite data with the necessary accuracy and cost-effectiveness. In the global (but also regional or even national) perspective, the common drawbacks of available automatic information procedures are as follows: (1) the necessity to collect expensive training sets, (2) the necessity of expensive ad hoc parameter setting and tuning, (3) the necessity of expensive computational infrastructures, and (4) the necessity of specialized input information not available globally. Thus, so far only time-expensive manual or semiautomatic operational procedures are available.

8.3 ON THE COMPLEXITY OF HR/VHR HUMAN SETTLEMENT DESCRIPTION

Some of the methodological challenges related to automatic image information retrieval from HR/VHR input data are related to computational complexity issues. Moreover, automatic recognition and analysis of urban areas and human settlement in general have always been considered a thematically complex issue due to the heterogeneous physical characteristics of the urban fabric and the complexity of the use context of the derived information. Assuming an automated image information retrieval process P starting from raw data x in input and generating information

I about human settlement as output $I = P(x)$, the total complexity O_{tot} of P can be described as the product of the complexity of input data O_{data} by the complexity of the thematic information O_{thema}:

$$O_{tot} = O_{data} \times O_{thema}.$$

The thematic and data complexities are described in Sections 8.3.1 and 8.3.2.

8.3.1 DATA COMPLEXITY

The O_{data} includes two main components: the data volume and the data (in)consistency. The data volume complexity increases quadratically with increasing input spatial detail. This practically means that assuming the mapping of the same surface on Earth with the same temporal and radiometrical resolution, the processing of VHR input images with 0.5 m spatial resolution will be at least 1204 times computationally more expensive than 30 m resolution image data and 1 million times computationally more expensive than the use of 500 m resolution image data. Translating these numbers to time cost, a given $P(x)$ designed for delivering information about a given area after 1 h of processing of Landsat 30 m resolution data would require more than 50 days using 50 cm resolution data as input. Similarly, a $P(x)$ designed for providing information after 1 h of 500 m resolution data in a specific area of the globe would require more than 114 years running on 50 cm data reporting about the same area. These rudimentary examples clearly show that the volume of data and related computational complexity are major issues in large area analysis using HR/VHR input data. While 2–3 orders of operational complexity magnitude can eventually be solved by brute force on multiple CPU environments as computer clusters or clouds, 6 orders of magnitude clearly show the need of a paradigm shift in the image processing algorithms applied during the image information extraction workflow. O_{data} also includes another important complexity component related to the spatial inconsistency of the data: in particular, the number of operations needed to fix spatial inconsistencies in the input data. Also, from this point of view, VHR imageries are challenging because of their intrinsic spatial inconsistency: even accurate processing of stereo pairs cannot reach subpixel root mean square error (RMSE) positional error, assuming a pixel size of 0.5 m. Because of the capacity to collect off-nadir image data of the VHR platforms, the apparent displacement of image pixels is even increasing due to panoramic and parallax distortions. Unfortunately, these effects are more evident in above-ground urban targets as in the case of rooftops of buildings that are strategic targets in remote sensing of urban areas. In practice, these facts lead to an expected apparent displacement of the targets in the order of several tens of pixels, assuming 0.5 m spatial resolution, tall buildings, and usual off-nadir data collection ranges. This has a direct bearing in increasing the complexity of reference data collection and in decreasing the expected accuracy and repeatability of the image information retrieval tasks, especially in the frame of monitoring activities.

8.3.2 THEMATIC COMPLEXITY

The O_{thema} factor can be described as made up of two main components: the heterogeneity of the physical characteristics of the human settlement and the complexity of the use of the information about settlements. From the point of view of the physical composition, the areas occupied by human settlements can be defined as the place of the heterogeneity. Almost all the possible materials and surfaces, including artificial and organic materials, can be reported as belonging to the settlement theme in the same places of the world. Moreover, in many cases we may observe that several different materials can be used for settlement components belonging to the same thematic class (i.e., clay tiles, corrugated metal, grass, concrete, plastic, bitumen, stone, for a building's roof) and at the same time identical materials can be used for making settlement components belonging to different thematic classes (i.e., the same stones can be used for paved roads and building roofs). From this point of view, the distinction among the settlement components based only on the study of materials would have strong limitations. The spatial scale of variability of these different materials is typically around a few meters (Small 2001). This situation is described here as *internal spatial heterogeneity* of the Earth's surfaces covered by settlement areas. Furthermore, because the settlements are also often made up of the same materials present in the surrounding natural areas, they are not distinguishable from the natural or agricultural areas, if we take into account only the characteristics of the materials or the surfaces. Typical examples are unpaved roads and bare soil, but also many roof materials (as clay tiles, or stone tile) and again bare soil or rocks. A very typical example is also the vegetated open spaces in settlements, private green surfaces in residential areas, or parks and other public recreational surfaces present in many settlements. This situation is described here as *mimetism* between the settlement theme and the surrounding areas.

The increase in resolution of HR/VHR input data amplifies the characteristics of internal spatial heterogeneity and mimetism of the settlement areas. From the thematic point of view, it is now well known that increased spatial resolution of sensors leads to increased spectral variability of the thematic classes: this is due to the changed scale of observation of the image information or *targets*. Changing the scale of observation to 0.5 m spatial resolution may reveal that an apparently simple *building roof* may become a multilevel universe made of gutters, chimneys, water tanks, windows, terraces, and even trees of roof gardens. Illumination incidence angle, surface slope, shadows, spatial pattern of the elements of these surfaces (e.g., tiles), and their degree of obsolescence may change their spectral reflectance/absorption characteristics dramatically. As a result, the spectral variability of the class *building roof* will also increase, which will weaken the inferential models for recognizing *building roofs* from image data. Attempts to classify the problem into subproblems by recognizing the different elements separately lead to an explosion in the number of target classes and their specific instances. This typically leads to the degradation of the model generality and applicability across different scenes and/or different geographical places and to the explosion of the cost needed for the collection of reference data for training and testing purposes. A fully automatic processing chain is required for reproducible, cost-effective, and sustainable image information retrieval in the conditions addressed by this study focusing on large areas with HR/VHR spatial resolution. Nevertheless, increasing input resolution and input inconsistency/variability typically decreases the stability of the inferential models translating image data in thematic information and dramatically increases the number of free parameters to be tuned. Moreover, more detail typically calls for more expensive training and testing reference data collection, which conflicts with the necessity of minimizing human intervention in the classification process. It is worth noting that all these mechanisms may show additional multiplicative effects on the whole computational complexity of image information retrieval.

Finally, we discuss the complexity in dealing with information related to human settlements. Human settlements display the highest concentration of human artifacts and functions, and they are typically the most valuable part of the territory for human societies. In fact, settlements typically show the maximum concentration of human activities and man-made structures and objects. Urban areas also show the maximum stratification of different functions in the same place. Settlements show overlapping economic interests among social groups as well as with and among public authorities. Urban areas typically also show the highest social stratification and differentiation that is often coupled with historical development and identity that define priorities and agendas of different social groups. All these multiple stratifications of interests and functions produce a remarkable multiplicity of views, equally valid but not coincident of the settlement "fact" or geographical physical entity. This explains why human settlement is hardly reducible to a unique descriptive scheme or "ontology," while for other kinds of earth surfaces such as "forests" or "corn fields" such reduction can eventually be relatively more successful. This is because in these other cases, it is easier to reach a consensus on a common descriptive scheme defining priorities and semantic hierarchies. Different functions and priorities also define a multiplicity of possible actions on settlements, as is the case of different planning and management strategies. The required information on settlements can therefore diverge remarkably in scales, semantics, priorities, and timeliness, by varying the social actor, the public authority, or the given point of view defining a specific "user." We describe this situation as *complexity of the use context* of the information regarding the human settlement. No unique description can be fully effective, but all descriptions have the same or comparable importance.

8.4   POSSIBLE SOLVING STRATEGY

As we have discussed earlier, automatic detection and analysis of human settlements using HR/VHR input image data must address two main challenges related to complexity of input data and complexity of the thematic information to be extracted. While input data complexity can be described in terms of data volume and (in)consistencies, thematic complexity can be described in terms of heterogeneous and mimetic physical characteristics of the targets to be described and complexity of the user definition or ontology of the settlement fact. The strategic methodological choices defined in the GHSL production system design make sense in the context of the challenges mentioned earlier, particularly specific data complexity and thematic complexity mitigation measures. Concerning data complexity, the main mitigation measures can be carried out in three steps: (1) definition of new optimized processing chains allowing textural and morphological multiscale image analysis with complexity growing linearly with the data volume and independently from the number of scales; (2) optimization of the feature space by drastic evidence-based reduction of the assumptions, linking image-derived features with information contents; and (3) definition of a new data representation schema based on a multiscale discrete field of image descriptors (DFID) including interscale mechanisms that are able to minimize spatial and thematic inconsistencies. Concerning thematic complexity, the main mitigation measures can be deployed in two steps: (1) adoption of a new classification schema based on multiple semantic abstraction levels in order to minimize the impact on the automatic processing chains of complex (heterogeneous, inconsistent, rapidly obsolete) user requirements about information contents and (2) optimization of semantic interoperability by drastic reduction of the abstraction embedded in the "level 0" or basic information level extracted during the GHSL production (level 0 is described in the Section 8.4.3).

8.4.1 FEATURES

The image-derived features used for the first GHSL operational production test of June 2012 can be classified under two main image descriptors: textural and morphological. Textural descriptors were calculated using a rotation-invariant anisotropic composition of gray-level co-occurrence matrix (GLCM) contrast textural measurements following the so-called PANTEX methodology (Pesaresi et al. 2008, 2011), also applied in experimental sets simulating global processing of HR/VHR data (Pesaresi et al. 2011; Ouzounis et al. 2012). Morphological descriptors were calculated by a multiscale morphological decomposition schema, which is an evolution of the Derivative of the Morphological Profile (DMP) methodology (Pesaresi and Benediktsson 2001) in the Characteristic-Saliency-Level (CSL) model. The CSL was introduced with the purpose of optimization of the computational complexity and of the disk space requirements for storage of image-derived features in massive multiple-scene input data scenarios (Pesaresi et al. 2012b). The selection of the features suitable for the experiment was done based on (1) computational sustainability for massive HR/VHR data processing and (2) evidence-based reasoning linking the image-derived features with the presence of the image information of interest.

From the point of view of image-derived features, the BU areas detectable in the images can be defined as "all the areas in the image showing high local density of high-contrast square corners and showing structures with a size in a specific range [s1...s2]." The first part of the statement is treated by textural descriptors (corners detection) and the second part by multiscale morphological analysis (scale-space decomposition). The designed inferential model tries to translate the observed physical characteristics of the BU areas in the minimal sufficient number of image descriptors. The adoption of textural contrast measurements is justified with the observation of the human settlements as determined by locally heterogeneous materials having a high probability of reflecting heterogeneous quantities of energy in specific wavelengths. Moreover, BU structures usually cast shadows and this amplifies the expected heterogeneity of reflectance detected by remote sensors in correspondence to BU areas. Furthermore, humans show a strong preference for building settlements made of objects with square corners, and the strong correlation between image spatial patterns showing high density of square corners and the presence of buildings can thus be measured (Pesaresi et al. 2011; Ouzounis et al. 2012). Finally, the expected sizes of BU structures can be estimated by analysis of the spatial characteristics of built-up areas as in Small (2001).

8.4.2 DATA REPRESENTATION

Data representation plays an important role in the GHSL production system design. The proposed approach is based on DFID, in analogy to the discrete fields approach in physics, where complex phenomena difficult to be modeled in a deterministic way at the micro scale show more stable statistical behavior if summarized at the macro scale. Accordingly, radiometric, textural, and morphological (shape) descriptors calculated at the geometry (resolution, coordinate of the origin, projection) of the input imagery are aggregated to the geometry of the global discrete field by analytical projective mapping transforms. The scale of the discrete field, and thus the size of the cells, is defined by the spatial resolution of the input data and their spatial uncertainty. This includes the implicit generalization that may be introduced by specific image descriptors, for example, the window size and the structuring element size in the textural and morphological descriptors, respectively. The entities that are classified are the cells of the discrete field organized in tiles in order to optimize the I/O through raster database operations. These cells are described by the image descriptors inherited from images processed at the original resolution, but then summarized at the size of the discrete field cell. Figure 8.3 depicts the proposed DFID concept. Inside and between the discrete fields, which may have different scales, various information collection and distillation processes may be discerned. In the proposed approach, the most important ones are related to aggregation and summarization from fine scale to coarse scale and, symmetrically, learning and classification from coarse scale to finer scales. The first kind of information processes can often be formalized by deductive and deterministic processing chains. On the other hand, the second kind of information processes is typically based on inductive statistical chains, for example, the learning techniques used for selecting the best thresholds in the image features for classification purposes.

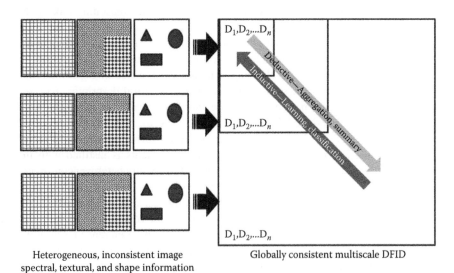

Heterogeneous, inconsistent image spectral, textural, and shape information

Globally consistent multiscale DFID

FIGURE 8.3 The general DFID concept. (From Pesaresi, M. et al., *IEEE J. Select. Top. Appl. Earth Observ. Remote Sens.*, 6(5), 2102, 2013.)

With respect to the pixel-oriented image analysis methodology, the proposed DFID provides a more consistent approach for integrating heterogeneous image descriptors in the same classification task, in particular for integrating textural and morphological (shape) multiscale descriptors into traditional radiometric descriptors that have a "natural" representation at the scale of the pixel.

With respect to other image analysis methodologies based on image segmentation and classification, as the so-called object-oriented image analysis (OBIA) (Bruzzone and Carlin 2006; Blaschke et al. 2008), the proposed DFID method has several key advantages: (1) explicit management of spatial uncertainty, (2) stabilization of the inferential models, (3) possibility of a second-level pattern analysis of the results, and (4) reduction of memory requirements.

The DFID method allows complete and consistent management of the spatial uncertainty embedded in the image information at any scale, which is one of the main drawbacks of OBIA methods assuming spatial uncertainty always negligible with respect to the image pixel size. As discussed before, these conditions are illusory in real HR/VHR image data processing scenarios, and they may become completely misleading in case of thematic information targets having a size comparable with the expected input spatial uncertainty and apparent observable displacement. With regard to the operational constraints discussed here, the image information retrieval methodologies based on preliminary image segmentation steps face computational problems in postprocessing when filtering and reaggregating spurious image segments resulting from misplacement of input data collected by different sensors and/or the same sensor at different times.

Moreover, the DFID method stabilizes the inferential information extraction model by first aggregating several image object/region instances in the same cell and then taking the classification decision based on the whole aggregated attributes. This is compatible with some well-established machine learning methodologies, for example, the "bag-of-words" approach. The increase in the number of instances makes the statistical inference more stable. At the level of the cell, omission and commission recognition errors may compensate, reducing the whole error rate. The OBIA paradigm instead takes the classification decision typically at the level of the single object/region, which is more risky from the statistical point of view: potential errors made at this point are thus directly propagated in the subsequent inferential steps.

Furthermore, DFID allows computationally efficient multiscale pattern analysis of the image information retrieval results. The same mathematical tools allowing pattern analysis on lattice or raster structures can be translated to the analysis of DFID.

Finally, DFID significantly reduces the memory required for storage of image-derived information and consequently the I/O efficiency of the whole image information retrieval workflow. Internal estimations show a reduction in memory requirements of 1 to 2 orders of magnitude for comparable image information (detail) stored using the DFID and OBIA approaches.

8.4.3 Classification Schema

In Pesaresi and Ehrlich (2009), a critical review of the standard LU/LC paradigm was made from the perspective of global mapping of human settlements. Based on observations of internal consistency and external adequacy aspects of the LU/LC paradigm, an alternative approach was proposed, structured with a modular abstraction model. In particular, three basic abstraction levels were identified as useful for the study of human settlements: <*level0*> where the basic settlement components are detected, <*level1*> where they are characterized by their physical characteristics and patterns, and <*level2*> where more abstract use of the settlement space by the population is inferred. The degree of generality of the inferential model translating image data in information can be described as inversely proportional to the degree of abstraction of the target information. Moreover, typically higher semantic abstraction is correlated with the need of higher integration with image-external information sources. In the first GHSL experimental test reported here, only a partial implementation of the first two semantic levels was made. In particular, only BU structures were detected at <*level0*> and they were characterized at <*level1*> on the basis of their morphological characteristics (size criteria).

The basic information contents of the current version of GHSL rely on the definition of BU structure (building) and BU areas: they are necessary for a quantitative description of human settlements using HR and VHR remotely sensed data inputs (Pesaresi and Ehrlich 2009). BU areas are the spatial generalization of the notion of building defined as "areas (spatial units) where buildings can be found." The working definition of BU structure (building) used in this experiment setting is as follows:

> Buildings are enclosed constructions above ground which are intended or used for the shelter of humans, animals, things or for the production of economic goods and that refer to any structure constructed or erected on its site.

This working definition is adapted from the data specification on buildings delivered by the Infrastructure for Spatial Information in Europe (INSPIRE),* taking into account the specific GHSL constraints and user requirements. In particular, as opposed to the INSPIRE definition, the GHSL definition does not include the underground building notion for obvious limitations associated with the input data.

Moreover, the GHSL notion does not impose the permanency of the BU structure on the site as INSPIRE does, following the classical topographic mapping tradition. The GHSL notion of BU structure is more inclusive and includes temporary human settlements for refugees or internally displaced people (IDP) camps.

Finally, in contrast to INSPIRE, the GHSL repository includes BU areas falling in the "slum" or informal settlement concept: the area of a city characterized by substandard housing and squalor and lacking in tenure security, also called "shanty town," "squatter settlement," and so on.

It is worth noting that the GHSL definition is only partially fitting with other similar available definitions already popular in the RS community such as the USGS "Urban or Built-up areas,"[†] "Impervious Surfaces" (Lu and Weng 2006), "Urban Soil Sealing,"[‡]

---

* INSPIRE Infrastructure for Spatial Information in Europe, D2.8.III.2 Data specification on building—Draft guidelines, INSPIRE Thematic Working Group Building 2012. http://inspire.jrc.ec.europa.eu/documents/Data_Specifications/INSPIRE_DataSpecification_BU_v3.0.pdf.

[†] http://landcover.usgs.gov/urban/umap/htmls/defs.php.

[‡] http://www.eea.europa.eu/articles/ urban-soil-sealing-in-europe.

and CORINE "Artificial Surfaces."[*] Compared to these LU/LC definitions, the GHSL classification schema is more general and does not assume any embedded urban/rural dichotomy (BU structures are mapped independently if they fall under "rural" or "urban" area definitions) and is more focused on quantitative support to crisis management and risk and disaster mitigation activities requiring detailed mapping of buildings, population, and their vulnerabilities with a multiscale approach. Furthermore, the GHSL classification scheme with its simplification and reduction of the embedded abstraction was designed to facilitate the semantic interoperability and multidisciplinary across-application sharing of data and results. This includes the sharing of data between different agencies (UN, WB, EC) working in similar areas, but not necessarily sharing exactly the same abstract definitions (Pesaresi and Ehrlich 2009).

8.5 GHSL PROCESSING WORKFLOW

8.5.1 INPUT IMAGE DATA AVAILABLE

The satellite and airborne data used in the first GHSL experiment were acquired with optical sensors with a spatial resolution of 10 m or higher in order to allow detection of single buildings or groups of buildings. The data are hosted in the Community Image Data (CID)[†] portal. The CID portal is a web portal to search and access remote sensing data and derived products hosted at JRC for a variety of applications.

In this study, we use in total 11,438 panchromatic and multispectral satellite data sets from SPOT 2 and SPOT 5, RapidEye, CBERS-2B, QuickBird-2, GeoEye-1, and WorldView 1 and WorldView 2. In addition, airborne data sets covering the whole of Guatemala were available as RGB imagery. The data set under test covers parts of Europe, South America, Asia, and Africa for a total mapped surface of more than 24 million square kilometers. The input data volume is estimated in the order of $4.00 + E12$ picture elements (pixels), stored in approximately 30 TB of disk space taking into account the various number of bands, bit depth, and compression formats applied in the available input scenes.

The different data sets cover a wide range of spatial resolutions from 0.5 m airborne data sets to 10 m of the SPOT 2 sensor. Radiometrically the entire visible and near-infrared part of the spectrum is covered by the test with wide panchromatic bands and up to eight multispectral bands of WorldView 2. In addition, some data sets consist of pan-sharpened multispectral images with the spatial resolution of the panchromatic band.

Around 50% of the VHR input data used in this experiment was available only in lossy data compression format: in particular JPEG, MrSid, and ECW formats were used in input during this experiment. It is worth noting that these formats introduce artifacts both in the radiometric and in the structural (texture, shape) image information descriptors, thus introducing robustness challenges in the whole image processing workflow.

---

[*] http://www.eea.europa.eu/publications/ COR0-landcover.
[†] http://cidportal.jrc.ec.europa.eu/.

Concerning the geocoding, the input quality condition was highly heterogeneous: the expected RMS absolute positional accuracy ranged from 3 to 5 m of orthorectified data, 25 m of raw VHR data, and up to 40 km in the CBERS 2B case.

The available input image data was collected in arbitrary and heterogeneous seasonal conditions, with arbitrary and heterogeneous sun/sensor elevation and azimuth parameters. In some 20% of the input data no precise information about collection parameters was available, especially in case of large mosaics of VHR input data made by third parties.

8.5.2 GLOBAL REFERENCE DATA

Several additional data sets were used in the workflow as ancillary data. For the orthorectification of some of the satellite data we used the TerraColor* as a reference layer. This is an orthorectified global imagery base map at 15 m spatial resolution built primarily from Landsat 7 satellite imagery. The Open Street Map[†] (OSM) data were used to extract a HR land–sea border. During the processing, LR global data sets are used for reference purposes. One of the data sets is urban class of the MODIS 500 Land Cover Type product (Schneider et al. 2010). In addition, LandScan (2008 and 2010)[‡] HR global population data sets were used.

The LR reference data were used for learning and consistency checking purposes before and after the classification steps, respectively. During the learning, LR reference data substitute a manual training set collection by a new interscale learning mechanism detailed in Pesaresi et al. (2013). Also, LR reference information contributes to global consistency checking and optimization of several alternative outputs done at the end of the image information extraction workflow, during the mosaic and integration phase. During the global consistency checking, the active visual training collection loop is activated. Accordingly, with the proposed DFID paradigm, no deterministic masking of HR GHSL information is done using LR information as input. Only statistical inferential chains are admitted from LR to HR information scales. Consequently, the GHSL output is considered as information extracted from HR/VHR input image data. The only deductive mask applied was the land mask derived from OSM data, assuming these data were produced at the same or finer scale than the GHSL output. Consequently, the seawater versus land dichotomy was not extracted from the input imagery by the proposed image information extraction workflow, but instead derived from an external source assumed as suitable for the purpose.

8.5.3 GENERAL WORKFLOW

There are four main ingredients of the workflow characterizing the GHSL experiment discussed here: (1) the input image data, (2) the reference set, (3) a preprocessing chain, and (4) a processing chain (Figure 8.4).

---

* http://www.terracolor.net.
[†] http://www.openstreetmap.org/.
[‡] Copyright by UT-Battelle, LLC, operator of Oak Ridge National Laboratory under Contract No. DE-AC05-00OR22725 with the United States Department of Energy.

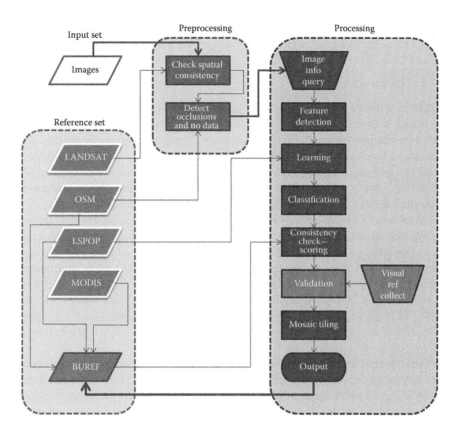

FIGURE 8.4 The workflow applied during the first operational GHSL production test of June 2012.

The reference set has the crucial function to support the optimization of the spatial and thematic consistency during the GHSL production. The workflow requires the existence of a global BU reference layer (BUREF) driving the automatic learning, classification, and information aggregation operations. This layer contains the best estimation of built-up presence at global scale at each iteration of the GHSL production system. At the first iteration, the best estimation was produced by two globally available data sets, LSPOP and MODIS500, at 1 km spatial resolution (Pesaresi et al. 2013). Independent of this initial choice, the whole system is designed to have an incrementally evolutionary approach: the output of any given image information extraction run/experiment, if passing a validation and consistency check, will contribute to improve (thematic accuracy, spatial/temporal completeness) the available BUREF layer. The expectation is that this retroaction mechanism will contribute toward a stepwise enhancement of the overall reliability and completeness of the GHSL output.

The preprocessing module basically performs two functions: (1) checking and optimizing the spatial consistency of the input image data and (2) checking and flagging eventual occlusions and no-data areas in the same images. The spatial consistency is optimized by using an available reference set having 15 m spatial resolution and an expected RMS spatial tolerance of around 20 m, while the occlusions and no-data areas are detected by an internal recognition mechanism.

8.5.4 LEARNING APPROACH

In this experiment, a new interscale learning and classification paradigm was introduced with the objective of allowing a fully automatic processing chain for heterogeneous and not calibrated input data set. The general idea behind this new approach is to move the calibration step from the input data—where it is placed classically—to the image-derived features (descriptors) before the actual classification. The general objective is to stabilize as much as possible the classification parameters against the complex input data used in this experiment.

The classical methodologies for standardization of image-derived features rely on observation of the statistical distribution of the values of the features in the specific scene. This strategy was tested but rejected during the experiment design: it provided unstable results in data-processing scenarios involving multiple-scene and heterogeneous input data. In particular, scene-relative standardization approaches assume homogeneous (or at least comparable) distribution of land cover classes in each scene. This condition was largely violated in the discussed experiment set, where fully "urban" scenes were processed together with scenes with only a few buildings in some remote rural areas.

In the proposed approach, the HR image-derived descriptors are rescaled through learning procedures that use LR globally available information layers as reference. Of course, a correlation between the HR image descriptors and the LR global reference layers must be assumed. The role of the LR reference information layers is to increase the consistency and comparability of HR classification outputs produced from heterogeneous HR/VHR sources. It is worth noting that in the proposed approach this objective is achieved exclusively by the learning procedures described in this section; no a priori masking of HR data is performed with the LR reference information.

Image-derived features that are standardized with respect to an explicit objective function can be used for a fully automatic classification chain. Consequently, the advantage of the proposed methodology is that the collection of training samples can be avoided. This is particularly important for the whole experiment and in particular for testing the sustainability of global HR/VHR image information retrieval. Nevertheless, it is evident that between HR image descriptors and LR reference layers there is a scale gap that may introduce geospatial generalization issues. In order to mitigate the scale gap effects, three different modalities of learning and classification are implemented in the experiment: (1) *adaptive learning*, (2) *meta-learning*, and (3) *discovery*.

In the "adaptive learning" modality, the system optimizes the decision thresholds in the input features using a given reference layer. The "meta-learning" modality is used to study the behavior of these decision thresholds in the set of scenes processed and to detect regularities, for example, typical thresholds for a given sensor in specific regions. The output of the meta-learning is then exploited during the "discovery" modality that can be activated in order to have the chance of recovering image information lost because of errors (incompleteness, inconsistencies) in the reference data or different scale generalization of the image-derived information and in the available reference data. In practice, adaptive learning optimizes consistencies between the image information under processing and the reference data, while the meta-learning and discovery modes proceed to a more tentative image information recognition phase where reference data is not available with the necessary thematic, spatial precision.

The typical workflow combining the three modalities is as follows: (1) run adaptive learning in all available scenes and classify the outputs by matching with the reference and the amount of available reference data, (2) run meta-learning in the set of successfully classified scenes with available reference data, and (3) run discovery mode in the set of scenes failing the learning and classification at the point (1).

8.5.5 VALIDATION STRATEGY

The strategy designed for dealing with GHSL validation is based on active linking of two different measurement sets: (1) *accuracy measures* using visually collected reference samples and (2) *consistency measures* using LR global reference sets. The strategy is inspired by the artificial intelligence "active learning" approach, also called "optimal experimental design" in the statistics literature, translated to the global multiscene evaluation case (Settles 2009). In particular, an iterative process exploiting interscale information for systematic comparison is established in order to maximize the impact of a minimized number of visually labeled samples. Assuming similar thematic contents and globally consistent behavior of the coarse-scale reference data, the expectation is that the low agreement areas are the most interesting for optimization of the visual labeling activities. In these areas, we concentrate on both the errors of the information under test and the errors or generalization effects of the coarse-scale reference data. During the iterative process, the integration of the samples visually labeled at the iteration n in the global reference data used at each iteration $n + 1$ would stepwise increase the overall reliability of the reference data and then the overall reliability of the derived quality measurements. Moreover, this mechanism will decrease the probability to select the same areas as priority to be visually analyzed.

8.6 DESCRIPTION OF THE FIRST GHSL RESULTS

Figure 8.5 shows the GHSL output on the characterization of BU structures in East London, UK. The color encodes the estimated size (surface) of BU structures in the range of 10–15,000 m^2 from blue (small) to red (large). Note how the modular abstraction model proposed for the classification schema as an alternative to the classical LU/LC model shows some interesting results in the specific area. BU structures detected and characterized at the abstraction <*level0*> and <*level1*>, respectively, of the classification schema show interesting behavior in the specific case if they are generalized by spatial aggregation. Local densities of <*level1*> descriptors seem to be correlated with a more abstract <*level2*> describing the dominant use of the settlement surfaces, for example, residential (blue) versus industrial commercial (orange-red).

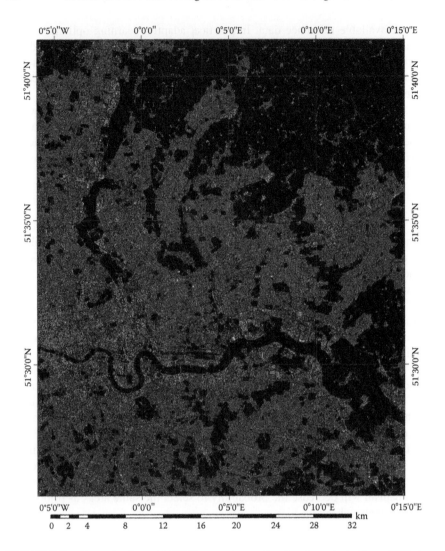

FIGURE 8.5 (See color insert.) GHSL output reporting the characterization of BU structures in East London, UK. The color encodes the estimated size (surface) of BU structures in the range of 10,000–15,000 m² from blue (small) to red (large). Note how relations between the dominant size of BU structures and the use of the settlement areas can be made.

The quality of the automatic GHSL output has been evaluated in Pesaresi et al. (2013) from the perspective of accuracy measurements and global consistency analysis. In this framework, the performance of the same automatic image information retrieval workflow has been benchmarked according to different criteria including (1) external sources used during the learning phase; (2) different sensors, input bands, and resolution; and (3) different geographical regions. The accuracy of the whole GHSL output produced during the experiment was estimated by using a total of approximately 95,000 and 700,000 samples of BU and NBU classes, respectively, collected by visual interpretation of HR/VHR input images. The collection of these samples was based on a random systematic grid approach as defined in the GHSL reference data collection protocol at scale 1:50 K (Pesaresi et al. 2012a). According to this procedure, an estimation of a total accuracy of more than 90% was assessed. These results are consistent with other more specific tests done on the same GHSL output. In particular, the GHSL output of 628 SPOT satellite scenes covering the major urban agglomerations in Europe was systematically compared with the HR European Soil Sealing Layer produced by the European Environment Agency.* The test provided a 90.8% ± 3.9% average agreement rate between the two sources (Pesaresi and Halkia 2012). In Brazil, a stratified random sampling procedure and visual reference data collection was applied to evaluate the GHSL output of more than 3000 input CBERS scenes (Kemper et al. 2013). The assessment provided an average agreement rate of 94% ± 6%. Finally, a systematic comparison between the GHSL output of 2,288 input CBERS scenes and the land cover of China derived from Landsat data was performed (Lu et al. 2013). This test provided an average 98.13% ± 5.6% agreement rate in the best of the benchmarked parameter sets.

By comparing the output of the GHSL produced from HR/VHR input image data against the available representations of human settlements produced by using moderate- or low-resolution input data, some observations can be made. HR/VHR images allow the recognition and classification of single BU structures according to their morphological characteristics: in the first GHSL experiment, a classification based on the estimated size of BU structures was implemented but of course many others can be designed starting from the basic information about single BU structures. The analysis of single BU structures is evidently impossible using input image data with a pixel size approaching the size of BU structures or even exceeding it by 1–2 orders of magnitude, as in the case of LR image data. The possibility to describe single BU structures is an evident value added of processing the HR/VHR image data for analysis of human settlement. The description of the morphological characteristics of single BU structures may contribute toward understanding the BU area fraction in larger pixel sizes and consequently may contribute to understanding the bias functions in the BU areas detected using different input sensors. Moreover, the characterization of single BU structures may contribute toward automatic inferential models recognizing more abstract semantic information layers, for example, different uses (industrial/residential) of the settlement surfaces, their physical vulnerability to specific hazards, the quality of their living conditions, and formal/informal patterns.

By observing the areas of major disagreement between the GHSL output processed during the June 2012 experiment and the available global LR reference data, some interesting phenomena can be reported. In particular, in some areas of the globe, the disagreement is generated by interaction between (1) the specific settlement pattern, (2) the spatial generalization rule embedded in the LR reference data, and (3) the criteria used for extraction of the settlement information from the LR image data. Figure 8.6 shows the detection and characterization of BU structures in the GHSL

---

* EEA fast track service precursor on land monitoring—Degree of soil sealing 100 m. http://www.eea.
europa.eu/data-and-maps/data/eea-fast-track-service-precursor-on-land-monitoring-degree-of-soil-
sealing-100m-1.

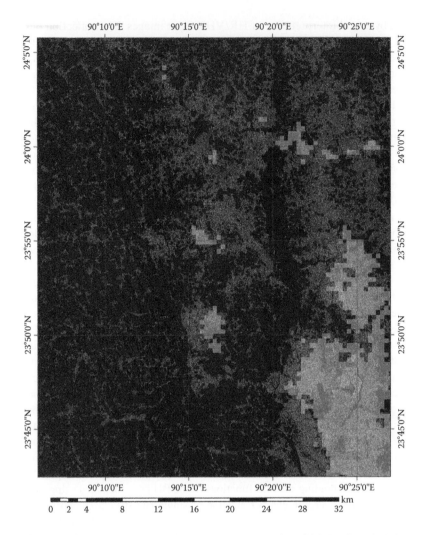

FIGURE 8.6 **(See color insert.)** GHSL output over the city of Dhaka, Bangladesh, overlaid with urban areas using moderate-resolution input data; white squared areas indicate the "urban areas" class of the MODIS 500 source, while colored dots are built-up structures recognized by the GHSL workflow.

output over the city of Dhaka, Bangladesh, compared with urban areas detected using moderate-resolution input data. In the image, white squared areas are derived from the "urban areas" class of the MODIS 500 source. The large disagreement between the two representations of the human settlement and in particular about the estimation of the density of BU structures in the north and in the west of the city is noticeable. Any impact, exposure, or risk spatial modeling using the two different information sources in input would presumably output very different results. Figure 8.7 displays a zoom in on the north-west part of the city of Dhaka showing large disagreement

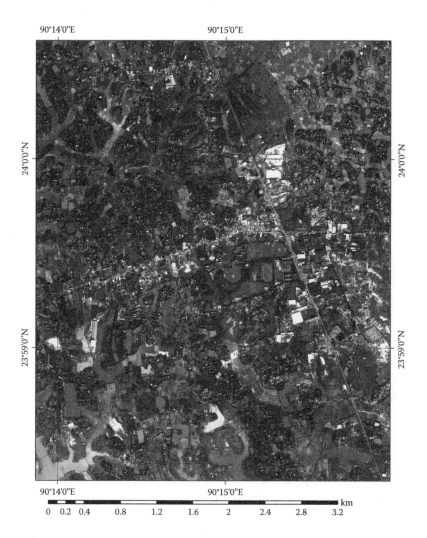

FIGURE 8.7 Zoom in on the north-west part of the city of Dhaka showing large disagreement between the GHSL and the MODIS outputs. The map shows the image input data generating the GHSL output. The settlement pattern shows heterogeneous residential, rural, and industrial use with scattered spatial arrangement. Note that in the area a significant positive correlation between the presence of dwellings and the presence of trees can be observed.

between the GHSL and the MODIS outputs. The map shows the HR image input data generating the GHSL output. The settlement pattern shows heterogeneous residential, rural, and industrial use with scattered spatial arrangement. Note that a significant positive correlation between the presence of dwellings and the presence of trees can be observed in the area. The two factors of spatial scattered patterns and the presence of a large fraction of stable and strong vegetated surfaces spatially correlated to the presence of BU structures may contribute toward explaining the underestimation of these areas in classification using LR input images. Similar considerations can be

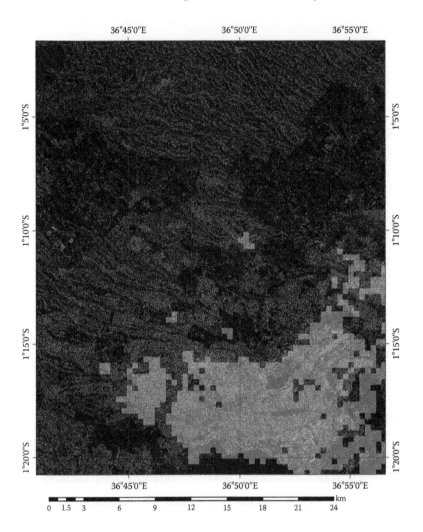

FIGURE 8.8 **(See color insert.)** GHSL output over the city of Nairobi, Kenya, compared with urban areas detected using moderate-resolution input data; white squared areas are derived from the "urban areas" class of the MODIS 500 source while colored dots are built-up structures recognized by the GHSL workflow.

extracted by observing Figure 8.8 showing the detection and characterization of BU structures in the GHSL output over the city of Nairobi, Kenya, compared with urban areas detected using moderate-resolution input data. In the image, white squared areas are derived from the "urban areas" class of the MODIS 500 source. A large disagreement between the two representations of the human settlement is noticeable in the north of the city. Figure 8.9 shows a zoom in on the northern part of the city of Nairobi where the disagreement between the GHSL and the MODIS outputs can be observed. The map shows the HR image input data generating the GHSL output. The settlement pattern displays a relatively dense rural settlement that is spatially

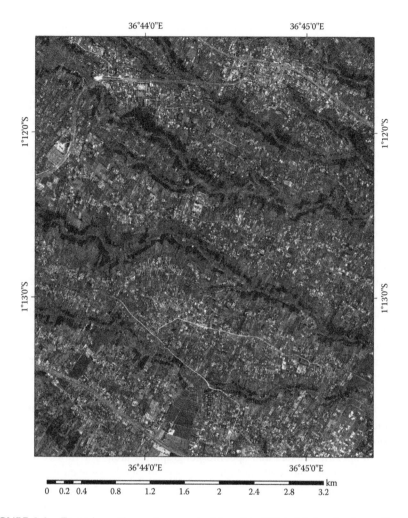

FIGURE 8.9 Zoom in on the northern part of the city of Nairobi showing large disagreement between the GHSL and the MODIS outputs. The map shows the image input data generating the GHSL output. The settlement pattern shows relatively dense rural settlement spatially organized along the crest lines of the hilly region.

organized along the crest lines of the hilly region. In this case, probably the major factor playing a role in the underestimation of BU areas using LR sensors is the relative small size of the single BU patches, together with their spatial scattering.

8.7 CONCLUSIONS

A proof-of-concept of the possibility to build a new GHSL derived from HR and VHR optical remotely sensed data was presented. The test involved 24.3 million square kilometers of test area spread across four continents, automatically mapped with the image data collected by a variety of optical satellite and airborne sensors with a spatial resolution ranging from 0.5 to 10 m. In this mapped area, the total number of people living in 2010 was estimated to be 1,268,448,973 (LandScan). It is the largest known test of automatic image classification involving such kind of image input. Several imaging modes were tested including panchromatic, multispectral, and pan-sharpened images. A new multiscale framework was introduced, integrating the automatic image information retrieval with global available geoinformation layers derived from other satellite sensors or GIS modeling. For the first time, the capability of automatic information extraction from remotely sensed data at a detailed scale in global realistic scenarios and the capacity to control the global consistency of the output both spatially and thematically were demonstrated. The robustness of the adopted image features was tested globally with a high variety of input data quality including extremely challenging "worst-case" scenarios as data lossy compression, pan-sharpening and data warping operations, large seasonal changes, and low signal/noise quality sources (CBERS-2B). New multiscale morphological and textural image feature compression and optimization methods were introduced, together with new learning and classification techniques allowing the processing of HR/VHR image data using LR reference data.

The validation of the automatic results by a visual inspection protocol provided an accuracy rate of more than 90%. These results are consistent with other independent validation campaigns testing the same classification output with comprehensive reference data available in Europe (Pesaresi and Halkia 2012), Brazil (Kemper et al. 2013), and China (Lu et al. 2013). The average agreement between the automatic HR output generated by the experiment and the available LR representation of the urban areas was estimated at 91.5%. Because of the comprehensive and systematic approach of the experiment, a comparative study across HR/VHR sensors, bands, and different geographic areas can be made using precisely the same image information extraction methodology and a consistent global reference layer.

The observation of the anomalies in the global agreement ratio will focus the attention on specific sensors and specific geographical areas for further analysis, validation campaigns, and methodological improvements. In some cases, this analysis shows that the available LR data underestimate significantly the presence of BU structures if compared with HR data. During the GHSL experimental run, this phenomenon was reported in several large regions in all the four processed continents and it is presumably mostly correlated to the fact that the settlements in these regions are spatially organized in relatively small patches. Because the large majority of the global human settlement footprint surface consists of small- and medium-sized BU patches, a significant change in the global estimation of this surface is expected from the analysis of HR/VHR data as it is conveyed by the proposed GHSL methodology.

REFERENCES

Bartholome, E. and A. Belward. 2005. GLC2000: A new approach to global land cover mapping from earth observation data. *International Journal of Remote Sensing 26*, 1959–1977.

Baud, I., M. Kuffer, K. Pfeffer, and R. Sliuzas. 2010. Understanding heterogeneity in metropolitan India: The added value of remote sensing data for analyzing substandard residential areas. *International Journal of Applied Earth Observation and Geoinformation: JAG 12*(5), 359–374.

Benediktsson, J., M. Pesaresi, and K. Amason. 2003. Classification and feature extraction for remote sensing images from urban areas based on morphological transformations. *IEEE Transactions on Geoscience and Remote Sensing 41*(9 part I), 1940–1949.

Blaschke, T., S. Lang, and G. Hay (Eds.). 2008. *Object-Based Image Analysis: Spatial Concepts for Knowledge-Driven Remote Sensing Applications* (1st edn.). Berlin, Germany: Springer-Verlag.

Bruzzone, L. and L. Carlin. 2006. A multilevel context-based system for classification of very high spatial resolution images. *IEEE Transactions on Geoscience and Remote Sensing 44*(9), 2587–2600.

Center for International Earth Science Information Network - CIESIN - Columbia University, and Centro Internacional de Agricultura Tropical - CIAT. 2005. Gridded Population of the World, Version 3 (GPWv3): Population Density Grid. Palisades, NY: NASA Socioeconomic Data and Applications Center (SEDAC). http://sedac.ciesin.columbia. edu/data/set/gpw-v3-population-density.

Chaudhuri, D., N. Kushwaha, and A. Samal. 2012. Semi-automated road detection from high resolution satellite images by directional morphological enhancement and segmentation techniques. *IEEE Journal of Selected Topics in Applied Earth Observations and Remote Sensing 5*(5), 1538–1544.

Dobson, J., E. Bright, P. Coleman, R. Durfee, and B. Worley. 2000. LandScan: A global population database for estimating populations at risk. *Photogrammetric Engineering & Remote Sensing 66*(7), 849–857.

Ehrlich, D. and C. Bielski. 2011. Texture based change detection of built-up on spot panchromatic imagery using PCA. In *Proceedings of Joint Urban Remote Sensing Event (JURSE)*, Munich, Germany, pp. 81–84.

Elvidge, C., M. Imhoff, K. Baugh, V. Hobson, I. Nelson, and J. Safran. 2001. Nighttime lights of the world: 1994–95. *ISPRS Journal of Photogrammetry and Remote Sensing 56*, 81–99.

Kemper, T., X. Blaes, D. Ehrlich, F. Haag, and M. Pesaresi. 2013. On the feasibility to map the settlements of Brazil with the CBERS-2B satellite. In *Urban Remote Sensing Event (JURSE), 2013 Joint*, Sao Paulo, Brazil, April 21–23, 2013, pp. 78–82. doi: 10.1109/ JURSE#.2013.6550670.

Khoshelham, K., C. Nardinocchi, E. Frontoni, A. Mancini, and P. Zingaretti. 2010. Performance evaluation of automated approaches to building detection in multi-source aerial data. *ISPRS Journal of Photogrammetry and Remote Sensing 65*(1), 123–133.

Lin, C. and R. Nevatia. 1998. Building detection and description from a single intensity image. *Computer Vision and Image Understanding 72*(2), 101–121.

Lu, D. and Q. Weng. 2006. Use of impervious surface in urban land-use classification. *Remote Sensing of Environment 102*(1–2), 146–160.

Lu, L., H. Guo, M. Pesaresi, P. Soille, and S. Ferri. 2013. Automatic recognition of built-up areas in China using CBERS-2B HR data. In *Urban Remote Sensing Event (JURSE)*, 2013 Joint, Sao Paulo, Brazil, April 21–23, 2013, pp. 65–68. doi: 10.1109/ JURSE#.2013.6550667.

Niebergall, S., A. Loew, and W. Mauser. 2008. Integrative assessment of informal settlements using VHR remote sensing data: The Delhi case study. *IEEE Journal of Applied Earth Observation and Remote Sensing 1*(3), 193–205.

Ouzounis, G. K., V. Syrris, and M. Pesaresi. 2012. Multiscale quality assessment of global human settlement layer scenes against reference data using statistical learning. *Pattern Recognition Letters 34*(14), 1636–1647.

Pesaresi, M. and J. Benediktsson. 2001. A new approach for the morphological segmentation of high-resolution satellite imagery. *IEEE Transactions on Geoscience and Remote Sensing 39*(2), 309 –320.

Pesaresi, M., X. Blaes, D. Ehrlich, S. Ferri, L. Gueguen,, F. Haag, M. Halkia et al. 2012a. Quality control and validation. In *A Global Human Settlement Layer from Optical*

High Resolution Imagery—Concept and First Results, Chapter 8, JRC technical report. Ispra, Italy: Joint Research Centre. http://publications.jrc.ec.europa.eu/repository/handle/111111111/27402.

Pesaresi, M. and D. Ehrlich. 2009. A methodology to quantify built-up structures from optical VHR imagery. In P. Gamba and M. Herold (Eds.), *Global Mapping of Human Settlement Experiences, Datasets, and Prospects,* Chapter 3, pp. 27–58. Boca Raton, FL: CRC Press.

Pesaresi, M., D. Ehrlich, I. Caravaggi, M. Kauffmann, and C. Louvrier. 2011. Towards global automatic built-up area recognition using optical VHR imagery. *IEEE Journal of Selected Topics in Applied Earth Observations and Remote Sensing* 4(4), 923–934.

Pesaresi, M., A. Gerhardinger, and F. Kayitakire. 2008. A robust built-up area presence index by anisotropic rotation-invariant textural measure. *Journal of Earth Observation Applications* 1(3), 180–192.

Pesaresi, M., G. Huadong, X. Blaes, D. Ehrlich, S. Ferri, L. Gueguen, M. Halkia et al. 2013. A global human settlement layer from optical HR/VHR RS data: Concept and first results. *IEEE Journal of Selected Topics in Applied Earth Observations and Remote Sensing* 6(5), 2102–2131.

Pesaresi, M. and M. Halkia. 2012. Global human settlement layer and urban atlas integration: Feasibility report. JRC Scientific and Policy Report EUR 25328 EN, European Commission, Joint Research Centre, Brussels, Belgium.

Pesaresi, M., M. Halkia, and G. Ouzounis. 2011. Quantitative estimation of settlement density and limits based on textural measurements. In *Proceedings of 2011 Joint Urban Remote Sensing Event (JURSE 2011),* Munich, Germany, pp. 89–92.

Pesaresi, M., G. K. Ouzounis, and L. Gueguen. 2012b. A new compact representation of morphological profiles: Report on first massive VHR image processing at the JRC. In *Proceedings of SPIE 8390, Algorithms and Technologies for Multispectral, Hyperspectral, and Ultraspectral Imagery XVIII,* Baltimore, MD, May 8, 2012, p. 839025.

Schneider, A., M. Friedl, and D. Potere. 2010. Monitoring urban areas globally using MODIS 500m data: New methods based on urban ecoregions. *Remote Sensing of Environment* 114(8), 1733–1746.

Settles, B. 2009. Active learning literature survey. Computer Sciences 1648, University of Wisconsin-Madison, Madison, WI.

Shettigara, V., S. Kempinger, and R. Aitchison. 1995. Semi-automatic detection and extraction of man-made objects in multispectral aerial and satellite images. In A. Gruen, O. Kuebler, and P. Agouris (Eds.), *Automatic Extraction of Man-Made Objects from Aerial and Space Images,* Monte Verita, pp. 63–72. Basel, Switzerland: Birkhauser.

Sirmacek, B. and C. Unsalan. 2011. A probabilistic framework to detect buildings in aerial and satellite images. *IEEE Transactions on Geoscience and Remote Sensing* 49(1), 211–221.

Small, C. 2001. Multiresolution analysis of urban reflectance. In *Remote Sensing and Data Fusion over Urban Areas, IEEE/ISPRS Joint Workshop 2001,* Rome, Italy, pp. 15–19.

Unsalan, C. and K. Boyer. 2004. A system to detect houses and residential street networks in multispectral satellite images. In *Proceedings of the 17th International Conference on Pattern Recognition, 2004 (ICPR 2004),* Paris, France, Vol. 3, pp. 49–52.

Zhu, C., W. Shi, M. Pesaresi, L. Liu, X. Chen, and B. King. 2005. The recognition of road network from high-resolution satellite remotely sensed data using image morphological characteristics. *International Journal of Remote Sensing* 26(24), 5493–5508.

Section III

Urban Observation, Monitoring, Forecasting, and Assessment Initiatives

9 Spatial Dynamics and Patterns of Urbanization

The Example of Chinese Megacities Using Multitemporal EO Data

Hannes Taubenböck, Thomas Esch,
Michael Wiesner, Andreas Felbier,
Mattia Marconcini, Achim Roth, and Stefan Dech

CONTENTS

9.1 INTRODUCTION

Our world is becoming urban. Although more than 50% of the global population already lives in urbanized areas, the United Nations (2011) foresees that by 2050, the entire expected global population growth of 2.3 billion people will be completely absorbed by urban areas. This massive process of urbanization produces immense spatial differences in stage, dynamics, or patterns across continents, cultural areas, or regions. For instance, Europe, North America, and Japan have been urban societies for decades; China is today experiencing a dynamic process of transformation toward an urban society, and developing countries (especially of Asia and Africa) have only recently entered this transformation process.

In China, this ongoing transformation—from a rural to an urban society—has been extremely dynamic in recent years. Thirty years ago, less than 20% of all Chinese people lived in cities (UN 2011). Today already more than 50% live in urban areas, thus transforming China into an urban society. With this development, China has the largest urban population in the world (Chen 2007). Figure 9.1 shows that since 1970, the global urban population has grown by 263%; in China the urban population has grown by 465%, considerably exceeding the global dynamics. Projections for China even estimate 250 million new and additional urban dwellers by 2025 (UN 2011).

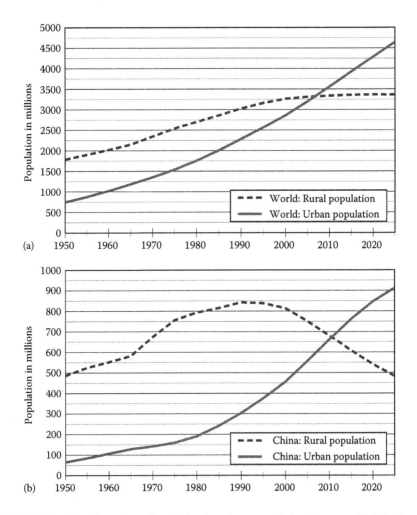

FIGURE 9.1 Transformation of societies based on population figures—Global (a) and Chinese (b) development since 1950 and prospects until 2025. (From United Nations, *World Urbanization Prospects–The 2009 Revision*, United Nations, New York, 2011.)

One result of this massive process of urbanization is the formation of so-called megacities, referring to the largest category of more or less individual monocentric city areas or regions. The United Nations (2003) define megacities quantitatively as a conurbation having more than 10 million inhabitants. In the year 1970, there were just two megacities in the world: New York City (United States) and Tokyo (Japan). Since then, their number has increased markedly to 28 throughout the world, and it is projected that this number will even rise to 37 by 2025 (Figure 9.2). With 16 (or 17 if we want to classify Istanbul as an Asian megacity) Asia features the largest number of megacities. China accounts for four megacities, namely, Shanghai, Beijing, Guangzhou, and Shenzhen, which are the study sites of this chapter. It is even expected that the number of megacities in China will rise to seven by 2025, with the cities of Chongqing, Wuhan, and Tianjin also gaining megacity status (Figure 9.4). Shanghai is projected to become the third largest city on the planet with respect to population (UN 2011).

The massive dynamics and large regions of urban sprawl often strain the ability to govern, organize, and plan new settlements, and it is often a difficult task to even document and measure what has already happened. In most cases, a large amount of spatial, quantitative, and qualitative data and information on urban sprawl exist. However, these data sets are seldom easily accessible, available in digital format, or unrestricted. These data sets are only in a few cases stored centrally, complete, consistent, standardized, substantially documented, and available as time-series or up-to-date, not to mention being comparable with data sets of other cities (Taubenböck et al. 2012).

In the last decades, Earth observation (EO) sensors developed to a stage where global maps of urban areas have been made possible at low spatial resolution from 300 m to 2 km (Potere and Schneider 2009). Several global land cover data sets include the thematic class "urban" (e.g., Bartholome and Belward 2005; Bontemps et al. 2011; Elvidge et al. 2001; Schneider et al. 2005), but their comparatively coarse resolution is insufficient to represent spatial variation of settlement patterns, especially in low density and scattered built-up areas of peri-urban or rural landscapes. Presentation, comparison, and evaluation of the various available global "urban" data sets are discussed by Potere and Schneider (2009). Currently, new EO initiatives aim at improving the geometric accuracy level for global urban mapping based on optical (Pesaresi et al. 2011, 2012) or radar data (Esch et al. 2012, 2013; Gamba and Lisini 2013; Miyazaki et al. 2012); however, to date a multitemporal component is still absent.

Multitemporal EO studies analyzing spatial urban growth at medium resolution (here defined as 10–100 m) have been conducted for megacities such as Dhaka in Bangladesh (e.g., Griffiths et al. 2010) or Tokyo in Japan (e.g., Bagan and Yamagata 2012) as well as for large multinuclei city regions (e.g., Sexton et al. 2013), mega-regions (Taubenböck et al. 2014), or for cities of different sizes (e.g., Yuan et al. 2005). Taubenböck et al. (2012) present a study systematically monitoring all current megacities across the globe. Four change detection examples taken from this study (namely, the megacities Delhi [India], Tehran [Iran], Kolkata [India], and Rio de Janeiro [Brazil]) are illustrated in Figure 9.3. The figure exemplifies spatial urban growth across the globe and its varying dynamics and dimensions as well as their evolving pattern differences at the urban footprint level at approximately decadal intervals since the 1970s.

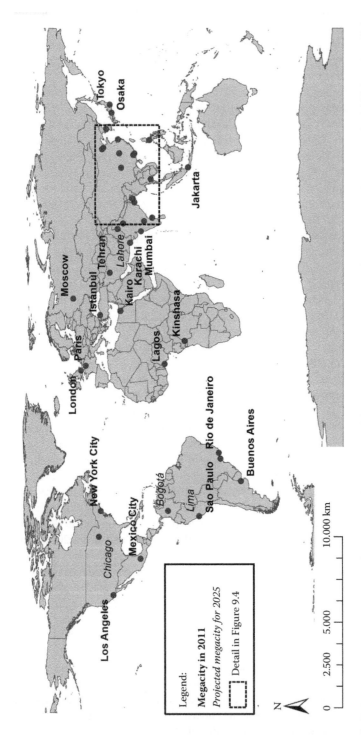

FIGURE 9.2 Global distribution of current and future megacities. (From United Nations, *World Urbanization Prospects–The 2009 Revision*, United Nations, New York, 2011.)

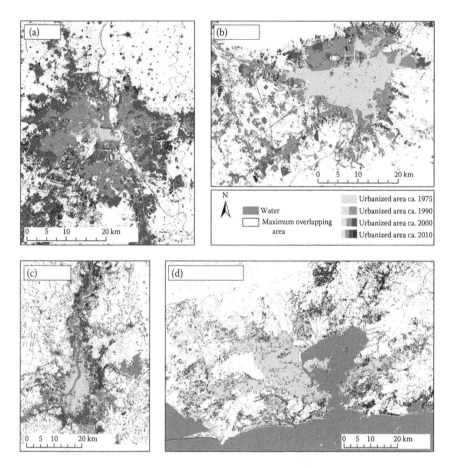

FIGURE 9.3 (See color insert.) Change detections derived from multitemporal remote sensing data: examples—(a) Delhi, India; (b) Tehran, Iran; (c) Kolkata, India; and (d) Rio de Janeiro, Brazil—for global megacity monitoring since the 1970s. (From Taubenböck, H., Esch, T., Felbier, A., Wiesner, M., Roth, A., and Dech, S., *Remote Sens. Environ.*, 117, 162, 2012.)

Angel et al. (2005) systematically compared 90 cities, including some Chinese cities (e.g., Guangzhou) at two time steps (i.e., 1990 and 2000). Beyond this, several studies measuring and comparing urban patterns have been presented (e.g., Civco et al. 2009; Jat et al. 2008; Ji et al. 2006; Schneider and Woodcock 2008; Taubenböck et al. 2009, 2011b). With a spatial focus on China, for example, multitemporal Landsat data are used to monitor spatial urbanization for the megacities Guangzhou (Ma and Xu 2010; Yu and Ng 2006), Beijing (Wu et al. 2006; Xie et al. 2007), and Shanghai (Chen 2007; Yin et al. 2011). Multitemporal SPOT data are used by Cheng and Masser (2003) for monitoring growth in the future megacity of Wuhan in the 1990s. One study analyzed the spatial pattern of the megacity Shanghai based on gradient analysis. Kong and Nakagoshi (2006) also apply gradient analysis for the evaluation of urban green spaces and their patterns in Jinan. Seto and Fragkias (2005) compared four cities in the Pearl River Delta (PRD) (i.e., Guangzhou,

Zhongshan, Dongguan, and Shenzhen) using multitemporal remote sensing data from 1988 to 1999 for inter- and intracity analysis.

However, a systematic approach of the urban remote sensing community toward global, continental, national, or large area and long-time monitoring of spatial urbanization is still absent. The Group on Earth Observation (GEO) is coordinating international efforts to build a Global Earth Observation System of Systems (GEOSS). This initiative was launched in response to calls for action by the 2002 World Summit on Sustainable Development, aiming at exploiting the growing potential of remote sensing data to support decision making. With the topic "SB-04 Global Urban Observation and Information" as information for societal benefits within GEO, the political significance of this topic becomes obvious (GEO 2013). However, this goal is also a commitment for the EO community to overcome the problem of delivering only case studies or stopping at the stage of scientific proof of concepts or methodologies. Rather, these scientific advancements need to be systematically applied to deliver new and value-added information to answer key geographical questions (as shown, with many examples, in this book [Weng 2014]).

This chapter deals with the following main aims contributing to the GEO framework as well as to existing studies using remote sensing to map and monitor urbanization systematically:

- Maximizing the time period for monitoring the spatial effects of urbanization based on EO data to a span of almost 40 years (~1975–2011)
- Mapping and monitoring systematically and consistently today's four megacities in China
- Measuring and comparing spatial patterns for derivation of new insights on the spatial effects of urbanization

9.2   STUDY SITES: CHINESE MEGACITIES

China is widely viewed as a case of underurbanization, and urbanization has lagged rather behind its industrialization as a consequence of restricting policies concerning migration toward cities before the economic reform (Chan 1996; Chang 1994; Chang and Brada 2002; Liu et al. 2003; Zhang and Zhao 2003). With the unprecedented economic growth, however, the country has been witnessing a dramatic growth in urbanization since 1978 when the economic reforms began. Its direct contribution is evident as the population growth of cities in China is fuelled in large measure by rural–urban migration (Heikkila et al. 2007). China prepares, for example, by building new towns from scratch (Taubenböck, 2013). One other result of this process is that today China accounts for the highest concentration of megacities on our planet: Beijing, Shanghai, Guangzhou, and Shenzhen are the megacities of today, and the number of megacities in China is expected to grow to seven by 2025 with Chongqing, Tianjin, and Wuhan being added to the list (UN 2011; Figure 9.4).

All Chinese megacities have undergone rapid urbanization. Beijing is the capital and the center of politics, economics, and culture in China; Shanghai is the economic center of eastern China and has a large industrial structure, is a transportation hub, and includes a service industry. These two cities are the largest cities in China today.

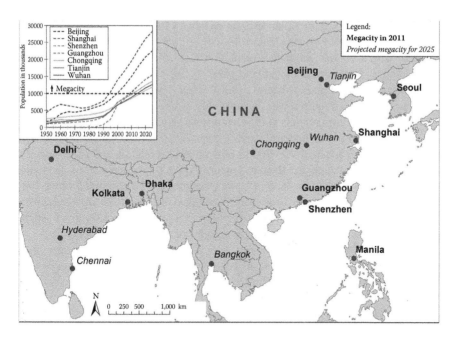

FIGURE 9.4 Focus map (compare Figure 9.2 for global distribution of current and future megacities): Location and population development of current and future Chinese and East Asian megacities.

Both turned into megacities as early as in the 1990s. Shanghai as the largest city is projected to have more than 28 million inhabitants by 2025.

The megacities Guangzhou and Shenzhen are located within the Pearl River Delta at about 100 km distance from each other. They are characterized by export-oriented industries as well as financial and other service industries. Both of them became megacities very recently. Shenzhen has experienced a remarkable population development since urbanization started comparatively late in the early 1990s but with incredible dynamics.

All these four megacities and their vicinity city clusters accounted for about 20% of the total gross domestic product in China in 2005 (Chan and Yao 2008), proving the importance of these urban hubs on the national as well as global economy. The future megacities of Tianjin, Chongqing, and Wuhan show very similar population developments with expectations to cross the megacity line around 2015. Figure 9.4 illustrates the location and the population development since the 1970s of both the current four megacities and the three future megacities. Apart from this detail, Figure 9.4 gives an overview of current and future megacities in East Asia, not shown in detail in Figure 9.2.

9.3 DATA

Earth observation (EO) data allow for consistent, multitemporal and large area data for monitoring spatial processes of urbanization. Maximizing the time period for monitoring spatial effects of urbanization limits the range for commercially or freely available EO data sets. Today's large extents of the megacities—e.g., the dimension of the core Shanghai metropolitan area in east–west direction roughly spans over 50 km—require remote sensing data sets with a large swath width for full areal coverage. Moreover, the geometric resolution must allow for the delineation of urbanized areas from nonurbanized land cover.

In this framework, Landsat data (e.g., GLCF 2013) are an obvious cost-effective choice fulfilling these requirements. Accordingly, we use for our analysis data from the Landsat multispectral scanner (MSS) featuring a geometric resolution of 60 m and a spectral resolution of four spectral bands (green, red, and two near-infrared bands), from the thematic mapper (TM) with 30 m geometric resolution and seven spectral bands, and from the enhanced thematic mapper (ETM) also with 30 m geometric resolution. Providing data sets since 1972, this program proved essential for long-time monitoring and maximizing the time span. With a field of view of 185 km, the Landsat satellites are able to survey the large metropolitan areas of the study sites. In our study, we consider data from the approximate time-points of 1975, 1990, and 2000.

In addition to these medium-resolution optical data sets, we also take very high-resolution radar data into account. In particular, we employ data from the two German satellites of TerraSAR-X (TSX) and TanDEM-X (TDX), which have acquired two coverages of the entire landmass of the world for the years 2011 and 2012. While the primary goal is to provide a global digital elevation model at 12 m resolution (Huber et al. 2010), this immense data source is also being used to classify urbanized areas globally (Esch et al. 2012, 2013a,b). The standard mode for classification of the urban footprint is the "stripmap" single polarized mode with approximately 3 m spatial resolution, which provides the data set for the latest time step in our study. All technical details of the data sets used have been presented by Taubenböck et al. (2012).

9.4 METHODS

Our general workflow presented here includes two separate processes: (1) *Classification*, intending to derive change detections mapping spatial urban growth from the multitemporal EO data sets for the megacities and (2) *Spatial metrics*, intending to quantify, analyze, and compare the classified spatial urban patterns of the megacities.

9.4.1 CLASSIFICATION OF MULTITEMPORAL AND MULTISENSORAL REMOTELY SENSED DATA

For optimizing the outcome of the monitoring of spatial urbanization based on multisensoral EO data sets, we apply a backdating chronological workflow, starting the classification procedure with the latest data set, the TSX/TDX data from 2011. With this conceptual idea, we aim at reducing the effect of the multisensor and multiscale approach because this data set features the highest spatial resolution, and its related urban footprint classification results support the classification of Landsat data with lower geometric resolutions. The schematic chronological processing chain is displayed in Figure 9.5.

In the proposed classification procedure, we start our analysis with the latest available data set, namely, the TSX/TDX stripmap data having the highest geometric resolution. Hence, we first classify very high resolution (VHR) synthetic aperture radar (SAR) images using a pixel-based approach (Esch et al. 2012).

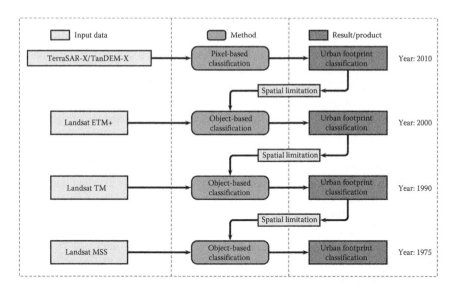

FIGURE 9.5 Processing chain from the multisensoral satellite data to the multitemporal urban footprint classification products.

Originally, an object-based approach had been developed for radar data; however, this approach was transferred to a pixel-based approach without losing accuracy but significantly improving processing time. The proposed methodology includes a specific preprocessing of the original intensity SAR data followed by an automated, threshold-based image analysis procedure.

The preprocessing aims to provide additional texture information for classification to highlight highly textured image regions, typically representing highly structured, heterogeneous built-up areas (thus, taking advantage of specific characteristics of urban SAR data showing strong scattering due to double bounce effects in these areas). In particular, the preprocessing focuses on the analysis of local speckle characteristics in order to provide this texture layer (referred to as "speckle divergence"). The analysis of the local image heterogeneity by means of the coefficient of variation is an established and straightforward approach to define the local development of speckle in SAR data. Highly textured landscapes such as urban landscapes showing a heterogeneous mix of objects within small areas lead to an increase of directional, non-Gaussian backscatter. Hence, the texture for such landscapes typically results in comparably high values. This information is used along with the original intensity information to automatically extract the urbanized areas, based on a fully unsupervised image analysis technique. The main concept of this approach is a two-stage procedure: first, a set of optimal thresholds for every specific scene is automatically determined by making use of the Jensen–Shannon divergence. These thresholds are then used to train a one-class classifier, which is based on support vector data description (SVDD) following principles of support vector machines (SVM). More details of this methodology are presented in Esch et al. (2013a,b). The result is a binary mask delineating "urbanized" from "nonurbanized" areas, a so-called urban footprint classification.

For backdating our chronological workflow, we use this urban footprint classification derived from TSX/TDX data from the year 2011 to support the classification of urban areas for the year 2000. Due to the lower geometric resolution of the Landsat data and the related problem of mixed pixels, we integrate the urban footprint classification from the year 2011 into our classification approach to reduce the possible areas for classifying urban areas in 2000 to the particular spatial extension. Thus, we classify urbanized areas in the Landsat data only if the later time step confirms an urban location (referred to as spatial limitation in Figure 9.5). We base our multitemporal classification approach on the hypothesis that the megacities of interest are basically expanding even if we miss out few areas of negative growth, if they exist at all. First, with this idea, we aim to minimize one of the primary obstacles to urban land cover classification from optical data sets: the diversity and spectral heterogeneity of urban reflectance often leading to misclassifications (Small, 2005). Second, if we start with the Landsat MSS data set as baseline approach, we would limit the classification inaccuracies due to the lower spatial and spectral resolutions of these data sets. This backdating workflow is consistently continued with the integration of the urban footprint of the particular previous time step as spatial limitation for classification until the 1970s data set of the MSS data, as is schematically presented in Figure 9.5.

The classification of the various Landsat scenes is based on an object-oriented classification procedure (Abelen et al. 2011). An object-based classification approach complies with the goal of an urban footprint classification, defined in detail below as an abstract spatial approximation of urbanized vs. nonurbanized areas as it reduces the so-called salt and pepper effect in pixel-based classifications of optical data sets.

The first step is a multiresolution segmentation, a bottom-up region-merging technique starting with one pixel objects for areas within the urban footprint information available from 2011. The second step is a hierarchical thematic classification using decision trees allowing mapping of four different thematic classes, namely, "water," "vegetation," "undeveloped land," and "urban area." Based on a feature selection study (Abelen et al. 2011), a set of features is suggested for classification and systematically applied to all scenes. This feature selection naturally cannot reflect the complete spectral diversity of urban reflectance; the selected features have to be seen as examples to reduce possible classification features, thus assuring methodological consistency when different operators are applying the algorithm to many different data sets, such as for the four megacities in this study. The thematic classes are identified hierarchically, starting with classes of significant separability from other classes and ending with those of lower separability, based on suggested features such as the Normalized Difference Vegetation Index (NDVI). Although the classification procedure is consistent, the definition of the thresholds used in the decision tree is scene dependent; this means, the procedure depends on the user experience and the time used for calibration, if reference data are available. With this concept, we aimed at a straightforward classification approach being consistent, traceable, and transparent for a large variety of optical Landsat scenes at different times and parts of the world.

This concept already includes a change detection strategy due to its backdating integration of the previous urban footprint layer; thus, spatial urban growth over time can be identified. The details of this procedure have been presented by Taubenböck et al. (2012).

9.4.2 Spatial Analysis of Urban Patterns

The resulting multitemporal classifications of the four megacities in China reveal complex spatial urban patterns. For a systematic spatial analysis (including the temporal development and comparison of these patterns), spatial metrics have recently become relevant for urban analysis. These metrics have been adopted from landscape ecology, where they are referred to as landscape metrics (Gustafson 1998; O'Neill et al. 1988).

In general, spatial metrics can be defined as quantitative and aggregate measurements derived from digital analysis of thematic categorical maps showing spatial heterogeneity at a specific scale and resolution (McGarigal and Marks 1994; McGarigal et al. 2002). A wide variety of indices have been developed to characterize and quantify the landscape structure and pattern, some of which describe the proportion of the landscape with a particular land cover class; the size, number, and perimeter of each land cover patch; and the complexity of the shape of the patch (McGarigal et al. 2002). In our study, we apply metrics to analyze and compare (1) the measured dimension of spatial growth, (2) the dominance of the urban core, (3) the fragmentation of the urban landscape, and (4) the spatial concentration of urban patches for the four current Chinese megacities.

1. Urban growth is calculated using both absolute and relative measures. In particular, absolute measures allow for a spatial comparison of the cities, whereas relative measures allow for a comparison in growth dynamics. For the latter approach, we take the extent of urbanized areas of each individual city in the 1970s as reference (we classify the urban extension at this particular time step as 100%) and calculate the percentage of urban growth with respect to this baseline.

2. The largest patch index (LPI) at the landscape level aims to quantify the percentage of total landscape area comprised by the largest patch (for mathematical details, see Table 9.1). The LPI equals 100 when the entire landscape consists of one single patch and approaches zero as the largest patch becomes increasingly small (McGarigal et al. 2002; Taubenböck et al. 2009). Accordingly, we apply this indicator as a measure to describe the dominance of the urban core.

3. Patch density (PD) is a fundamental indicator for measuring the fragmentation of a settlement pattern. The PD, defined as the number of urban patches per area (in our case the area of interest remains constant), is a measure of discrete urban areas in the landscape (for mathematical details, see Table 9.1). PD is expected to increase during periods of rapid urban nuclei development (sprawl) but may decrease if urban areas expand and merge into a continuous urban fabric (McGarigal et al. 2002; Seto and Fragkias 2005).

4. The mean Euclidean nearest neighbor distance (ENN) is used as a measure of patch dispersion (for mathematical details, see Table 9.1); indeed, smaller mean distances imply a fairly compact distribution of patches across landscapes, whereas large distances imply a distribution of patches tending to a dispersed pattern.

TABLE 9.1

Mathematical Details of the Applied Spatial Metrics

Metric	Formula	Units	Range
Largest Patch Index (LPI)	$LPI = \dfrac{\sum\limits_{j=1}^{n} max(a_{ij})}{A} \cdot (100)$	Percent	$0 < LPI \le 100$
Patch Density (PD)	$PD = \dfrac{n_i}{A} \cdot (10,000) \cdot (100)$	Number per 100 ha	PD > 0 (constrained by cell size)
Euclidean Nearest-Neighbor (ENN) Distance	$ENN = h_{ij}$	Meters	ENN > 0 (without limit)

Source: McGarigal, K. et al., FRAGSTATS v4: Spatial pattern analysis program for categorical and continuous maps, Computer software program produced by the authors at the University of Massachusetts, Techn. Bericht, University of Massachusetts, Amherst, MA, http://www.umass.edu/landeco/research/fragstats/fragstats.html.

Note: a_{ij}, area (m²) of patch ij; A, total landscape area (m²); n_i, number of patches in the landscape of patch type (class) i; h_{ij}, Distance (m) from patch ij to nearest neighboring patch of the same type (class), based on patch edge-to-edge distance, computed from cell center to cell center.

To ensure spatial comparability for all four megacities, we apply a standardized area of interest for the analysis of settlement patterns. The area of interest consists of a square with an edge length of 50 km centered in the downtown area of the respective city (e.g., for Beijing, we opt for the Tiananmen Square and for Shanghai, we choose the People's Park). In this way, we ensure spatial comparability for the city regions of interest, as for example, artificial administrative borders do not allow for consistent spatial analysis and comparability across cities.

9.5 RESULTS

In the following discussion, we present experimental results on the mapping of urban growth using EO data, its accuracy, and a related discussion on the term "urban footprint." Moreover, we quantify and compare spatial expansion and the evolving patterns of megacities in China offering an interpretation from a physical point of view on the process of urbanization.

9.5.1 URBAN FOOTPRINT PRODUCT AND ITS ACCURACY

In the following discussion on experimental classification results we consistently used the terms "urban area" or "urban footprint," which are also commonly employed in literature. These terms basically refer to the spatial extent of urbanized areas on a regional scale; however, this is a spatially and thematically fuzzy definition as this can inconsistently refer to either a functional area, the density of population, buildings, or the built-up area. The latter definition goes along with Cahn (1978), who understood the term "urban footprint" as the land directly occupied by a particular physical man-made structure.

From a remote sensing perspective, the latter definition is an obvious choice as functional information or population densities are not directly displayed in the remote sensing data. Thus, we adopt the idea of Cahn (1978) and define a "settlement mask" for the definition of the urban footprint as a combination of buildings, streets, and other impervious surfaces (Taubenböck et al. 2011a). The basic concept behind classifying a pixel in an optical data set as "urban" is as follows: Pixels are categorized as "urban," if the corresponding area is dominated by built environment, which includes human-constructed elements, roads, buildings, runways, and industrial facilities. When open land (e.g., vegetation, bare soil) dominates a pixel, these areas are not considered urban, even though they may function as urban space (Potere et al. 2009). The urban footprint retrieved from radar data is slightly different with respect to the physical characteristics of the input data and the algorithm; however, due to the generalization of the urban seeds, the basic concept of the urban footprint is similar.

On a regional scale, the urban footprint product provides insight into dimension, directions, patterns, and location of a city. However, the "urban footprint" should not be misunderstood as a spatially and thematically exact measurement of individual urban objects, and their detailed spatial arrangements, but as abstract delineation of the physical man-made structures of cities (Taubenböck et al. 2012).

Properly evaluating the accuracy of the classification results is essential for assessing the performances of the proposed approach. To this aim, we checked the accuracy by a visual verification process overlaying the 2011 classification result from TSX/TDX data to high-resolution optical satellite data available in Google Earth. Using a random distribution of 200 check points per megacity, the accuracy has been calculated for every individual city (Table 9.2). For the older time steps of the Landsat classifications, VHR optical data sets, such as from Google Earth, are not available. Thus, we compare the check points to the original Landsat data due to missing reference data sets. Here, we also use a random distribution of 200 check points per time step and city; the results are reported in Table 9.2 in terms of percentage of overall accuracy and the kappa coefficient.

TABLE 9.2

Accuracy Assessment of the Various Urban Footprint Products

| City | Landsat MSS | | Landsat TM | | Landsat ETM+ | | TerraSAR-X | |
	Overall Accuracy (%)	Kappa Coefficient	Overall Accuracy (%)	Kappa Coefficient	Overall Accuracy (%)	Kappa Coefficient	Overall Accuracy (%)	Kappa Coefficient
Beijing	78.3	0.69	86.0	0.79	91.1	0.80	84.4	0.77
Shanghai	72.7	0.60	78.8	0.74	82.5	0.78	87.2	0.71
Guangzhou	81.2	0.72	84.9	0.77	88.2	0.83	88.1	0.81
Shenzhen	66.7	0.58	76.9	0.71	84.7	0.72	79.4	0.69

The aforementioned method for accuracy assessment followed an approach presented in Congalton (1991). Naturally, this evaluation implies errors due to subjective misclassifications by the person in charge and the visual verification process is also dependent on the respective reference data. For 2011, VHR optical data sets from Google Earth allow an evaluation on the highest geometric level although the classification result of the urban footprint is an abstract, more regional assessment of urbanized areas, defined earlier as settlement mask. Thus, because of this difference in geometric resolutions between classification result and reference data sets, errors can be detected in great detail; thus, the assessment of accuracy is often lower than visual comparisons to lower resolution Landsat data. However, although this reveals subjectivity in the accuracy assessment, we see that, on average, the overall accuracy came up to values constantly over 80% and high kappa coefficients. Lower accuracies have been identified for the classification results obtained using Landsat MSS data; however, due to the backdating chronologic classification workflow, their low spatial resolution did not significantly influence the classification accuracy. In general, the urban footprint classifications provide high and stable accuracy values appropriate for spatial urban settlement analysis.

9.5.1.1 Change Detection of Chinese Megacities

The main result of this study is a consistent monitoring of the four current Chinese megacities—Beijing, Shanghai, Guangzhou, and Shenzhen. The change detection result displays their spatiotemporal evolution and their respective spatial patterns since the 1970s. Figure 9.6 allows a first visual comparison of dynamics, dimensions, and patterns. The areas of interests for each megacity are held constant with a 50 km edge length; thus, the analysis covers an area of 2500 km².

At first glance, it becomes obvious that the dimension of spatial urbanization of China's current megacities has been massive since the 1970s; its magnitude, however, does not show an obvious similar trend. Also, the resulting urban patterns do not show a characteristic structure.

Beyond this, the chosen area of interest, 2500 km², is generally appropriate for the time step 1975 or 1990; however, the current extent of the two largest megacities, Beijing and Shanghai, prove that the transformation from urban to suburban to peri-urban and finally to rural patterns is fluid and spatially not distinct. This brings up the question, whether comparing settlement patterns of cities based on standardized and thus comparable spatial entities is appropriate. It brings up the research question regarding where an appropriate spatial border of today's megacities would be for consistent comparisons.

The problem tackled here becomes obvious when looking at the urban footprint classified for the extended region of the megacity Shanghai and its peri-urban and rural surroundings (Figure 9.7—with the urban core of Shanghai located in the east of the scene). The spatial transition of urban to peri-urban to rural patterns is floating within these complex, highly urbanized settlement patterns surrounding Shanghai. Thus, defining an appropriate spatial entity for systematic comparison of cities is a complex problem since it is unclear without administrative borders (which are artificial and thus not appropriate for spatial cross city comparisons). Thus, we chose a standardized spatial entity for a systematic comparison of cities.

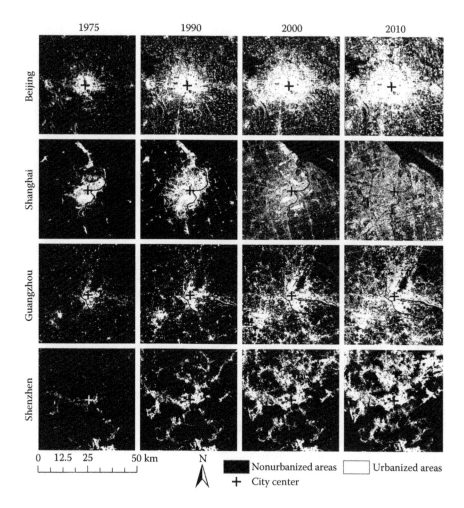

FIGURE 9.6 The change detections for the megacities Beijing, Shanghai, Guangzhou, and Shenzhen within 50 km² around the urban centers since the 1970s based on multitemporal Landsat and TerraSAR-X data.

However, these results, based on the standardized areas of interest, allow identifying, localizing, and quantifying the dimension and pattern of urban sprawl over time, and because consistent data and methods are applied, comparisons across cities are feasible.

9.5.1.2   Measuring and Comparing Spatial Patterns of Chinese Megacities

In general, spatial urbanization in China's megacities is highly dynamic. However, spatial urban expansion across China's megacities of today is not and has not been analogical. For evaluation, we calculate absolute and relative spatial growth displayed in Figure 9.8. Basically, the results show that Beijing has—within the 50 km² area—the largest urban extension, followed by Shanghai, Guangzhou, and Shenzhen. This does not match with the population figures as Shanghai has the largest population

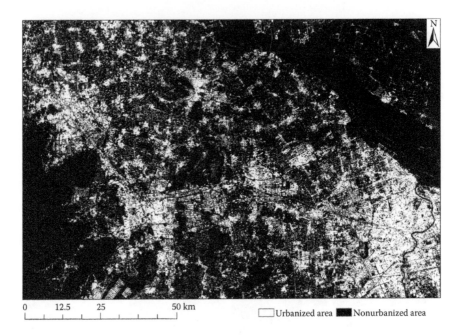

0   12.5   25   50 km
$\llcorner_____\lrcorner$ ☐ Urbanized area   ■ Nonurbanized area

FIGURE 9.7   The urban footprint of megacity Shanghai and the urban, peri-urban, and rural settlement patterns in the megacity's surroundings.

in China (UN 2011). One could conclude that the change detection in two dimensions measured here with EO data neglects urban growth in its third dimension as Shanghai is known to have already a higher concentration of high-rise buildings than, for example, New York City. Thus, we assume Shanghai's floor space index is significantly higher than the one in Beijing.

The Guangzhou region closes spatially the gap to Shanghai regarding urban extension. This may be related to a spatial double structure of the area. The megacity of Guangzhou has another urban hub, the city of Foshan, nearby. This is clearly visible in the urban footprint classifications for 1975 and 1990, with Guangzhou's urban center in the area's core and the additional urban center Foshan in the southwest of the area. Today, the two cities basically form one coalescent multinuclei large area city region.

Using relative measures, it becomes obvious that megacity Shenzhen experienced the highest spatial dynamics with an extent in 2011 almost 12 times greater than in 1975; overall, all Chinese megacities show immense spatial urban expansion rates since the 1970s: Guangzhou (8 times), Beijing (7 times), and Shanghai (6 times). In relation to global relative growth in megacities, the most explosive spatial growth is happening in Asia (Taubenböck et al. 2012). The four considered Chinese cases exhibit a spatial behavior similar to that of other Asian megacities such as Mumbai, India, or Jakarta, Indonesia (both show today a spatial extent, which is about 8 times larger than their spatial extent of 1975), Manila, Philippines (7.5 times) or Seoul, South Korea (5.5 times), confirming the statement. Moreover, the growth of Chinese megacities exceeds by far examples from other continents, such as Mexico City, Mexico (about 3 times), Cairo, Egypt (about 4.5 times) or London, Great Britain (about 2 times).

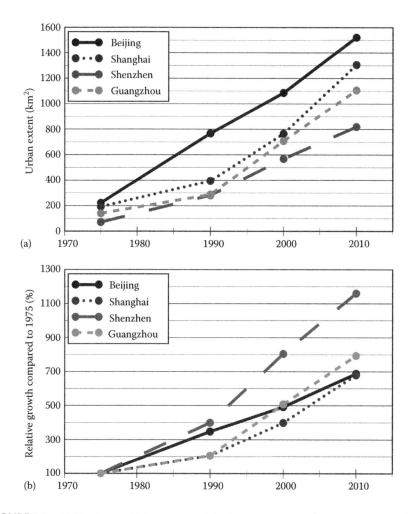

FIGURE 9.8 (a) Absolute and (b) relative spatial urban growth in the four Chinese megacities Beijing, Shanghai, Guangzhou, and Shenzhen based on the change detection using EO data.

In any case, megacity Shenzhen can be seen as an exceptional example from a global as well as a continental perspective because of its remarkable spatial dynamics. The remarkable spatial dynamics correspond to the high dynamics of population development. The city of Shenzhen with only about 50,000 inhabitants in the 1970s turned into a city with more than 10 million inhabitants within only about 35 years (UN 2011).

For an analysis and comparison of evolving settlement patterns, we apply the three different landscape metrics described in Section 9.4.2 to provide a spatial perspective on the urban cores, on landscape fragmentation, and on concentration of urban patches in the four Chinese megacities.

In general, the LPI is rising for all megacity regions (Figure 9.9). Beijing and Shanghai show the highest values during the entire monitoring period. With an

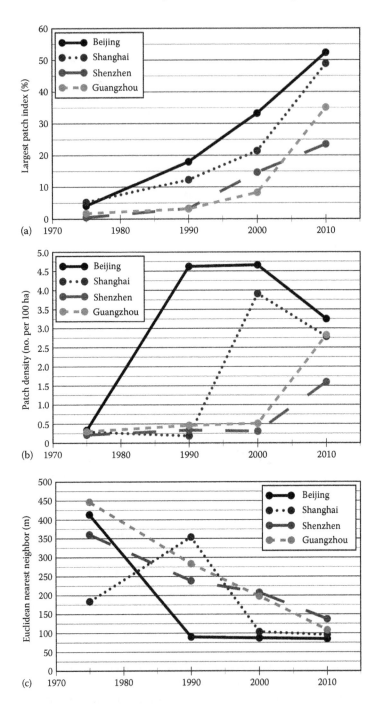

FIGURE 9.9 Landscape metrics—(a) largest patch index (LPI), (b) patch density (PD) and (c) Euclidean nearest neighbor (ENN)—for analysis and comparison of settlement patterns in the four Chinese megacities.

LPI of a value around 50, the urban core is dominating the urban pattern of both megacities. In comparison, Guangzhou experienced a dynamic increase of the LPI only after the year 2000. The spatial double structure, because of the proximity of the other large city of Foshan next to the megacity resulted in a low dominance of the urban core until 1990. However, the urban expansion and, in succession, the spatial coalescence of the settlement patterns of both cities led to this enormous increase in LPI in recent years. Shenzhen shows the lowest LPI value. This is due to its location in a very hilly and fragmented coastal area. The influence of orographic and related environment conditions channeled the settlement pattern development. The result is a highly complex pattern without any typical characteristics of concentric growth around the urban core.

The PD, which is a measure of fragmentation, shows for the two largest cities—Beijing and Shanghai—a similar development with a temporal shift of about 10 years: a rise in fragmentation because of basically concentric noncoalescent urban sprawl in peripheral regions (Figure 9.9). After the year 2000, PD in both megacity regions decreased significantly; the decrease can be interpreted as a redensification and, thus, compacting of urbanized areas. It is interesting to note that Guangzhou and Shenzhen (both cities having about half the population of Beijing and Shanghai today) show the same characteristic trend of an increasing fragmentation, which was true of Beijing and Shanghai 10–20 years earlier. This raises the question whether a similar settlement pattern development can be expected for these two cities.

Regarding ENN, the general trend for the megacities are decreasing values over time, a logical result due to urban expansion and sprawl while keeping the area of interest stable (Figure 9.9). However, it is remarkable that the complex structure of Shenzhen exhibits the highest ENN values measured at all four megacities, whereas the more mono- and concentric urban patterns of Beijing and Shanghai show lowest ENN values. Deviation from the general temporal trend is only measured for Shanghai from 1975 until 1990. There, urban sprawl happened in a pattern with comparatively large distances from the dominating urban core. In urban planning, this spatial process is often described as a leap-frog development, which causes the rising ENN values. However, after 1990, immense redensification processes in the peripheral urban areas engage the settlement pattern development in Shanghai within the general trend.

Overall, we can state that similar trends in spatial urban growth for megacity regions in China are observed. Regarding settlement patterns, similarities as well as significant differences have been highlighted.

9.6 CONCLUSION

With respect to the main goals defined in the introduction, this chapter underlines the capabilities of remote sensing to (1) enable long-time multitemporal approaches (almost 40 years) on a medium geometric resolution for large areas such as all current Chinese megacities, (2) enable to systematically map and monitor large areas such as megacities with consistent data and methods with sufficient accuracy (accuracies basically higher than 80%), and (3) provide value addition to remote sensing classification results by using spatial metrics to measure and compare the urban patterns with the aim to derive new insights on the spatial effects of urbanization. In our case, similarities and differences regarding the dimension and patterns of urban expansion have been evaluated and discussed. With these capabilities, EO provides a unique data source to significantly contribute to mapping, characterizing, and understanding physical processes in urban environments.

However, it must be critically stated that remote sensing data and results also have clear limitations. First, the classification accuracy especially when dealing with multitemporal and multisensoral data sets is limited. Nevertheless, with accuracies of constantly over 80% for the urban footprints, an objective basis is given for monitoring, assessing, and evaluating the process of urbanization, at least at a regional scale. Second, two-dimensional urban sprawl is mapped with the multitemporal approach based on multisensoral remote sensing data. Thus, when using these two-dimensional data sets, it must be clear that an important aspect of urban expansion is neglected: the expansion toward the third dimension, for example, especially in China, often large-area high-rise buildings replace low-rise buildings. Although high-resolution laserscan or stereo data may be highly precise data sets for three-dimensional mapping of urban morphology (see e.g., Rottensteiner et al. 2007; Sirmacek et al. 2012), large area 3D monitoring approaches using EO data for entire megacity regions or beyond are still absent. Taubenböck et al. (2013) approach this topic by large-area digital surface models from Indian Cartosat-1 stereo data, classifying Central Business Districts (CBDs) for entire mega city areas. Third, although this chapter contributes—together with many similar studies—to the approach for a more comprehensive global documentation and deeper knowledge on spatial urbanization processes at the regional scale, an organized combination with other spatial data sets enabling continuative studies with a more global perspective is nonexistent; even clear political leadership to do so is absent. And, without claiming to be complete, fourth, the physical monitoring of urbanization is only one perspective of the analysis of the ongoing urbanization processes. The development of multidisciplinary thinking, data integration, and method development holds immense potential for new insights on this complex issue of global change—a task that is still underrepresented.

As discussed, within these critical comments, an enormous potential for future (urban) research using EO data becomes visible. EO provides unique possibilities for consistent analysis of settlements on global, continental, regional, or local scale. And the possibilities to do so are constantly increasing due to constantly increasing availability of and accessibility to modern remote sensing technologies, which open up new opportunities for a wide range of urban applications. We must make use of them in a more structured and organized way. Now.

ACKNOWLEDGMENTS

The authors would like to thank the TerraSAR-X Science Team for providing the SAR data for this study. Furthermore, we would like to thank Leopold Lesko for his great support.

REFERENCES

Abelen S, Taubenböck H, and Stilla U (2011): Interactive classification of urban areas using decision trees. In: Stilla, U., Gamba, P., Juergens, C., Maktav, D. (eds.) *JURSE 2011—Joint Urban Remote Sensing Event*, Munich, Germany, April 11–13, pp. 373–376.

Angel S, Sheppard S C, and Civco D L (2005): *The Dynamics of Global Urban Expansion*, Washington, DC: Transport and Urban Development Department, the World Bank, p. 102.

Bagan H and Yamagat Y (2012): Landsat analysis of urban growth: How Tokyo became the world's largest megacity during the last 40 years. *Remote Sensing of Environment*, 127, 210–222.

Bartholome E and Belward A S (2005): GLC2000: A new approach to global land cover mapping from Earth observation data. *International Journal of Remote Sensing*, 26, 1959–1977.

Bontemps S, Defourny P, Van Bogaert E, Arino O, Kalogirou V, and Ramos Perez J J (2011): GLOBCOVER 2009 Products description and validation report. Université catholique de Louvain (UCL) and European Space Agency (ESA), Vers. 2.2, 53pp. Available at http://due.esrin.esa.int/globcover/LandCover2009/GLOBCOVER2009_Validation_Report_1.0.pdf (accessed January 13, 2014).

Cahn R (1978): *Footprints on the Planet: A Search for an Environmental Ethic*, New York: Universe Books, p. 277.

Chan C K and Yao X (2008): Air pollution in megacities in China—A review. *Atmospheric Environment*, 42(1), 1–42.

Chan K W (1996): Post-Mao China: A two-class urban society in the making. *International Journal of Urban and Regional Research*, 20(1), 134–150.

Chang H G and Brada J C (2002): China's urbanization lag during the period of reform: A paradox. Working Paper. University of Toledo and Arizona State University. Available at http://genechang.net/upload/lagurban4.pdf (accessed January 13, 2014).

Chang K S (1994): Chinese urbanization and development before and after economic reform: A reappraisal. *World Development*, 22(4), 601–613.

Chen J (2007): Rapid urbanization in China: A real challenge to soil protection and food security. *Catena*, 69, 1–15.

Cheng J and Masser I (2003): Urban growth pattern modelling: A case study of Wuhan city, PR China. *Landscape and Urban Planning*, 62(4), 199–207.

Civco D L, Chabaeva A, and Parent J (2009): KH-series satellite imagery and Landsat MSS data fusion in support of assessing urban land use growth. *SPIE Europe Remote Sensing*, 747801–747812.

Congalton R G (1991): A review of assessing the accuracy of classification of remotely sensed data. *Remote Sensing Environment*, 37(1), 35–46.

Elvidge C, Imhoff M L, Baugh K E, Hobson V R, Nelson I, Safran J, Dietz J B, and Tuttle B T (2001): Nighttime lights of the world: 1994–95. *ISPRS Journal of Photogrammetry and Remote Sensing*, 56, 81–99.

Esch T, Marconcini M, Felbier A, Roth A, Heldens W, Taubenböck H, Huber M, Schwinger M, Müller A, and Dech S (2013): Urban footprint processor—Fully automated processing chain generating settlement masks from global data of the TanDEM-X mission. *IEEE Geoscience and Remote Sensing Letters*, 10(6).

Esch T, Taubenböck H, Roth A, Heldens W, Felbier A, Thiel M, Schmidt M, Müller A, and Dech S (2012): TanDEM-X mission: New perspectives for the inventory and monitoring of global settlement patterns. *Journal of Selected Topics in Applied Earth Observation and Remote Sensing*, 6, 22.

Gamba P and Lisini G (2013): Fast and efficient urban extent extraction using ASAR wide swath mode data. *IEEE Journal of Selected Topics in Applied Earth Observations and Remote Sensing*, 6(5), 2184–2195.

GEO (2013): Group on Earth observation. SB-04 Global urban observation and information. http://www.earthobservations.org/index.shtml (accessed January 13, 2014).

GLCF—Global Land Cover Facility (2013): University of Maryland. Available at http://glcf. umd.edu/data/ (accessed January 13, 2014).

Griffiths P, Hostert P, Gruebner O, and van der Linden S (2010): Mapping megacity growth with multi-sensor data. *Remote Sensing of Environment*, 114, 426–439.

Gustafson E J (1998): Quantifying landscape spatial pattern: What is the state of the art? *Ecosystems*, 1(2), 143–156.

Heikkila E J (2007): Three questions regarding urbanization in China. *Journal of Planning, Education and Research*, 27, 65–81.

Huber S, Younis M, and Krieger G (2010): The TanDEM-X mission: Overview and interferometric performance. *International Journal of Microwave and Wireless Technologies*, 1–11. DOI: 10.1017/S1759078710000437.

Jat M K, Garg P K, and Kahre D (2008): Monitoring and modelling of urban sprawl using remote sensing and GIS techniques. *International Journal of Applied Earth Observation and Geoinformation*, 10(1), 26–43.

Ji W, Ma J, Twibell R W, and Underhill K (2006): Characterizing urban sprawl using multistage remote sensing images and landscape metrics. *Computers, Environment and Urban Systems*, 30, 861–879.

Kong F and Nakagoshi N (2006): Spatial-temporal gradient analysis of urban green spaces in Jinan, China. *Landscape and Urban Planning*, 78(3), 147–164.

Liu S H, Li X B, and Zhang M (2003): Scenario analysis on urbanization and rural–urban migration in China. Interim Report IR-03-036. International Institute for Applied Systems Analysis. Available at http://webarchive.iiasa.ac.at/Admin/PUB/Documents/ IR-03-036.pdf (accessed January 13, 2014).

Ma Y and Xu R (2010): Remote sensing monitoring and driving force analysis of urban expansion in Guangzhou City, China. *Habitat International*, 34(2), 228–235.

McGarigal K, Cushman S A, and Ene E (2012): FRAGSTATS v4: Spatial pattern analysis program for categorical and continuous maps, Computer software program produced by the authors at the University of Massachusetts, Amherst, MA. Techn. Bericht, University of Massachusetts. Available at http://www.umass.edu/landeco/research/fragstats/fragstats. html (accessed January 13, 2014).

McGarigal K, Cushman S A, Neel M C, and Ene E (2002): FRAGSTATS: *Spatial Pattern Analysis Program for Categorical Maps*, Computer software produced by the authors at the University of Massachusetts, Amherst, MA.

McGarigal L and Marks B J (1994): FRAGSTATS manual: Spatial pattern analysis program for quantifying landscape structure. Available at http://ftp.fsl.orst.edu/pub/fragstats.2.0 (accessed January 13, 2014).

Miyazaki H, Shao X, Iwao K, and Shibasaki R (2012): An automated method for global urban area mapping by integrating ASTER satellite images and GIS data. *IEEE Journal of Selected Topics in Applied Earth Observations and Remote Sensing*, 6(2), 1004–1019. DOI: 10.1109/JSTARS.2012.2226563.

O'Neill R V, Krummel J R, Gardner R H, Sugihara G, Jackson B, Deangelis D L, Milne B T et al. (1988): Indices of landscape pattern. *Landscape Ecology*, 1, 153–162.

Pesaresi M, Blaes X, Ehrlich D, Ferri S, Gueguen L, Haag F, Halkia M et al. (2012): *A Global Human Settlement Layer from Optical High Resolution Imagery*, European Union, Luxembourg: Publications Office of the European Union, p. 121.

Pesaresi M, Ehrlich D, Caravaggi I, Kauffmann M, and Louvrier C (2011): Toward global automatic built-up area recognition using optical VHR imagery. *IEEE Journal of Selected Topics in Applied Earth Observations and Remote Sensing*, 4(4), 923–934.

Potere D and Schneider A (2009): Comparison of global urban maps. In: Gamba, P. and Herold, M. (eds.) *Global Mapping of Human Settlements: Experiences, Data Sets, and Prospects*, Boca Raton, FL: Taylor & Francis Group, pp. 269–308.

Potere D, Schneider A, Angel S, and Civco D L (2009): Mapping urban areas on a global scale: Which of the eight maps now available is more accurate? *International Journal of Remote Sensing*, 30(24), 6531–6558.

Rottensteiner F, Trinder J, Clode S, and Kubik K (2007): Building detection by fusion of airborne laser scanner data and multi-spectral images: Performance evaluation and sensitivity analysis. *ISPRS Journal of Photogrammetry and Remote Sensing*, 62(2), 135–149.

Schneider A, Friedl M, and Woodcock C (2005): Mapping urban areas by fusing multiple sources of coarse resolution remotely sensed data: Global results. *Proceedings of the Fifth International Symposium of Remote Sensing of Urban Areas*, Tempe, AZ, March 2005.

Schneider A and Woodcok C E (2008): Compact, dispersed, fragmented, extensive? A comparisons of urban growth in twenty-five global cities using remotely sensed data, pattern metrics and census information. *Urban Studies*, 45, 659.

Seto K C and Fragkias M (2005): Quantifying spatiotemporal patterns of urban land-use change in four cities of China with a time series of landscape metrics. *Landscape Ecology*, 20, 871–888.

Sexton J O, Song X P, Huang C, Channan S, Baker M E, and Townshend J R (2013): Urban growth of the Washington, D.C.-Baltimore, MD metropolitan region from 1984 to 2010 by annual, Landsat-based estimates of impervious surfaces. *Remote Sensing of Environment*, 29, 42–53.

Sirmacek B, Taubenböck H, Reinartz P, and Ehlers M (2012): Evaluation of automatically generated three-dimensional city models derived from remotely sensed data. *Journal of Selected Topics in Applied Earth Observation and Remote Sensing*, 5(1), 59–70.

Small C (2005): A global analysis of urban reflectance. *International Journal of Remote Sensing*, 26(4), 661–681.

Taubenböck H (2013): Die Ruhe vor dem Sturm: Chinas Weg in ein urbanes Zeitalter. *Standort*, 37(1), 56–61.

Taubenböck H, Esch T, Felbier A, Roth A, and Dech S (2011a): Pattern-based accuracy assessment of an urban footprint classification using TerraSAR-X data. *IEEE Geoscience and Remote Sensing Letters*, 8(2), 278–282.

Taubenböck H, Esch T, Felbier A, Wiesner M, Roth A, and Dech S (2012): Monitoring of mega cities from space. *Remote Sensing of Environment*, 117, 162–176.

Taubenböck H, Klotz M, Felbier A, Wegmann M, and Ludwig R (2011b): Spatio-temporal cross-city comparison using multi-sensoral remote sensing for Mexican cities. In: Stilla, U., Gamba, P., Juergens, C., Maktav, D. (eds.) *JURSE 2011—Joint Urban Remote Sensing Event*, Munich, Germany, April 11–13, pp. 81–84.

Taubenböck H, Klotz M, Wurm M, Schmieder J, Wagner B, Wooster M, and Esch T (2013): Central business districts: Delineation in mega city regions using remotely sensed data. *Remote Sensing of Environment*, 136, 386–401.

Taubenböck H, Wegmann M, Roth A, Mehl H, and Dech S (2009): Urbanization in India—Spatiotemporal analysis using remote sensing data. *Computers, Environment and Urban Systems*, 33, 179–188.

Taubenböck H, Wiesner M, Felbier A, Marconcini M, Esch T, and Dech S (2014): New dimensions of urban landscapes: The spatio-temporal evolution from a poly-nuclei area to a mega-region based on remote sensing data. *Applied Geography*, 47, 137–153.

United Nations (2003): *World Urbanization Prospects—The 2003 Revision*. New York: United Nations.

United Nations (2011): *World Urbanization Prospects–The 2009 Revision*, New York: United Nations.

Weng Q (2014): *Global Urban Monitoring and Assessment through Earth Observation*, Boca Raton, FL: Taylor & Francis/CRC Press.

Wu Q, Li H-Q, Wang R, Paulussen J, He Y, Wang M, Wang B, and Wang Z (2006): Monitoring and predicting land use change in Beijing using remote sensing and GIS. *Landscape and Urban Planning*, 78, 322–333.

Xie Y, Fang C, Lin G, Gong H, and Qiao B (2007): Tempo-spatial patterns of land use changes and urban development in globalizing China: A study of Beijing. *Sensors*, 7, 2881–2907.

Yin J, Yin Z, Zhong H, Xu S, Hu X, Wang J, and Wu J (2011): Monitoring urban expansion and land use/land cover changes of Shanghai metropolitan area during the transitional economy (1979–2009) in China. *Environmental Monitoring and Assessment*, 177, 609–621. DOI: 10.1007/s10661-010-1660-8.

Yu J X and Ng C N (2006): An integrated evaluation of landscape change using remote sensing and landscape metrics: A case study of Panyu, Guangzhou. *International Journal of Remote Sensing*, 27(6), 1075–1092.

Yuan F, Sawaya K, Loeffelholz B, and Bauer M (2005): Land cover classification and change analysis of the Twin Cities (Minnesota) Metropolitan Area by multitemporal Landsat remote sensing. *Remote Sensing of Environment*, 98, 317–328.

Zhang L and Zhao XBS (2003): Reinterpretation of China's underurbanization: A systemic perspective. *Habitat International*, 27, 459–483.

(a)   (b)

Typology	Color
Open set mid rise	Red
Compact low rise	Blue
Open set low rise	Green
Dispersed low rise	Yellow
Extensive low rise	Purple

(c)

FIGURE 3.1 Experimental results for the urban climate zone extraction in a small subsample of the XuZhou (People's Republic of China) scene: (a) the urban climate zone map to be compared with (b) a ground truth obtained by visual classification and superimposed on the original data set. Classes are identified by colors, according to the legends displayed in (c).

(a)   (b)

FIGURE 3.2 Climate zone extraction for the town of Xuzhou, People's Republic of China: (a) final results of the proposed segmentation and classification procedure and (b) detailed ground truth obtained by manually delineating and labeling individual blocks.

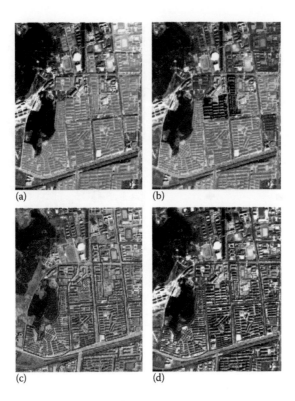

FIGURE 3.3 Example of problematic assignment of urban blocks to urban climate zones: (a, b) two different visual interpretations by remote sensing experts, to be compared with the block borders superimposed on (c) the panchromatic and (d) the color image of the same area.

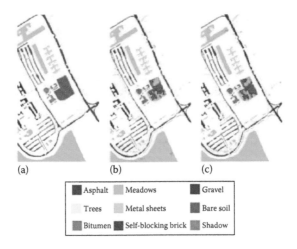

Asphalt	Meadows	Gravel
Trees	Metal sheets	Bare soil
Bitumen	Self-blocking brick	Shadow

FIGURE 3.4 Urban land use mapping results in Pavia, Italy: (a) ground truth for the hyperspectral ROSIS data sets, (b) mapping results using semisupervised classification but no unmixing information, and (c) mapping results jointly considering a semisupervised technique and unmixing information.

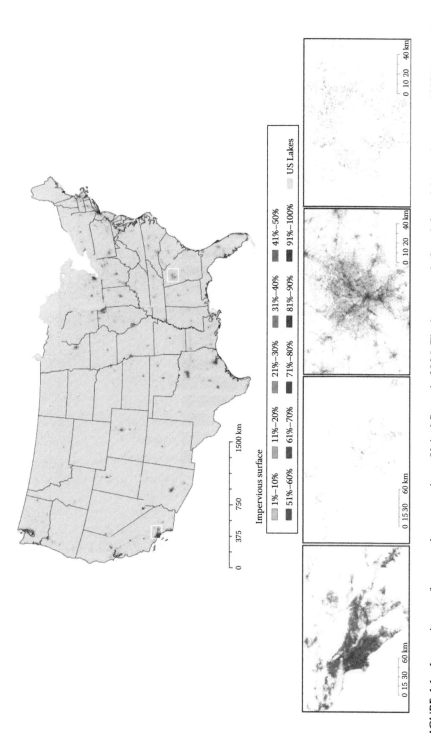

FIGURE 4.1 Impervious surface over the conterminous United States in 2006. The lower panels from left to right are the maps of 2001 impervious surface and 2006 new impervious surface in Los Angeles, California, and those in Atlanta, Georgia.

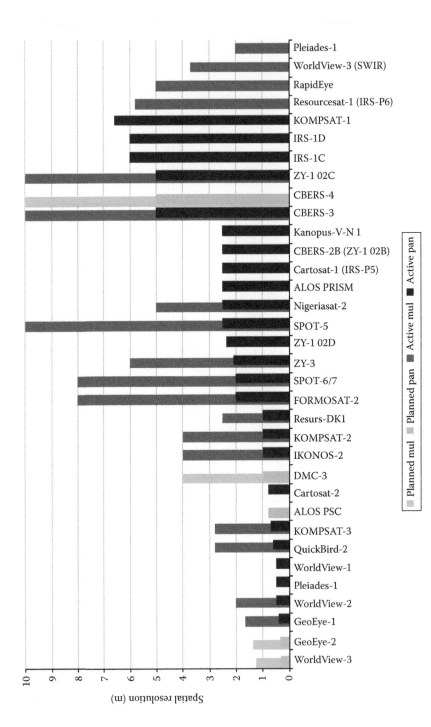

FIGURE 4.2 Pixel size (meters) of current and planned high spatial resolution satellite sensors.

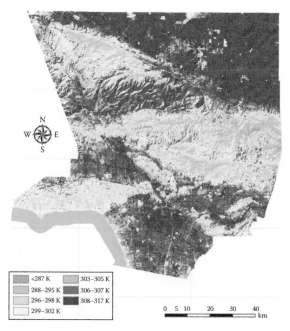

<287 K	303–305 K
288–295 K	306–307 K
296–298 K	308–317 K
299–302 K	

0 5 10   20   30   40
km

FIGURE 4.3 Mean annual surface temperature in Los Angeles determined by an unconstrained nonlinear optimization with the Levenberg–Marquardt minimization scheme. LST measurements of all available 115 Landsat-5 TM scenes between 2000 and 2010 were used for modeling by a sine function. (From Weng, Q. and P. Fu, *Remote Sens. Environ.*, 2014, 140, 267.)

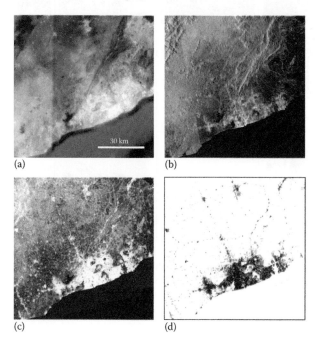

(a)   (b)

(c)   (d)

FIGURE 4.4 Optical data from Google Earth (a), TerraSAR-X amplitude image (b), calculated texture image (c), and urban footprint mask derived from combined classification of amplitude and texture (d).

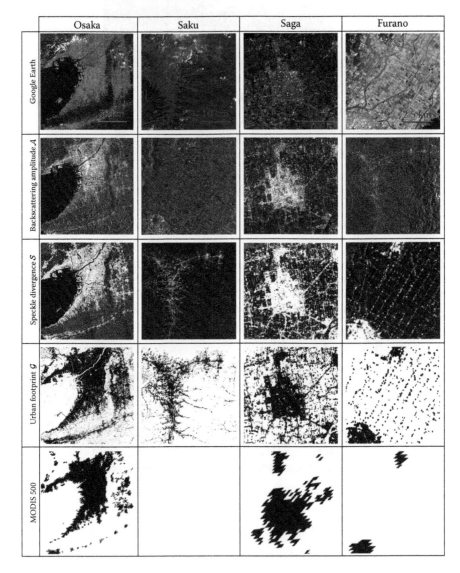

FIGURE 5.3 Optical data (from Google Earth), TanDEM-X mission backscattering amplitude $\mathcal{A}$, speckle divergence $\mathcal{S}$, urban footprint $\mathcal{G}$, and the corresponding MODIS 500 urban class map for the Japanese cities of Osaka, Saku, Saga, and Furano.

FIGURE 6.1 DMSP stable lights for the year 2010.

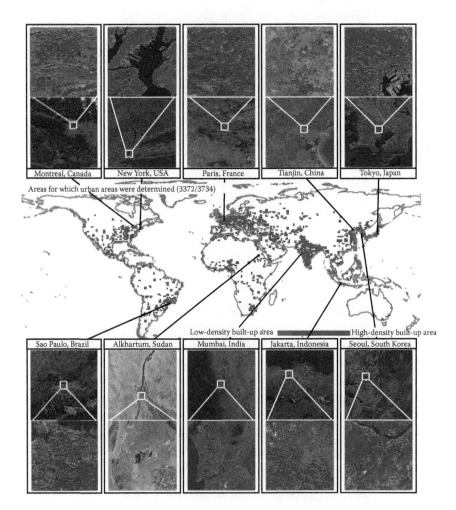

FIGURE 7.3 Examples of the results of the global urban area maps. Red pixels in the close-up figures indicate urban areas.

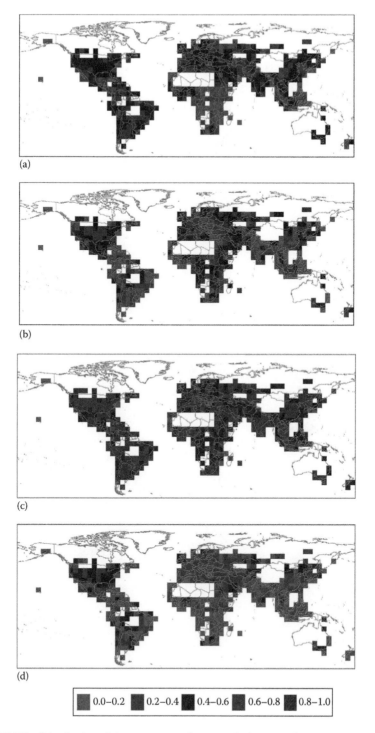

FIGURE 7.7 Distribution of the agreements between the integrated map and the MCD12: (a) User's agreement, (b) producer's agreement, (c) overall agreement, and (d) kappa coefficient of agreement.

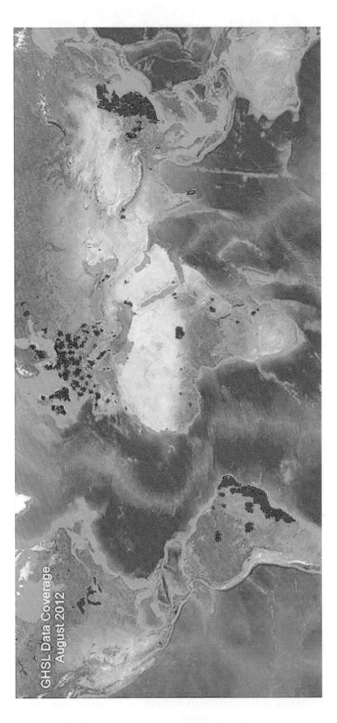

GHSL Data Coverage
August 2012

FIGURE 8.1 An overview of HR/VHR image data processed during the GHSL experiment. (From Pesaresi, M. et al., *IEEE J. Select. Top. Appl. Earth Observ. Remote Sens.*, 6(5), 2102, 2013.)

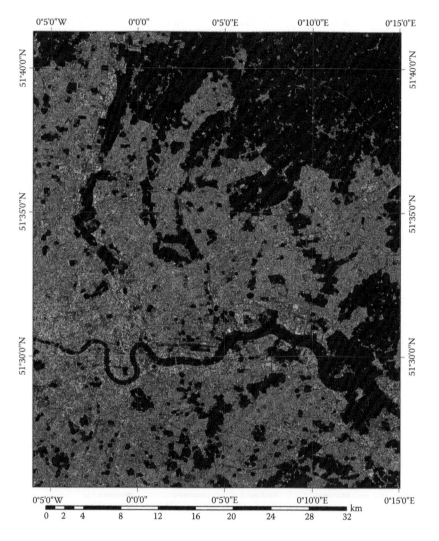

FIGURE 8.5 GHSL output reporting the characterization of BU structures in East London, UK. The color encodes the estimated size (surface) of BU structures in the range of 10,000–15,000 m² from blue (small) to red (large). Note how relations between the dominant size of BU structures and the use of the settlement areas can be made.

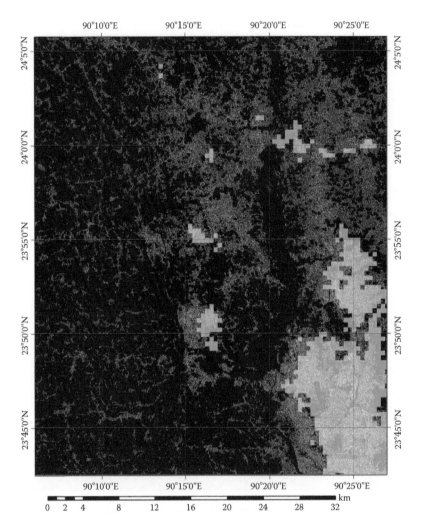

FIGURE 8.6 GHSL output over the city of Dhaka, Bangladesh, overlaid with urban areas using moderate-resolution input data; white squared areas indicate the "urban areas" class of the MODIS 500 source, while colored dots are built-up structures recognized by the GHSL workflow.

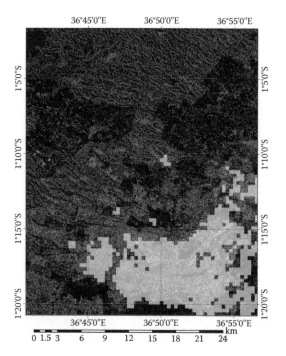

FIGURE 8.8 GHSL output over the city of Nairobi, Kenya, compared with urban areas detected using moderate-resolution input data; white squared areas are derived from the "urban areas" class of the MODIS 500 source while colored dots are built-up structures recognized by the GHSL workflow.

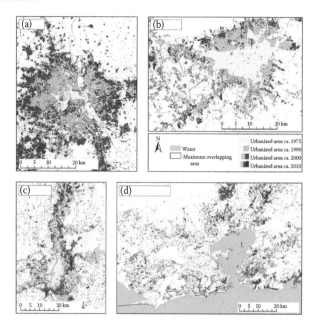

FIGURE 9.3 Change detections derived from multitemporal remote sensing data: examples— (a) Delhi, India; (b) Tehran, Iran; (c) Kolkata, India; and (d) Rio de Janeiro, Brazil—for global megacity monitoring since the 1970s. (From Taubenböck, H., Esch, T., Felbier, A., Wiesner, M., Roth, A., and Dech, S., *Remote Sens. Environ.*, 117, 162, 2012.)

FIGURE 10.8 Example of an operational refugee camp map. (Map, © 2013 European Commission.)

FIGURE 10.10 Aerial photograph of the Ardamata camp near Al Geneina, West Darfur. (Courtesy of WFP, © 2009 WFP.)

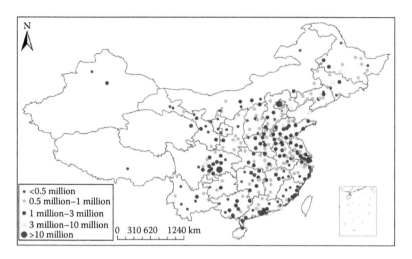

FIGURE 11.1 Spatial distribution of China's prefectural-level cities and municipalities grouped by 2010 population size.

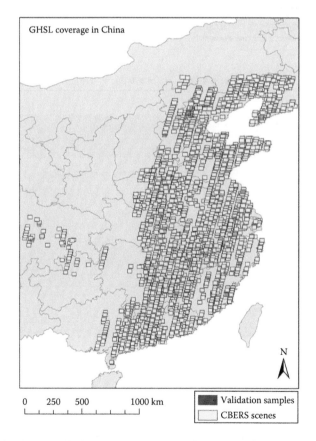

FIGURE 11.3 Geographic distribution of the CBERS HR input images processed during the experiment. Sample grids for visual validation of GHSL are marked in red.

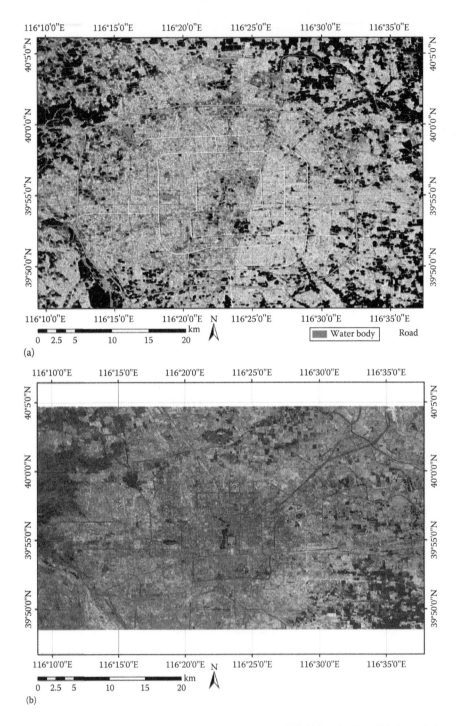

FIGURE 11.4 GHSL of the Beijing metropolitan area at 1:200,000 scale (a) and high-resolution images from Bing Maps aerial view (b).

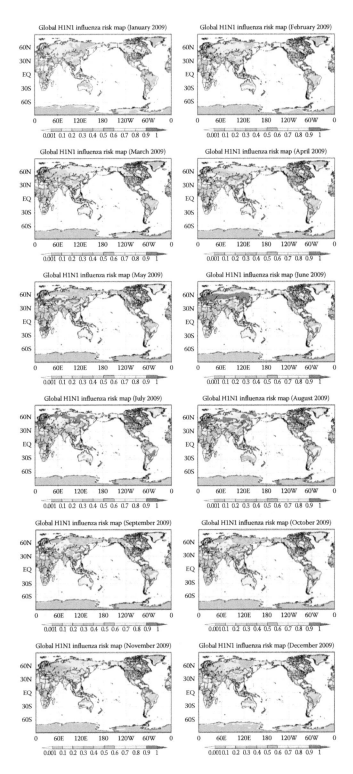

FIGURE 12.2 Global risk prediction maps for January through December. The warmer the color, the more suitable conditions were for virus survival, and the higher the risk predicted.

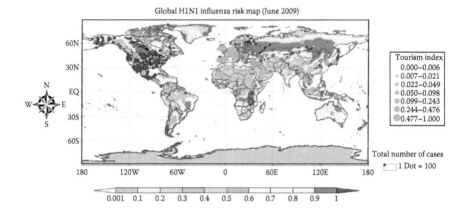

FIGURE 12.3 Tourism index—a useful indicator of the size of influenza outbreaks. The tourism index for each country (0–1) is superimposed on the predicted disease risks (0–1) for June. Pink circles depict the total number of confirmed cases worldwide reported until May 24, 2009.

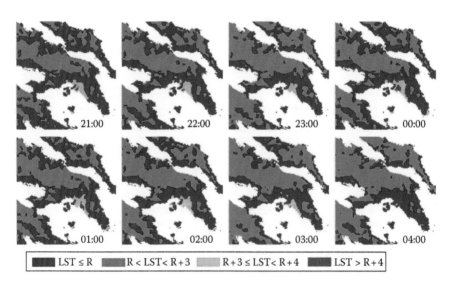

FIGURE 13.5 Hourly development and decay of SUHI patterns for a summer day (July 16/17, 2009); time is in UTC. The reader is advised to focus on the orange and red patterns that represent the SUHI patterns at different times.

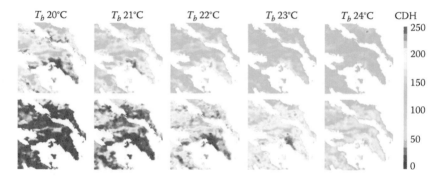

FIGURE 13.8 Cooling degree hours in degrees for a single day for various base LSTs (T_b). Top row represents a typical day in July and the bottom row a hot day.

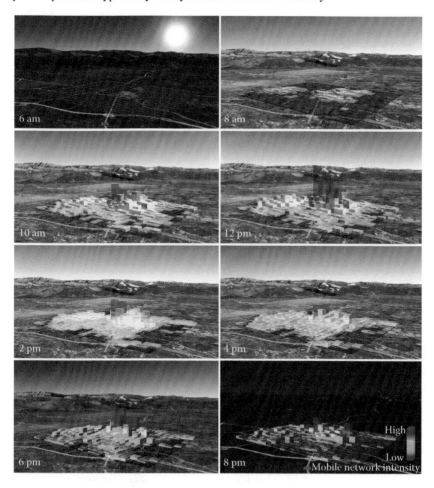

FIGURE 14.4 Collective human activity in the city (Udine, northern Italy)—spatio-temporal mobile communication activity on a typical working day as seen from a mobile network operator's perspective. (From Sagl, G. et al., From social sensor data to collective human behaviour patterns: Analysing and visualising spatio-temporal dynamics in urban environments, in Jekel, T., Car, A., Strobl, J., and Griesebner, G., eds., *GI-Forum 2012: Geovisualization, Society and Learning*, Wichmann Verlag, Berlin, Germany, 2012c.)

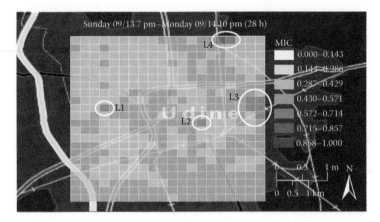

FIGURE 14.5 Measuring the strength of the relationship between adverse weather conditions and unusual human activity using the MIC. 0, no relationship; 1, functional relationship. (Modified from Sagl, G. et al., *Sensors*, 12, 9835, 2012a.)

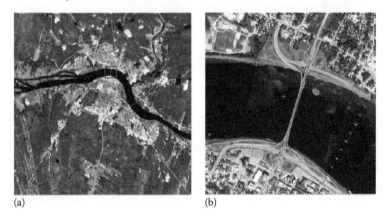

(a)   (b)

FIGURE 15.1 Depiction of the visual details in low- and high-resolution satellite images of same dimension, 512 × 512 pixels of downtown Fredericton, Canada: (a) Landsat TM 30 m MS and (b) QuickBird 2.44 m MS. (From Wuest, B.A., Towards improving segmentation of very high resolution satellite imagery, MScE thesis, Department of Geodesy and Geomatics Engineering Technical Report No. 261, University of New Brunswick, Fredericton, New Brunswick, Canada, 2008.)

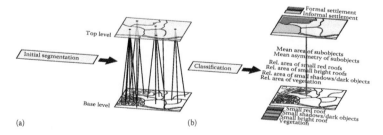

FIGURE 16.4 Object–hierarchical relationships after segmentation (parameterization, see Table 16.2) and classification between top- and base-level objects. For detailed class descriptions, see Table 16.3. Every top-level object relates to its subobjects in the base level and vice versa (a). Relationships to subobjects can be used for classification of superobjects (b).

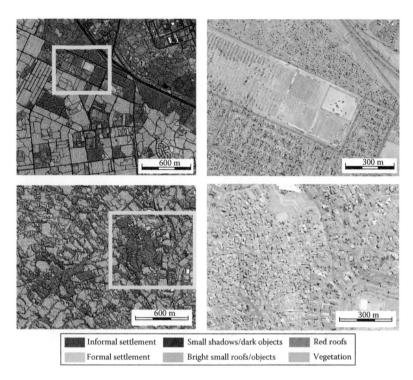

| Informal settlement | Small shadows/dark objects | Red roofs |
| Formal settlement | Bright small roofs/objects | Vegetation |

FIGURE 16.5 Segmentation and classification results for IKONOS (top) and QuickBird (bottom). Left: top segmentation level with respective classes. Right: base segmentation level with respective classes. Blue rectangle in the left images indicates the location of the right images.

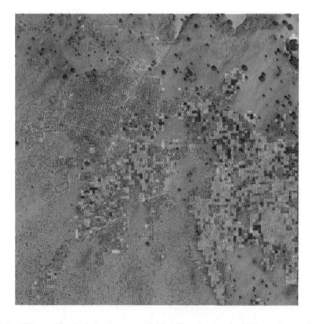

FIGURE 17.14 Generalized change map of Abu Suruj: new buildings (green), low to moderate change (yellow), and extensive change (red).

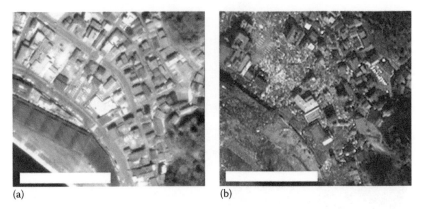

(a)   (b)

FIGURE 17.16   Test dataset: The pre-event image (© DigitalGlobe 2013) used for digitizing the ground truth information (a) and the postevent image (© GeoEye 2013) used for change detection (b). (Satellite images, courtesy of Google Earth.)

(e)   (f)

FIGURE 17.18   (e) reconstructed 3D building models, and (f) DSM with 5 m resolution generated from the original stereo pair.

FIGURE 18.1   ENVISAR ASAR image over part of Toronto.

FIGURE 18.2 ENVISAT MERIS image over Toronto.

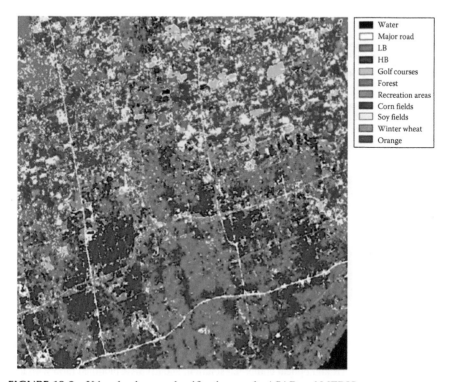

■	Water
□	Major road
▨	LB
■	HB
▨	Golf courses
▨	Forest
▨	Recreation areas
■	Corn fields
□	Soy fields
▨	Winter wheat
■	Orange

FIGURE 18.3 Urban land cover classification result: ASAR and MERIS.

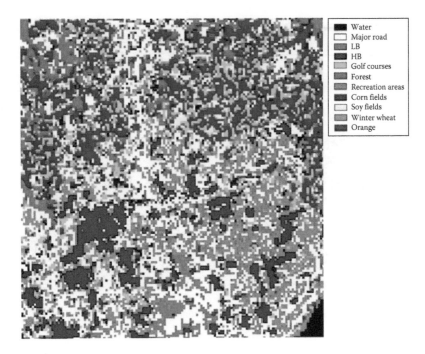

■	Water
□	Major road
■	LB
■	HB
■	Golf courses
■	Forest
■	Recreation areas
■	Corn fields
□	Soy fields
■	Winter wheat
■	Orange

FIGURE 18.4 Urban land cover classification result: MERIS.

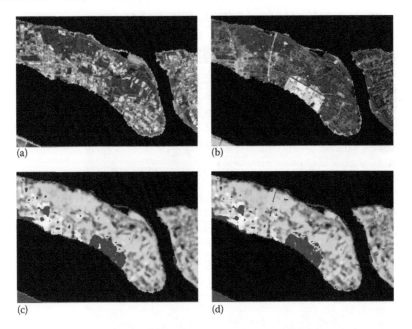

FIGURE 18.7 Detected changed areas in yellow, overlaid in a false color composite using (a) false color composite of 03.11.1999 Landsat image, (b) false color composite of 06.07.2008 Landsat image, (c) SAR and optical combined solution, and (d) SAR modified ratio with log normal solution.

10 Mapping and Monitoring of Refugees and Internally Displaced People Using EO Data

Thomas Kemper and Johannes Heinzel

CONTENTS

10.1 INTRODUCTION

The focus of this book is the monitoring and assessment of urban settlements. As such, it looks at the circumstances in which the majority of people around the globe are living. Apart from the remaining rural population, there is a considerable part of the global population that was forcibly displaced by natural disaster, conflict, or persecution. They are often forced to live in temporary or transitional settlements as refugees or internally displaced persons (IDPs, see more detailed definition in the next section). By the end of 2012, about 44 million people worldwide were considered as forcibly displaced due to conflict and persecution. They included 15.2 million refugees and 28.8 million IDPs (IDMC 2013).

A large part of refugees and IDPs are living in camps that are managed or supported mostly by national or international relief organizations. These camps may host several tens of thousands of refugees or IDPs, sometimes for many years. Although we may have some stereotyped ideas about refugee camps, there is a plethora of camp situations. One of the objectives of this chapter is to raise awareness about this often neglected part of the global population and at the same time to show how Earth Observation (EO) can help in improving its situation by providing up-to-date information to relief agencies and decision makers. Knowing how many refugees/IDPs there are is fundamental for planning and managing efficient relief operations. This chapter reviews the scientific literature to identify how EO may support the mapping and monitoring of refugees and IDPs with EO data and proposes a robust methodology that was already tested successfully in different camp situations.

10.2 SOME DEFINITIONS

This chapter discusses how EO helps in mapping and monitoring refugees and IDPs. In this context, it is important to understand the differences (and similarities) between refugees and IDPs.

The status of refugees is clearly defined by international law. The 1951 United Nations Convention Relating to the Status of Refugees, also known as the "Geneva Convention" defines, in Article 1.A.2, a refugee as "any person who: owing to a well-founded fear of being persecuted for reasons of race, religion, nationality, membership of a particular social group, or political opinion, is outside the country of his nationality, and is unable to or, owing to such fear, is unwilling to avail himself of the protection of that country" (UNHCR 2011). The definition was expanded by the Convention's 1967 Protocol to include persons who had fled war or other violence in their home country (UNHCR 2011).

The status of IDPs is legally less well defined. However, according to the *Guiding Principles on Internal Displacement* published by the UN Office for the Coordination of Humanitarian Affairs (UNOCHA 2004), "internally displaced persons are persons or groups of persons who have been forced or obliged to flee or to leave their homes or places of habitual residence, in particular as a result of or in order to avoid the effects of armed conflict, situations of generalized violence, violations of human rights or natural or human-made disasters, and who have not crossed an internationally recognized state border." It is important to note that IDP camps generally tend to have fewer relief resources available, making camp residents more dependent on locally available natural resources (UNEP 2007).

In this chapter, we focus our attention on mapping and monitoring of conflict and disaster-induced displacement of groups of refugees and IDPs. There are other forms of forced migration like environmental displacement, asylum seekers, and smuggled and trafficked people, which are not dealt with in this chapter.

10.3   UNIVERSE OF REFUGEE/IDP CAMPS

Millions of refugees and IDPs are living in all kinds of crisis situations. It is crucial to understand this variety and its implications in order to develop appropriate methods for the mapping and monitoring based on EO data. This section provides a typology of camps based on a number of parameters. The first set of parameters describes the framework conditions of the camp including the causes for its creation, the status of the displaced persons (refugee or IDP), the origin of the installation of the camp, and the time the camp already exists (Figure 10.1).

Such kind of information can only be derived indirectly from EO imagery; however, it has a strong impact on the camp. For example, refugees are more likely to be hosted in planned camps, while IDPs are often grouped in self-settled camps. Some parameters can be derived from the EO imagery (Figure 10.2). There is information related to the size of the camp. Large camps like in Darfur or at the Horn of Africa can be detected even with coarser-resolution EO data. For dispersed, fragmented IDP gatherings, like the example of Haiti discussed next or the IDPs spread in Mogadishu (Somalia), often very high spatial resolution and/or pre-event imagery are necessary to identify IDPs. Specific structures of the camp (fences, lined-up tents) usually assist in identifying if a camp is a planned or a self-settled area. Very high resolution (VHR) imagery helps in estimating the size and to some extent the type of dwellings in a camp. Finally, the surroundings of the camp provide information about the location, accessibility, and risks of the camp.

The following examples highlight some emblematic camp situations referring to the parameters discussed.

FIGURE 10.1   Framework conditions for a camp.

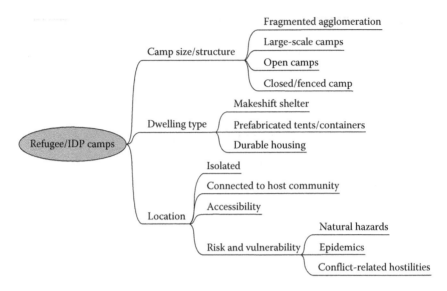

FIGURE 10.2   Parameters that can be derived from EO data.

10.3.1   SELF-ORGANIZED IDP CAMP AFTER A NATURAL DISASTER IN HAITI

The starting point for the creation of a refugee/IDP camp is always the (urgent) need for people to leave their homes in search of a secure and safe place. The main causes for the creation of camps are usually either natural disasters or conflicts. The occurrence of a disaster (e.g., earthquake, flood, technical accident) leads mostly to temporary displacement of population from the affected area. In the immediate aftermath, people tend to group in safe places as a survival strategy and return to their original habitat once the disruption is over, for example, after the floodwater has receded. There are also cases where people are permanently displaced, as in the case of the Chernobyl and Fukushima nuclear accidents. The large majority of these people continue to live as IDPs in the country of origin. Depending on the coping capacity of the country, the IDPs are moved as soon as possible into temporary shelters by the national civil protection or they remain, at least initially, in the disaster area mostly in self-organized camps. This is illustrated with an example of an IDP gathering in Port-au-Prince (Haiti) in Figure 10.3. In fear of the frequent aftershocks, the people started putting up makeshift shelters with plastic tarpaulins to protect themselves from sun and rain. At this point they were still entirely self-dependent without access to water, sanitation, and alimentation. Some days later, the entire area was densely covered with tarpaulins hosting 8,000–10,000 IDPs. This camp was functional until the end of 2012. In the immediate aftermath of the disaster, there was a need, first, to identify the locations where IDPs had gathered, and, second, to estimate the camp population to scale the emergency relief.

10.3.2   SELF-ORGANIZED IDP CAMP DURING A CONFLICT IN SRI LANKA

In violent conflict situations, people try to flee the fighting zone. Sometimes they are even deliberately targeted by the fighting parties. As in the case of disaster

FIGURE 10.3 IDP gathering in Delma/Saint-Martin (Haiti) 5 days after the catastrophic magnitude 7.0 Mw earthquake that struck Haiti on January 12, 2010. IDPs are seen gathered in open spaces, because their homes are destroyed or people are afraid of aftershocks. (Imagery, © 2010 Digitalglobe.)

refugees/IDPs, their first reflex would be to seek refuge in secure places as a survival strategy. In this context, usually groups offer more protection and hence the displaced gather in informal camps away from the fighting. Such a situation is depicted in Figure 10.4, where several tens of thousands of IDPs are seen gathered in an ad hoc manner in the so-called non–fighting zone (NFZ). The IDPs mostly put up tarpaulin sheds all over the NFZ, which were several times reduced in size by the fighting parties. The area resembled a big contiguous camp without any kind of coordination. IDPs were largely deprived of access to safe water, sanitation, alimentation, and health care from relief agencies. Toward the end of the conflict, access to the combat zone was strongly limited, and independent information was necessary to monitor the conflict, estimate the entrapped civilian population, and support the collection of evidence on alleged war crimes.

10.3.3 PLANNED IDP CAMP AFTER A CONFLICT IN SRI LANKA

The civilians who managed to escape the fighting in the NFZ and the civilians who were still in the NFZ after the end of combat were detained in a number of camps. The reasons given by the Sri Lankan government/military for not allowing the civilians to return to their homes were the existence of land mines and the need to identify Liberation Tigers of Tamil Eelam (LTTE) fighters whom they alleged were hiding among the civilians. The largest camp—Menik Farm—consisted of eight separate camps that hosted more than 220,000 IDPs (Figure 10.5). The camps were

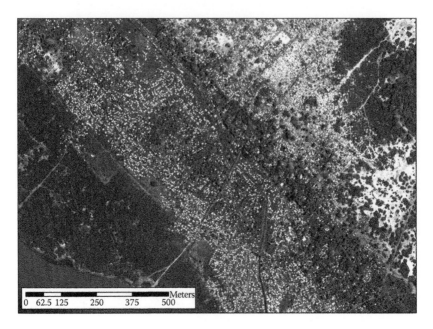

FIGURE 10.4 Mullaitivu (Sri Lanka). IDP tents in the designated "safe zone" for civilians trapped in the fight between the Sri Lankan military forces and the LTTE on May 10, 2009. (Imagery, © 2009 Digitalglobe.)

FIGURE 10.5 Close-up view of Menik Farm camp 1 (Sri Lanka) on June 26, 2009, where IDPs were detained after the end of the civil war in Sri Lanka. It is an example of a planned camp with clearly separated blocks and lined-up dwellings. (Imagery, © 2009 Digitalglobe.)

established by the military in an isolated location. Each camp was subdivided into rectangular blocks separated by driveways. Each block was occupied by clearly lined up tents of different sizes. Each camp zone and many blocks were fenced. IDPs were initially not allowed to leave the camp, and UN and humanitarian agencies had only limited access to the IDPs.

Sri Lanka was initially planning to keep the IDPs in the camps for up to 3 years. Following the pressure from the international community the camps were partially opened up and the IDPs slowly released. Since access to the camps was limited to international observers, EO could be used to verify the promises of the government.

10.3.4 PARTLY PLANNED CAMP IN A PROTRACTED IDP SITUATION IN DARFUR (SUDAN)

In the aftermath of a conflict, refugees and IDPs have to adapt to the situation in the camps—they are often trapped there for several years. Wherever possible, they try to improve their living conditions by generating income (offering labor, self-sustaining food production, trading) and improving their dwellings (building more resistant, spacious structures). The refugees and IDPs in the Darfur conflict that started in 2003 are in such a situation. Ten years after the start of the conflict, 1.4 million IDPs and 300,000 refugees still live in camps in protracted refugee/IDP situations. The refugees/IDPs find themselves in a long-lasting and intractable state of limbo. Their lives may not be at immediate risk, but their basic rights and essential economic, social, and psychological needs remain unfulfilled after years in the camps and/or exile (UNHCR 2004).

The change in the dwelling situation is clearly visible in Figure 10.6. The figure shows the original village of Jaffalo in the center with the typical Darfurian settlement structure: the dwellings are a combination of round thatched-roof huts and flat-roofed, mud-brick buildings; the properties are delimited by thorn hedges. The right part of Figure 10.6 shows the old Zamzam camp, where the IDPs have already adopted the traditional style, though more densely built up. The left part shows the extension of the camp, where the new arrivals are living in much more primitive conditions. In camp situations, as described for Zamzam with a continuous movement of the population, it is difficult to monitor the actual camp population size. EO could monitor the population without resorting to a cost- and labor-intensive and often dangerous population census.

Figure 10.6 also addresses an important aspect of refugee/IDP camps: the interaction with the host communities and the impact of the camp on the surrounding environment. The sudden influx of large numbers of IDPs or refugees—often outnumbering the host population—into a spatially limited area can place severe pressure on the local resources. In many cases this results in grave consequences for the local environment, leading to further deterioration in the socioeconomic and political, as well as sanitary, conditions within the host communities (Hugo 2008).

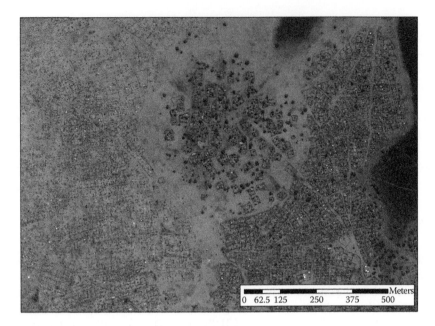

FIGURE 10.6 The village of Jaffalo in Darfur (Sudan) in the center surrounded by the Zamzam IDP camp on May 29, 2009. The right part of the image shows an old Zamzam camp, the left part the Zamzam extension. At this point, the camp hosted approximately 100,000 IDPs. (Imagery, © 2009 Digitalglobe.)

10.3.5 Refugee Camp in a Protracted Situation in an Urban Context in Lebanon

The oldest camps in the world are the Palestinian refugee camps, which were established after the 1948 Arab–Israeli war. Some of the camps are today entirely absorbed by the host cities. Figure 10.7 shows the Bourj el-Barajneh refugee camp in Beirut, Lebanon. It is entirely surrounded by dense urban fabric. It is only partly distinguishable from the neighboring building development. However, the building density is extremely high, with no open spaces observable.

The example of Bourj el-Barajneh clearly shows some of the limits of EO, particularly in urban environments. Often it is impossible to provide information purely from EO data, if it is not combined with additional information from the ground.

On the other hand, today's refugee/IDP reality is that only a minority now live in camps. Many of those forcibly displaced have moved to urban areas in search of greater security, including a degree of anonymity, better access to basic services, and greater economic opportunities. Today, approximately half of the world's estimated 10.5 million refugees and at least 4 million IDPs are assumed to be living in urban areas (Fielden 2008, UNHCR 2009, Metcalfe et al. 2011). But the lives of urban refugees and IDPs also present dangers: refugees may not have legal documents that are respected; they may be vulnerable to exploitation, arrest, and detention; and they can be in competition with the poorest local workers for the worst jobs.

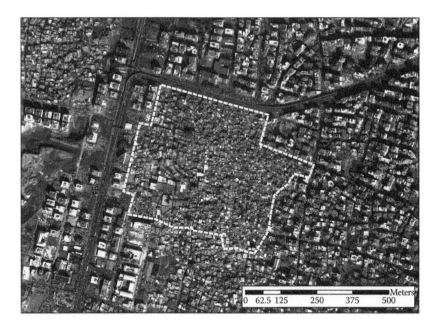

FIGURE 10.7   Bourj el-Barajneh refugee camp in Beirut (Lebanon) on February 10, 2010. This Palestinian refugee camp was established in 1948 and hosts more than 15,000 refugees. The white dashed line represents the approximate camp limits. (Imagery, © 2010 Digitalglobe.)

The examples described in this chapter are only a representative sample of the many situations encountered by refugees/IDPs every day. The next section provides an overview of how EO can support this struggle with up-to-date information.

10.4   SUPPORT OF EO DATA

In most relief operations, the key information needed for efficient planning and management is the location and number of the affected population. The size of the displaced population in a given site is crucial for an initial assessment to effectively allocate resources (food, shelter, health, etc.) and take measures to address added pressure on natural resources due to the arrival of the displaced population (Noji 2005). In particular, in early phases of a conflict, this information is impossible to obtain from the ground. This is because the location of the camps is not known and it is often impossible to access the area due to ongoing violence or due to the impact of the natural disaster. Even at later stages, obtaining up-to-date statistics remains difficult. A large number of approaches have been developed, but most methods to estimate the population are based on in situ assessments and require statistical sampling and analysis (e.g., Brown et al. 2001, Grais et al. 2006). In the first days or weeks of an emergency, such knowledge is mostly unavailable. In addition, humanitarian access is limited in many IDP situations due to ongoing violence or access limitations imposed by the impact of the disaster. Hence, the current guidelines of

UNHCR (2001) propose the use of Geographic Information Systems (GIS) and aerial and satellite imagery for estimation of the affected population based on visual interpretation and/or simple modeling of the population figures using a statistical sampling approach. Ehrlich et al. (2009) provide examples of how remote sensing with satellites can indirectly be used to estimate population.

10.4.1 DWELLING AND POPULATION ESTIMATION IN CAMPS

The initial studies in the scientific literature focus mostly on the characterization of refugee/IDP camps and their surrounding environment due to the limitations in the spatial resolution of satellites. Lodhi et al. (1998) analyzed the impact of Afghan refugees on the forest cover in the vicinity of their camp in Pakistan using Landsat data. Bjorgo (2000) used satellite photography from the Russian KVR-100 camera with a spatial resolution of 3.3 m to assess the environment around a Khmer refugee camp in Thailand and successfully correlated the camp area with census-based population estimates. The first study to propose an automated procedure to estimate the number of dwellings in a refugee camp was published by Giada et al. (2003). They tested different approaches to enumerate the dwellings based on statistical sampling with visual interpretation, supervised/unsupervised classification, object-based image analysis, and mathematical morphology, where the last produced the best results. Subsequently, more studies were published for the automated enumeration of dwellings in camps on the basis of sound precision rather than visual interpretation (Jenerowicz et al. 2010, Kranz et al. 2010, Tiede et al. 2010, Kemper et al. 2011a). Unfortunately, the studies could not provide population figures and validate their methods against ground population estimates. Checchi et al. (2013) tested the validity and feasibility of a satellite imagery–based analysis for rapid estimation of displaced populations and validated it against field-based population estimates. They concluded that in settings with clearly distinguishable individual structures, the remote imagery–based method has a good potential to be used operationally even with automated approaches. According to their findings, it may not work in settings with connected shelters, a complex pattern of roofs, or multilevel buildings. A particularly interesting aspect of the application of EO imagery is the fact that an area can be easily observed multiple times allowing monitoring of the situation. Lang et al. (2010) monitored the development of the Zamzam camp in Darfur with three time steps between 2004 and 2008. Kemper et al. (2011b) monitored the dismantling of Menik Farm camps in Sri Lanka using an advanced change detection methodology.

Today, EO is playing an important role in providing relief agencies with up-to-date information about the location, structure, and number of dwellings in a camp (Figure 10.8), including frequent updates of the situation. This is possible due to the enhanced capability of EO in terms of availability, extent, and timeliness of satellite imagery covering a certain crisis situation. However, it is necessary to have operational mapping centers with the appropriate know-how to provide the needed information. In the last few years, several initiatives have been developed with regard to operational capacities in this field. There are, just to give some examples, international initiatives such as the Emergency Mapping Service of the European

FIGURE 10.8   **(See color insert.)** Example of an operational refugee camp map. (Map, © 2013 European Commission.)

Copernicus Programme (http://emergency.copernicus.eu), UNOSAT (http://www. unitar.org/unosat), or national capabilities such as the German ZKI (Center for Satellite-Based Crisis Information; http://www.zki.dlr.de).

10.4.2   ASSESSING THE ENVIRONMENTAL IMPACT OF REFUGEE/IDP CAMPS

Refugees and IDPs have a major impact on the host communities and the surrounding environment (e.g., Black 1994, Jacobsen 1997). As such, environmental concerns related to IDP and refugee streams have become increasingly important, since they strongly impact postconflict recovery. The UNEP has recently started to address the environmental impacts of IDP and refugee camps through their postconflict environmental assessment (PCEA) activities. Primary assessments have been conducted in Liberia (UNEP 2006), Sudan (UNEP 2007), the Democratic Republic of the Congo (UNEP 2011b), and Rwanda (UNEP 2011c). However, despite the fact that such PCEAs are often carried out in areas that are remote and unsafe, they are mostly based on field surveys that are both time- and cost-intensive, and make little or no use of state-of-the-art remote sensing methodologies. There are, however, examples showing that this topic can be supported with EO data analysis. Lodhi et al. (1998) and Gorsevski et al. (2012) used medium-resolution Landsat data to assess the impact of conflict and refugees in Pakistan and Uganda, respectively. Hagenlocher et al. (2012) used VHR data to assess the environmental impact of an IDP camp in Darfur (Sudan). They developed a weighted natural resource depletion index that integrates selected land-use/land-cover target classes and their relative importance for human security and/or ecosystem integrity. In 6 years, they observed a noticeable decrease in the area covered by single shrubs and small trees, which is clearly linked to the presence of a growing number of IDPs.

Over the last few years, there have been significant advances in the usage of EO imagery to support refugee/IDP camp management. The following section will provide a more in-depth description of a robust approach for population estimation in camps.

10.4.3 ROBUST METHODOLOGY FOR ENUMERATION AND SUBSEQUENT POPULATION ESTIMATION WITH FIELD KNOWLEDGE

As described in the previous sections, there have been several approaches to automatically or visually estimate the number of dwellings in refugee camps. Most of them were tested only in one camp context and stopped at the level of dwelling enumeration. In the following sections, we describe a robust methodology that was tested in very different camp conditions (among them the situations described previously in this chapter) and in one case validated with census information. The goal is to develop a reliable and consistent method to estimate the total number of dwellings in refugee/IDP camps robust enough to provide sound figures under different environmental and camp conditions and to combine these with limited field data collection for the population estimation. With robustness, we refer to the needs of the field worker that need population figures (or at least dwelling numbers) accurate enough to base decisions on. They need not necessarily be very accurate, but should help in a good understanding of the error margin. This is exemplified with an analysis of the Al Salam camp in South Darfur (Sudan).

The approach (Figure 10.9) relies on an initial verbal characterization of the camp context to identify the searched dwellings and structures. This characterization is used for a visual interpretation of a representative sample, which is necessary for validation and quality control. It is also translated in terms of image processing transformations, enabling the automatic extraction of these dwellings. The intermediate result is subsequently refined to binary masks for direct counting. The same intermediate result can be used to estimate the number of structures by performing a regression against the visually interpreted samples (area-based modeling in Figure 10.9) to provide an additional measure of quality of the output.

10.4.3.1 Verbal Dwelling Characterization

Camp characterization is a crucial step in understanding the complexity of camp reality. Ideally, this phase is supported by field or aerial photography and a point of contact in the field that is able to describe specific features in the camp. For the Al Salam camp in South Darfur, the World Food Program (WFP) provided such field information.

On the basis of a comparison of the satellite image with field pictures (Figures 10.10 and 10.11), it can be concluded that dwellings correspond to a mixture of thatched-roof huts, traditional clay constructions with flat roofs, and other structures covered with bright materials (e.g., UN tent fabrics or plastic sheets).

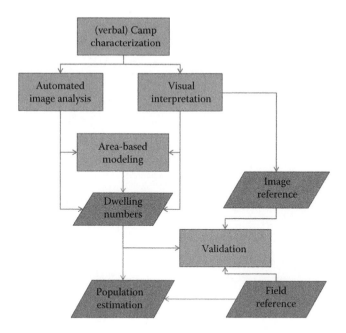

FIGURE 10.9   High-level workflow for remote sensing–based population estimation.

This led to the following characterization of dwellings visible in the satellite image:

- Most dwelling structures appear as either dark objects over a brighter background or bright objects over a darker background. Sometimes dwellings made of mud bricks display a flat roof hardly discernible from the surrounding ground (similar intensity levels).
- Disk-shaped dark structures correspond to either thatched-roof huts or foliated tree crowns.
- The minimum size of a dwelling structure can be set to 6 m² corresponding to a dwelling of 3 × 2 m, which is the smallest dwelling identified in the images. It is also reasonable to consider that a dwelling structure must be compact enough to contain a square of 2 × 2 m.
- Bright structures corresponding to objects above the ground display thin shadows in the direction opposite to the sun-azimuth angle.
- Dwellings are clustered within a series of rectangular compounds outlined by a fence made of clay walls, thorn hedges, or straw matting. These outlines are revealed in the imagery mainly through the shadows they cast so that dwellings appear at the interior or edge of an area surrounded by a thin dark rectangular outline. Note, however, that the fences contain one or more entrances so that these outlines are not closed.

Given the coarser spatial resolution of the multispectral imagery and given the fact that most dwelling structures have a size only slightly larger than the multispectral

FIGURE 10.10 **(See color insert.)** Aerial photograph of the Ardamata camp near Al Geneina, West Darfur. (Courtesy of WFP, © 2009 WFP.)

FIGURE 10.11 Typical example of the Al Salam camp structure comparable to the situation in Ardamata (Figure 10.10). (Imagery, © 2009 Digitalglobe.)

imagery, it is not possible to recognize individual dwellings on the multispectral images. Therefore, the identification of single dwellings was made based on panchromatic and pansharpened multispectral imagery. Moreover, the pansharpened multispectral imagery is useful for producing a vegetation mask so as to better discriminate thatched-roof huts from tree crowns in the panchromatic imagery (both appear as dark disk-shaped objects). By definition, dwellings that are located below trees are not visible from satellites operating in the visible wavelengths and are thus the source of underestimation of the real number of dwellings.

10.4.3.2 Dwelling Extraction

Dwelling extraction relies on the theory of mathematical morphology, which is a theory and technique for the analysis and processing of geometrical structures (Soille 2003). The image processing chain aims at automatically extracting dwelling structures from the panchromatic input satellite imagery. It has been designed by translating the previous verbal description (e.g., shape and minimum/maximum size of dwellings) into a series of morphological operators. The theory of morphological operators is explained in depth, for example, by Soille (2003).

Morphological area opening and closing (Cheng and Venetsanopoulos 1992, Meijster and Wilkinson 2002) suppress all bright and dark objects, whose area is below a given threshold value (and considering a given connectivity rule such as the 4- or 8-connectivity). The suppressed objects are then retrieved by computing the difference between the original and the transformed image (top-hat operation). The following top-hat images were calculated:

- Top-hat by 8-connected area opening with an area of 100 pixels. This operation highlights bright objects of less than 100 pixels. The threshold was chosen to exclude larger structures in the camp that are used as health centers, schools, or similar establishments.
- Top-hat by 8-connected area closing with an area of 100 pixels. This operation highlights dark objects of less than 100 pixels.

The union (i.e., point-wise maximum) of these two top-hat images corresponds to an image containing bright as well as dark structures smaller than 100 pixels. This image can be directly used to estimate the number of structures by performing a regression against visually interpreted samples (area-based modeling). The top-hat images can also be further processed in parallel so as to generate binary masks of bright and dark structures. The total number of structures is then used as an estimate of the total number of dwellings.

The applied processing chain is as follows:

- Filtering of the top-hat images using containment and size criteria: each structure must contain at least a square with a width of 4 pixels and have an area of at least 24 pixels. The first filter corresponds to an area opening by a 4 × 4 square structuring element and the second to an area opening with an area threshold value of 24 pixels.

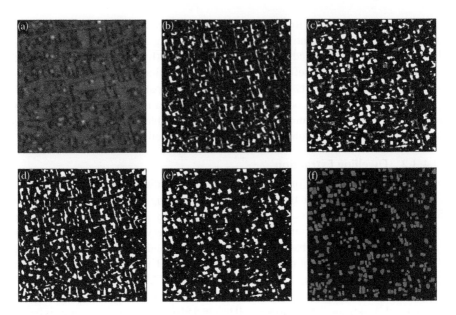

FIGURE 10.12   Example of the processing steps of the dwelling extraction: (a) input sample, (b) top-hat by area opening, (c) top-hat by area closing, (d) filtered top-hat by area opening, (e) filtered top-hat by area closing, and (f) bright (red) and dark (green) structures obtained by thresholding of filtered images (d) and (e).

- Threshold of the previous images for all intensity values greater than or equal to 4 for less contrasted structures corresponds mainly to intensity variations in the background.
- Union of the threshold images generates a binary image of filtered bright and dark structures.

Figure 10.12 illustrates the successive steps of this process. The resulting binary mask still contains isolated trees since they correspond to dark structures in the panchromatic image. They are masked out by computing the intersection of the central point (centroid) of automatically extracted structures with the mask of vegetation obtained by thresholding the Normalized Difference Vegetation Index (NDVI) image computed on the multispectral image and resampled at the resolution of the panchromatic image.

10.4.3.3   Random Sampling and Visual Interpretation

Visual interpretation is used for the quality control of the results. It is performed on randomly selected samples. The camp area was overlaid with a grid of 80 m by 80 m in order to randomly select a representative group of grid cells for visual analysis. Approximately 5% of the cells were selected for the random sample. An image interpreter was tasked to label each dwelling in a sampling cell with a point taking into account the minimum and maximum criteria defined earlier. Consequently, large structures (such as public facilities, storage, and shopping buildings) were not marked. These structures are characterized by a regular spatial arrangement and mostly dwelling structures of a larger size than usual. They are often surrounded by solid fences and to some extent separated from the densely situated dwellings. In addition, they are matched against the camp maps provided by WFP. The visual counting relies on the dwelling characterization described earlier. The distinction between tree crowns and thatched-roof huts could be achieved thanks to the use of a pan-sharpened false color composition highlighting the vegetated areas in red. The individual dwellings were stored as a single point in the database.

10.4.3.4   Regression Analysis

In order to have a greater reliance on the derived dwelling numbers, we used a statistic regression approach to find the relation between the visually interpreted number of dwellings and the automatically extracted structures for the random sample. Based on that function, the total number of dwellings for the entire camp is estimated. The reasons to add this indirect estimation are twofold. First, this area-based approach solves problems of attached buildings that are otherwise counted as single dwellings. Second, the regression model allows a sensitivity analysis that provides error margins for the dwelling numbers, which is highly relevant information for the user.

The union of the top-hat area opening and closing results (Figure 10.12b and c) is the basis for area estimation. In these gray-level images, nondwelling structures are still apparent (e.g., fence structures). Therefore, a fuzzy-logic min/max operator was applied to better distinguish between dwellings and nondwelling structures. Prior to area estimation, the NDVI was used to eliminate isolated trees. The area was then calculated as the sum of gray levels in each randomly selected cell as input for the regression model by relating this model to the visually counted number of dwellings. The stability of the regression model was tested using different numbers of samples for calibration and validation. The number of samples ranges from 10% to 90% with 10% intervals. The remainder of the samples was used for validation. Each model was repeated 500 times with a different random selection in each run. The final regression coefficients for the estimation of the total number of dwellings in the camp are the average of the coefficients derived in each step.

10.4.3.5   Field Data Collection

Remote sensing analysis provides the overall number of dwellings in the camp. This information is, however, not sufficient for relief agencies, who are interested in the population figures. In order to translate the number of dwellings, it is necessary to derive information on the average household size and the average number of dwellings per household. This is important especially because in protracted refugee/IDP situations, like Darfur, people use some shelters for cooking, sleeping, living, and possibly as stable for animals. For the survey in Al Salam, the camp was divided into homogeneous units based on visual interpretation of the camp densities. In each unit, two transects were randomly selected to collect the relevant information. According to this survey, a household in Al Salam camp on average consists of 5.72 persons living in 2.79 structures.

10.4.3.6   Results

The analysis of the Al Salam camp is based on a WorldView-2 image acquired on November 10, 2010. Although eight multispectral bands were available, only the

50 cm resolution panchromatic band was used. For the visual interpretation, 40 cells were randomly selected. In these cells, 1522 dwellings were visually detected and 1363 were mapped automatically. The scatterplot in Figure 10.13a shows the slight underestimation of the dwellings compared to the visual interpretation in the selected areas; nevertheless, the correlation coefficient is strong at 0.83. The area-based modeling using the regression model produced a mean error of 42 structures (6%) and a maximum error of 92 structures (11%) in the sensitivity analysis (Figure 10.13b).

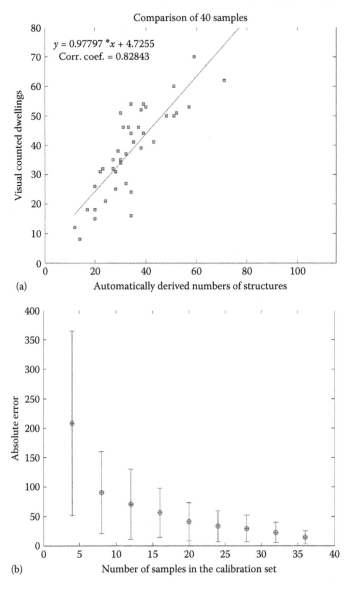

(a)

(b)

FIGURE 10.13 Analysis results of automated processing. Scatterplot of visually versus automatically detected dwellings (a) and sensitivity analysis of area-based modeling (b).

These values are similar to those obtained in earlier experiments (Jenerowicz et al. 2010, Kemper et al. 2011a,b).

Based on the direct counting of all dwellings in the camp, 17,394 dwellings were detected. The estimation of the camp population based on the automatically extracted number of dwellings (and the limited field data collection by WFP) matched the figures of a camp census conducted by WFP with an error of less than 1%. Due to the sensitivity of the figures, we are unable to reveal more detailed information.

10.5   CONCLUSION

This chapter described a particular part of the global population, namely, those displaced by natural disasters or conflicts that have to persevere in refugee or IDP camps, sometimes for many years. The focus was on how these camps are structured from a remote sensing point of view and how EO may help in identifying and monitoring the camps. We made an attempt to classify the camps according to some external framework conditions and parameters that can be derived from satellite imagery. If at least some of the parameters can be derived automatically from satellite imagery, it would be possible to undertake regular monitoring, which might help relief agencies like UNHCR or WFP, which are responsible for a large number of camps, to plan their resources better, because they may identify needs in certain camps at an early stage.

EO data is useful with regard to refugee/IDP camps in mainly two ways: population estimation and environmental impact assessment. The latter is only addressed by few publications and has definitively more potential given all the research that has been undertaken for monitoring the environment in general. More attention has been given to the population estimation in refugee/IDP camps, although most analyses stop at the enumeration of dwellings in refugee camps without turning it into population data, which is needed for camp management. This highlights the clear gap between the scientific community and the practitioners in the field that we have to overcome, if (automated) EO information extraction is to be used more operationally. The first step has been taken with operational units that are able to perform rapid population estimations with up-to-date imagery using manual and/or automated approaches and by interacting with practitioners in the field to obtain the necessary field information. This link needs to be strengthened in any case. We believe that if this is achieved, automated data analysis of EO data in combination with field data can provide rapid population estimations for scaling the relief needs in the early phases of a crisis. Moreover, at later stages regular monitoring will allow tuning of the resources to the actual needs.

ACKNOWLEDGMENT

We would like to express our gratitude to the World Food Program Office in Nyala, South Darfur (Sudan), for their support in field data collection. In particular, we would like to thank Pedro Matos for his enthusiasm and for the lively discussions we have had with him.

REFERENCES

Bjorgo, E. 2000. Using very high spatial resolution multispectral satellite sensor imagery to monitor refugee camps. *International Journal of Remote Sensing*, 21(3), 611–616.

Black, R. 1994. Forced migration and environmental change: The impact of refugees on host environments. *Journal of Environmental Management*, 42(3), 261–277. doi:10.1006/jema.1994.1072.

Brown, V., Jacquier, G., Coulombier, D., Balandine, S., Belanger, F., and Legros, D. 2001. Rapid assessment of population size by area sampling in disaster situations. *Disasters*, 25(2), 164–171.

Checchi, F., Stewart, B.T., Palmer, J.J., and Grundy, C. 2013. Validity and feasibility of a satellite imagery-based method for rapid estimation of displaced populations. *International Journal of Health Geographics*, 12, 4. doi:10.1186/1476-072X-12-4.

Cheng, F. and Venetsanopoulos, A. 1992. An adaptive morphological filter for image processing. *IEEE Transactions on Image Processing*, 1, 533–539.

Ehrlich, D., Lang, S., Laneve, G., Mubareka, S., Schneiderbauer, S., and Tiede, D. 2009. Can Earth Observation help to improve information on population? Indirect population estimations from EO derived geo-spatial data: Contribution from GMOSS. In: Jasani, B., Pesaresi, M., Schneiderbauer, S., and Zeug, G. (Eds.), *Remote Sensing from Space—Supporting International Peace and Security*, Springer, Berlin, Germany, pp. 211–237.

Fielden, F. 2008. Ignored displacement persons: The plight of IDP's in urban areas, UNHCR Research Paper 161. UNHCR, Geneva, Switzerland.

Giada, S., De Groeve, T., Ehrlich, D., and Soille, P. 2003. Information extraction from very high resolution satellite images over Lukole refugee camp, Tanzania. *International Journal of Remote Sensing*, 24(22), 4251–4266.

Gorsevski, V., Kasischke, E., Dempewolf, J., Loboda, T., and Grossmann, F. 2012. Analysis of the impacts of armed conflict on the eastern Afromontane Forest region on the South Sudan–Uganda border using multitemporal Landsat imagery. *Remote Sensing of Environment*, 118, 10–20. doi:10.1016/j.rse.2011.10.023.

Grais, R., Coulombier, D., Ampuero, J., Lucas, M., Barretto, A., Jacquier, G., Diaz, F., Balandine, S., Mahoudeau, C., and Brown, V. 2006. Are rapid population estimates accurate? A field trial of two different assessment methods. *Disasters*, 30(3), 364–376.

Hagenlocher, M., Lang, S., and Tiede, D. 2012. Integrated assessment of the environmental impact of an IDP camp in Sudan based on very high resolution multi-temporal satellite imagery. *Remote Sensing of Environment*, 126, 27–38. doi:10.1016/j.rse.2012.08.010.

Hugo, G. 2008. *Migration, Development and Environment* (IOM Migration Research Series, No. 35). UN, Geneva, Switzerland.

Internal Displacement Monitoring Centre. 2013. *Global Overview 2012. People Internally Displaced by Conflict and Violence*. IDMC, Geneva, Switzerland.

Jacobsen, K. 1997. Refugees' environmental impact: The effect of patterns of settlement. *Journal of Refugee Studies*, 10(1), 19–36.

Jenerowicz, M., Kemper, T., Pesaresi, M., and Soille, P. 2010. Post-event damage assessment using morphological methodology on 0.5m resolution satellite data. *Italian Journal of Remote Sensing*, 42(3), 37–47.

Kemper, T., Jenerowicz, M., Gueguen, L., Poli, D., and Soille, P. 2011a. Monitoring changes in the Menik Farm IDP camps in Sri Lanka using multi-temporal very high-resolution satellite data. *International Journal of Digital Earth*, 4(Suppl.1), 91–106.

Kemper, T., Jenerowicz, M., Soille, P., and Pesaresi, M. 2011b. Enumeration of dwellings in Darfur Camps from GeoEye-1 satellite images using mathematical morphology. *IEEE Journal of Selected Topics in Applied Earth Observations and Remote Sensing*, 4(1), 8–15.

Kranz, O., Gstaiger, V., Lang, S., Tiede, D., Zeug, G., Kemper, T., Vega Ezquieta, P., and Clandillon, S. 2010. Different approaches for IDP camp analyses in West Darfur (Sudan)—A status report. In: *Sixth International Symposium on Geoinformation for Disaster Management (Gi4DM)*, Turin, Italy. Available at http://www.gdmc.nl/zlatanova/Gi4DM2010/gi4dm/Pdf/p187.pdf (accessed March 22, 2013).

Lang, S., Tiede, D., Hölbling, D., Füreder, P., and Zeil, P. 2010. Earth observation (EO)-based ex post assessment of IDP camp evolution and population dynamics in Zam Zam, Darfur. *International Journal of Remote Sensing*, 31(21), 5709–5731.

Lodhi, M., Echavarria, F., and Keithley, C. 1998. Using remote sensing data to monitor land cover changes near Afghan refugee camps in Northern Pakistan. *Geocarto International*, 13(1), 33–39.

Meijster, A. and Wilkinson, M. 2002. A comparison of algorithms for connected set openings and closings. *IEEE Transactions on Pattern Analysis and Machine Intelligence*, 24(4), 484–494.

Metcalfe, V., Pavanello, S., and Mishra, P. 2011. Sanctuary in the city? Urban displacement and vulnerability in Nairobi. HPG Working paper. Available at http://www.internal-displacement.org/8025708F004BE3B1/(httpInfoFiles)/79CA6774ADC9B401C12579610058D6AD/$file/sanctuary-in-the-city-nairobi-sep2011.pdf (accessed March 22, 2013).

Noji, E.K. 2005. Estimating population size in emergencies. *Bulletin of the World Health Organ*, 83(3), 164.

Soille, P. 2003. *Morphological Image Analysis: Principles and Applications*, 2nd edn. Springer, Berlin, Germany.

Tiede, D., Lang, S., Hölbling, D., and Füreder, P. 2010. Transferability of OBIA rulesets for IDP Camp Analysis in Darfur. In: Addink, E.A. and Coillie, F.M.B.V. (Eds.), *GEOBIA 2010—Geographic Object-Based Image Analysis, Ghent University*, Ghent, Belgium, 29 June–2 July. *ISPRS* XXXVIII-4/C7, Archives ISSN No 1682-1777.

UN Environmental Programme. 2006. *Environmental Considerations of Human Displacement in Liberia: A Guide for Decision-makers and Practitioners*. UNEP, Nairobi, Kenya.

UN Environmental Programme. 2007. *Sudan—Post-Conflict Environmental Assessment*. UNEP, Nairobi, Kenya.

UN Environmental Programme. 2011a. *Livelihood Security: Climate Change, Migration and Conflict in the Sahel*. UNEP, Nairobi, Kenya.

UN Environmental Programme. 2011b. *The Democratic Republic Congo—Post-Conflict Environmental Assessment. Synthesis for Policy Makers*. UNEP, Nairobi, Kenya.

UN Environmental Programme. 2011c. *Rwanda—From Post-Conflict to Environmentally Sustainable Development*. UNEP, Nairobi, Kenya.

UN High Commissioner for Refugees. 2001. *Handbook of Emergencies*, 2nd edn. UNHCR, Geneva, Switzerland. Available at http://www.unicef.org/emerg/files/UNHCR_handbook.pdf (accessed March 22, 2013).

UN High Commissioner for Refugees. 2004. *Protracted Refugee Situations*. EC/54/SC/CRP.14. Available at http://www.unhcr.org/refworld/docid/4a54bc00d.html (accessed March 22, 2013).

UN High Commissioner for Refugees. 2009. *UNHCR Policy on Refugee Protection and Solutions in Urban Areas*. UNHCR, Geneva, Switzerland.

UN High Commissioner for Refugees. 2011. *Handbook and Guidelines on Procedures and Criteria for Determining Refugee Status under the 1951 Convention and the 1967 Protocol Relating to the Status of Refugees*. HCR/1P/4/ENG/REV. 3. UNHCR, Geneva, Switzerland. Available at http://www.unhcr.org/refworld/docid/4f33c8d92.html (accessed March 22, 2013).

UN Office for the Coordination of Humanitarian Affairs (UNOCHA). 2004. *Guiding Principles on Internal Displacement*, 2nd edn. UNOCHA, Geneva, Switzerland. Available at http://www.brookings.edu/~/media/Projects/idp/GPEnglish.pdf (assessed March 22, 2013).

11 Assessment of Fine-Scale Built-Up Area Mapping in China

Linlin Lu, Huadong Guo, Martino Pesaresi, Daniele Ehrlich, and Stefano Ferri

CONTENTS

11.1 INTRODUCTION

In view of its vast territory, frequent occurrence of extreme meteorological and geological events, and socioeconomic conditions, China is one of the most disaster-affected countries in the world. Frequent natural disasters, consisting of floods, droughts, earthquakes, forest fires, snow, typhoons, and marine disasters, have caused severe human and economic loss to the country. For example, in 2010, 7844 people were killed, 2.73 million houses were destroyed, 37.4 million hectares of croplands were damaged, and there was an economic loss of 534 billion RMB because of disastrous events (Ministry of Civil Affairs of the People's Republic of China 2011). Disaster (damage) risk quantifies the damages that an element at risk may suffer when a hazard strikes. It is an integral part of decision- and policy-making processes of national and local disaster risk reduction activities. Exposure is the collection of the elements at risk that are subject to potential losses (ISDR 2009). It is reported to be increasing due to population growth and urbanization and is a major factor in increased disaster risk in hotspot countries (ISDR 2009). Fine-scale, standardized exposure information is crucial for disaster risk models, and thus for quantitative disaster risk assessment (Bal et al. 2010).

Human settlement maps derived from remote sensing data are widely used to generate location information of physical elements exposed to risk (Ehrlich et al. 2010; Ehrlich and Tenerelli 2012; Guo 2010). Different fundamental abstraction levels have been identified for the study of human settlements derived from remote sensing images (Pesaresi and Ehrlich 2009). For instance, level 0 is where settlement components are detected, and level 1 indicates their physical characteristics and patterns. This chapter addresses specifically the mapping of built-up areas, which can be defined as "areas (spatial units) where buildings can be found."

With the development of high-resolution (HR) and very high-resolution (VHR) satellite sensors, some initiatives aimed at identifying built-up areas based on HR and VHR remote sensing data have been established at different scales. Some initiatives such as Monitoring Land-use/Cover Change Dynamics (MOLAND), the Urban Environmental Project, and the European Urban Atlas provide geo-information on urban areas for specific cities (Esch et al. 2012). Coordination of Information on the Environment (CORINE) Land Cover maps urban areas with improved geometric and thematic detail in Europe (EEA 2010). Several initiatives were designed specifically for urban area mapping at the global scale. In the ASTER Global URban Area Map (AGURAM) project, an automated classification method was developed to produce urban area maps at 15 m resolution by integrating satellite images taken by the Visible and Near-Infrared Radiometer of Advanced Spaceborne Thermal Emission and Reflection radiometer (ASTER/VNIR) and GIS data derived from existing urban area maps (Miyazaki et al. 2013). The German Remote Sensing Data Center (DFD) of the German Aerospace Center (DLR) has applied a fully automated processing system that detects and extracts built-up areas to produce an urban footprint mask from global TerraSAR-X and TanDEM-X imagery (Esch et al. 2010, 2012). An automatic production framework has been implemented by the Joint Research Centre (JRC), European Commission (EC), in order to produce a global human settlement layer (GHSL) from the large volume of optical HR images (Pesaresi et al. 2013).

The China–Brazil Earth Resources Satellite (CBERS)-2B panchromatic imagery at 2.36 m spatial resolution covering mainly the eastern part of China was processed using the JRC work flow to generate GHSL products, which can match the 1:50,000 scale human settlement specifications. The objective of this chapter is to assess the GHSL products generated with CBERS panchromatic data. It is structured as follows:

The 10 m GHSL products are illustrated and compared with several available reference datasets covering representative settlement patterns of China visually.

In order to quantitatively assess the quality of the GHSL product, validation results are calculated by comparing the classification results with manual interpretation of sampled CBERS scenes.

A benchmark experiment is conducted between GHSL at three scales and the moderate resolution imaging spectroradiometer (MODIS) 500 m urban layer with Chinese land use data.

The improvement of fine-scale built-up area identification and quality measurements are analyzed and discussed.

11.2 BACKGROUND

As the most populous country, China has been experiencing unprecedented urbanization with rapid economic growth over the past two decades (Deng et al. 2008). The 2010 population census shows that nearly half of the Chinese population lives in urban areas, an increase of 13.5% since 2000 (National Bureau of Statistics of China 2011). The spatial distribution of China's prefectural-level cities and municipalities grouped by 2010 population size is illustrated in Figure 11.1. Ma et al. (2012) used coarse spatial resolution data, including nighttime data derived from the Defense Meteorological Satellite Program's Operational Linescan System (DMSP/OLS) to measure urbanization dynamics for prefectural-level cities and municipalities from 1994 to 2009. The combination of medium and coarse spatial resolution images was used for settlement mapping at the regional scale by Lu et al. (2008). Liu et al. (2003) produced a national land use/cover database with medium-resolution satellite data including Landsat and CBERS data at the national scale, and these data were used to successfully simulate population distribution (Mao et al. 2012). Taubenböck et al. (2012) analyzed the last 40 years of growth in three megacities consisting of Beijing, Shanghai, and Guangzhou with Landsat and TerraSAR-X data. Yang et al. (2011) extracted HR settlement and population information for local studies. The ZY-3 satellite, launched in 2012, is China's first civil HR satellite, producing 1:50,000 scale map products and updates of 1:25,000 scale maps with settlements as an essential information layer (Tang and Xie 2012). However, efforts are still lacking in creating finer-scale built-up area products with HR remote sensing data over large areas at the national scale.

The first effort to produce a 10 m resolution GHSL covering China was reported by Lu et al. (2013). An image information query (I2Q) system to produce GHSL was developed and implemented by the JRC (Pesaresi et al. 2013). The I2Q system can

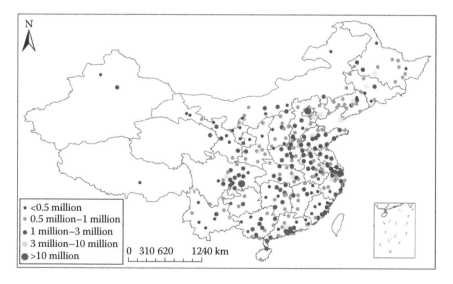

FIGURE 11.1 **(See color insert.)** Spatial distribution of China's prefectural-level cities and municipalities grouped by 2010 population size.

detect and characterize built-up areas based on the average size (scale) of built-up structures from a set of optical remotely sensed imagery having spatial resolutions in the range of 0.5–10 m. The workflow of the I2Q system includes fully automatic image information extraction, which is based on multiscale textural and morphological image feature extraction, generalization, and mosaic (Pesaresi et al. 2013). In this study, a built-up area was extracted with the JRC I2Q system with CBERS-2B HR data as the main input. The CBERS-2B satellite was launched by the CBERS program on September 19, 2007. The HR data are captured with an HR CCD camera onboard at a single spectral band (0.5–0.8 µm) with a swath of 27 km and resolution of 2.36 m. A total of 3810 scenes acquired in 2009 and 2010 were downloaded at a volume of approximately 306 GB. The following sections assess the quality of GHSL_10, GHSL_50, and GHSL_200 products representing GHSL at 10, 50, and 200 m, respectively.

11.3 QUALITY ASSESSMENT

11.3.1 REFERENCE DATASET

The reference dataset used in the quality assessment included three kinds of land use or urban classification products, a vector dataset, and validation samples collected by visual interpretation. A summary of the reference dataset is listed in Table 11.1, and a collection of visual interpretations is introduced in Section 11.3.2.

Landuse2000 data were extracted from Landsat Thematic Mapper (TM) images using a land use/land cover classification scheme appropriate for the 1:100,000 scale (Liu et al. 2003). To produce this map, a total of 508 TM images collected in 1999–2000 covering the entire Chinese territory with supplements of CBERS-1 CCD data in invalid areas were manually interpreted and classified. The final product had an

TABLE 11.1
Reference of Urban or Urban-Related Maps

Abbreviation	Map	Definition of Urban Feature	Resolution
Landuse2000	Chinese Landuse 2000 (Liu et al. 2003)	Three raster layers, ild51, ild52, and ild53, corresponding to the classes "urban settlement," "rural settlement," and "infrastructures," respectively	1 km
MOD500	MODIS 500 m 2002 (Schneider et al. 2009)	Areas dominated by built environment (>50%), including nonvegetated, human-constructed elements, with minimum mapping unit >1 km	500 m
Glob2009	Globcover 2009 (Arino et al. 2010)	Artificial surfaces and associated areas (urban areas >50%)	300 m
Urban2010	Urban vector map (Wang et al. 2012)	Urban built-up areas of 663 cities in China in 2010	1:50,000

average positional accuracy higher than 50 m after orthorectification. The thematic accuracy was higher than 90% compared with field validation. The original format of the land use data was raster files at 1 × 1 km resolution using the cell-based percentage breakdown encoding method. Every file represented a land cover type, and the value of each grid cell in the raster file corresponded to the area of the type of land use in the grid cell. This study used three raster layers, ild51, ild52, and ild53, respectively, corresponding to the classes "urban settlement," "rural settlement," and "infrastructures." The union of all three reference classes was taken as a reference for matching the semantics of the GHSL output. Specifically, the reference estimation of the built-up percentage was derived by the formula y = (ild51 + ild52 + ild53)/10,000.

The MOD500 data are a global distribution of urban land use at 500 m spatial resolution using remotely sensed data from MODIS with a supervised decision tree classification algorithm (Schneider et al. 2009, 2010). MOD500 maps the 500 × 500 m cells as built-up if more than 50% of the land is covered by built-up structures. An accuracy assessment based on sites from a stratified random sample of 140 cities shows that the new map has an overall accuracy of 93% at the pixel level.

Glob2009 is a 300 m global land cover map derived by an automatic and regionally tuned classification of a time series of global Medium Resolution Imaging Spectrometer Instrument (MERIS) Fine Resolution mosaics for the year 2009 (Arino et al. 2010). The global land cover map counts 22 land cover classes defined with the United Nations Land Cover Classification System. The map projection is a Plate-Carrée (WGS84 ellipsoid). The overall accuracy weighted by the class area reaches 67.5% using 2190 points globally distributed and including homogeneous and heterogeneous landscapes.

The Urban2010 data are urban built-up areas mapped with manual interpretation of Landsat TM/ETM + data from 2008 to 2010. All of the 663 cities were manually interpreted. For accuracy analysis, 5% of the cities were randomly selected based on HR imagery on Google Earth. The area differences were less than 10% with the same mapping standard.

11.3.2 VISUAL VALIDATION

Visual validation was conducted on the original CBERS panchromatic imagery. The original CBERS data were warped with coordinates using affine transformation and the nearest-neighbor resampling algorithm in ArcGIS9.3. The reference data collection includes the following steps: (1) collection of spatial samples by a systematic grid procedure, (2) development of a stratified sampling schema, and (3) interpretation of each sample by visual inspection of the corresponding part of the image.

The surrounding was defined in this study by assuming an output target nominal scale of 1:50,000 and calculating an admitted displacement error with a radius of 25 m. This gave a generalization grid of 50 × 50 m cells. A stratified sampling schema of CBERS scenes was applied to represent scenes with different building densities. The building density of each scene was calculated with the Landuse2000 built-up density index. Image scenes of different built-up densities including low, medium, and high density were selected. For each scene, about 72 grids were delineated. For each selected grid, 16 cells were interpreted. Given the adopted dichotomic protocol, the interpreter was asked to check if the specific subsample was intersecting a visible building in the image with four possible coded answers: yes, no, not sure, and no data available. A classification of "not built-up" or "0" indicates that the cell does not contain any building, while "built-up" or "1" means the cell contains one building or part of a building.

The standardized GHSL output was the continuous built-up presence index. For each 50 × 50 m cell, the maximum built-up index was extracted. The maximum built-up index can be dichotomized as built-up or not built-up by selecting a threshold. For the threshold selection, an optimization model can be formulated to test the recognition performances of all the possible thresholds in the available range if training samples are collected (Pesaresi et al. 2011). In this study, the threshold was selected by human interaction, visually comparing the GHSL output with the original image. A threshold of 190 was chosen to differentiate built-up and not built-up in the final GHSL product. A confusion matrix produced by comparing the polygons generated by the threshold of the built-up presence index and the reference dataset was produced after manual interpretation.

11.3.3   BENCHMARK EXPERIMENT SETTING

Although a visual validation protocol is highly accurate, it is also very time consuming. The samples interpreted with the adopted protocol covered only a small fraction of the territory, and the complete collection of each satellite scene used in the experiment would require several years of visual reference data collection for an image interpreter. To overcome this issue, a systematic evaluation of the results of the automatic image information retrieval was carried out and the Landuse2000 data were used as a reference.

The classes considered in this benchmarking exercise include built-up areas representing cities (class 5.1), villages (class 5.2), and infrastructure (class 5.3). The information was made available as density information at a grid with 1000 × 1000 m cells and will be referred to herein as Ref_51, Ref_52, and Ref_53. For this analysis, we used Ref_51 and Ref_52. We also combined classes 51, 52, and 53 into a single file referred to as Ref_Tot.

The information sources benchmarked include the MOD500 urban layer and the built-up information from the GHSL. The 500 × 500 m MODIS cells were recoded to 1000 × 1000 m grid cells corresponding to that of the reference information and herein will be referred to as MODIS. The GHSL information makes available the built-up area information at resolutions of 10, 50, and 200 m. For consistency with the reference, the GHSL information was recoded to 1000 × 1000 m, referred to herein as GHSL_10, GHSL_50, and GHSL_200. The GHSL built-up area products do not cover the entirety of the territory of eastern China. CBERS data were available only in part, and the resulting GHSL covers the territory discontinuously. A comparison with the reference and MODIS was conducted only for the area for which GHSL data were available, as shown in Figure 11.2.

Two types of comparisons were carried out. One was based on the linear regression between the continuous values in the reference and target layers and one on the agreements between binary maps. The production of binary maps is addressed in the next paragraph. The comparison based on linear fit regression aimed to provide an

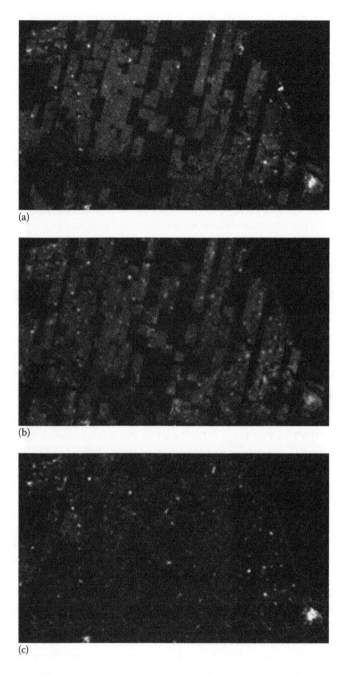

FIGURE 11.2   Reference (a), GHSL (b), and MODIS (c) datasets used in the analysis. Bottom right is the city of Shanghai.

in-depth understanding of the density produced by MODIS and GHSL layers and those that better match the values in the reference. The comparison of binary maps aimed to provide an agreement on the extent of built-up land as areal units. In fact, a percentage computed on an actual binary built-up layer allows derivation of an areal measure that can be expressed in hectares or square kilometers. The areal built-up estimate also requires computing the relative omission and commission errors based on agreement measures.

The binary maps were produced by defining a threshold within the density continuum of target and reference data layers. For the Ref_total map and the Ref_51 map, we assigned a threshold of 0.5 corresponding to 50% of the cell covered by built-up land. Cell values were assigned a value of 0 when density was less than 50% and 1 when cells had a density larger or equal to 50%. For the Ref_52 map (villages), the selected threshold was 10%, which means cells with a built-up density greater than 10% were assigned a value of 1. The target datasets were also assigned a threshold analogous to the reference data to which they were compared.

11.3.4 QUALITY MEASUREMENT

For quantitative evaluation of the results, the following notations were adopted. "True positive" (Tp) and "true negative" (Tn) samples are samples detected as "built-up" (BU) and "not built-up" (NBU), respectively, by the automatic recognition procedure and then confirmed by visual inspection. "False positive" (Fp) samples were samples classified as BU by automatic recognition and labeled as NBU by visual inspection. Correspondingly, "false negative" (Fn) samples were samples classified as NBU by automatic recognition and labeled as BU by visual inspection. Then, given a total number N of samples, we defined the overall accuracy agreement as $A = (Tp + Tn)/N$, the producer's accuracy as $PA_{bu} = Tp/(Tp + Fn)$, $PA_{nbu} = Tn/(Tn + Fp)$, and the user's accuracy as $UA_{bu} = Tp/(Tp + Fp)$, $UA_{nbu} = Tn/(Tn + Fn)$. For this benchmarking experiment, we only used the BU accuracy as a quality measurement, defined as $Tp/(Tp + Fp)$.

11.4 RESULTS AND DISCUSSION

11.4.1 OVERVIEW OF THE FINE-SCALE CHINESE GHSL

Most of the cities with a high population density are distributed in eastern China (see Figure 11.1). The CBERS imagery processed in the GHSL production experiment covers most of the populated areas in eastern China (Figure 11.3). A total of 2288 CBERS panchromatic scenes cover 1.67 million km², which is about 18% of Chinese territory. Validation samples are randomly, evenly distributed over the whole dataset. The GHSL of the Beijing metropolitan area and corresponding HR imagery are illustrated in Figure 11.4. It can be seen that various building patterns in the metropolitan areas, including departments, commercial buildings, and airports, are recognized as built-up with high values. By contrast, green spaces, croplands, and water bodies have low values.

There are a variety of settlement patterns over the large territory of China. These patterns as detected through the GHSL procedure in three Chinese cities are

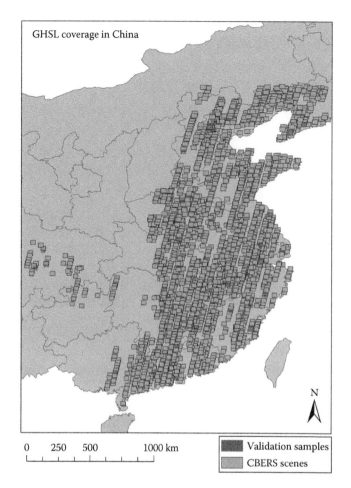

FIGURE 11.3 **(See color insert.)** Geographic distribution of the CBERS HR input images processed during the experiment. Sample grids for visual validation of GHSL are marked in red.

depicted at 1:50,000 scale in Figure 11.5. The cities range in size from 3,000 km² (Longquan in Zhejiang Province) to 16,000 km² (Beijing, the capital city). The GHSL data covering all cities were compared with available reference datasets. Comparing the images visually, Glob2009 underestimates the extent of metropolitan areas of a city. Small villages have been identified in Landuse2000 data because it is based on the Landsat data, which has much finer spatial resolution than MODIS data. Landuse2000 and GHSL show that urbanization occurred in all the cities from 2000 to 2010. The GHSL products draw the exact picture of the metropolitan area and also the surrounding small villages and towns. The improvement of spatial details is highly visible. Overall, these observations demonstrate the diversity of global urban maps and the improvement of GHSL products.

In this study, the GHSL product was compared with global urban maps. In disaster risk assessment tasks, the national land cover/land use databases are often used as a

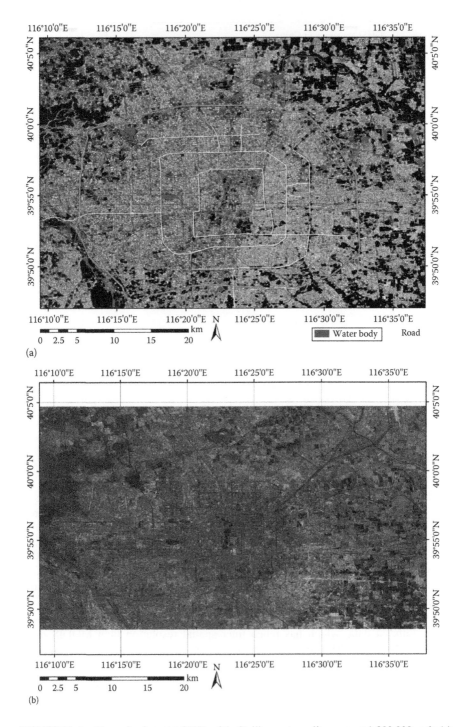

FIGURE 11.4   **(See color insert.)** GHSL of the Beijing metropolitan area at 1:200,000 scale (a) and high-resolution images from Bing Maps aerial view (b).

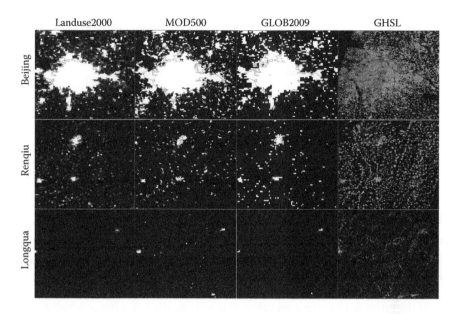

FIGURE 11.5 Comparison of GHSL with other global urban maps over various settlement patterns covering different parts of China.

source for exposure information extraction. The settlement information from national databases at different scales including 1:250,000, 1:100,000, and 1:50,000 should also be acquired and integrated for the GHSL quality assessment at the next step.

11.4.2 ACCURACY ASSESSMENT

Quantitative validation of the quality of the 10 m GHSL output against visual reference data relies on validation samples covering sample output (ca. 50 of the 2000 scenes). The samples were collected from the same 2.5 m resolution CBERS-2B panchromatic imagery used for automatic information retrieval. A total of 59,680 samples were collected using the GHSL reference data collection protocol (Pesaresi et al. 2013). The total ground surface processed employing this visual interpretation protocol was over 149 km². In the sampling sets, 41,946 samples including 4,701 built-up and 37,245 not built-up samples were valid. Others were samples falling in areas outside the scene and clouds or other occlusions and sensor failures.

The results of the quantitative analysis for the accuracy assessment are presented in Table 11.2, which shows the estimation of the commission and omission errors and the producer and user accuracies obtained by matching the results of GHSL classification with the reference dataset. The overall accuracy of the built-up and not built-up classes were 87.04%. The accuracy for the not built-up class was higher than 85%. The producer's accuracy for built-up areas was 74.68% while the user's accuracy was only 32.68%. It is similar to the result of the benchmarking experiment, which might be attributed to the resolution as discussed in the following section.

TABLE 11.2

Commission and Omission Errors and Producer's and User's Accuracy

Output Class	Prod. Acc	User Acc.
Built-up	0.7468	0.3268
Not built-up	0.8800	0.9781

Note: Overall accuracy = 0.8704.

11.4.3 GHSL Benchmark

The benchmark experiment described in Section 11.3.3 provides the following datasets for comparison. The linear regression between the continuous values in reference and target layers is listed in Table 11.3, and a comparison of binary maps is given in Table 11.4.

For the linear fit of continuous values (Table 11.3), the following conclusions can be observed:

- For Ref_51, MODIS and GHSL_50, both obtained the highest correlation value of 0.57, GHSL_200 and GHSL_10 also obtained relatively high values.
- For Ref_52, the correlation with the three GHSL layers was in general lower than for class 51.
- For MODIS, the linear correlation was insignificant.

TABLE 11.3

Linear Correlation of Density Values

Comp.	Reference	Target	Linear Correlation
1	Ref_51	MODIS	0.57
2		GHSL_200	0.52
3		GHSL_050	0.57
4		GHSL_010	0.51
5	Ref_52	MODIS	0.12
6		GHSL_200	0.33
7		GHSL_050	0.36
8		GHSL_010	0.35
9	Ref_Tot	MODIS	0.35
10		GHSL_200	0.46
11		GHSL_050	0.49
12		GHSL_010	0.44

TABLE 11.4

Agreement Measures between Binary Built-Up Layers

Comp.	Reference Dataset	Target Dataset	Bu_Acc
1	Ref_51_r_0.5	MOD_500_r0.5	0.58
2		GHSL_200_r0.5	0.81
3		GHSL_050_r0.5	0.69
4		GHSL_010_r0.5	0.03
5	Ref_52_r_0.1	MOD_500_r0.1	0.08
6		GHSL_200_r0.1	0.71
7		GHSL_050_r0.1	0.64
8		GHSL_010_r0.1	0.27
9	Ref_Tot_r_0.5	MOD_500_r0.5	0.37
10		GHSL_200_r0.5	0.65
11		GHSL_050_r0.5	0.70
12		GHSL_010_r0.5	0.02

- For Ref_Tot, the highest correlation in absolute terms was obtained with GHSL_50 and the second highest with GHSL_200. GHSL_010 performed unexpectedly poorly.
- For the binary comparison (Table 11.4), the best agreement was for GHSL_200 with the second best being GHSL_50 for Ref_51. MODIS ranked third, while GHSL_010 obtained a low accuracy value.
- For Ref_52, GHSL_200 resulted in the highest accuracy. MODIS showed very low agreement.
- For Ref_Tot, GHSL_50 and GHSL_200 resulted in high accuracy values and GHSL_10 and MODIS resulted in extremely low accuracy values.

In summary, all GHSL scales rank better than MODIS for the linear fit. MODIS ranked best together with GHSL_50 for estimating cities (Ref_51), MODIS ranked worst for estimating villages (Ref_52), and GHSL_50 ranked best for estimating villages (Ref_52). GHSL_50 was also the best estimator of the combination of cities and villages (Ref_Tot). For the binary comparison, MODIS compared well only for cities (Ref_51) and with values lower than those of GHSL_200. GHSL 200 showed a high degree of agreement in both city (Ref_51) and village estimation (Ref_52). GHSL_50 resulted in the best agreement with the combination of cities and villages (Ref_Tot). GHSL_010 did not perform as well for all three classes.

The disagreement between GHSL_010 and the reference might be caused by the different observation scales of GHSL and that of the reference (Landsat) at 10 m resolution. The GHSL at 50 and 200 m resolution obtained high accuracy values in both continuous and binary value comparison. The aggregation of GHSL from 10 to 50 and 200 m increased the agreement between GHSL and the reference. GHSL at 200 m resolution performed better for binary values, while GHSL at 50 m resolution was better for continuous values. This may be attributed to the use of a global threshold, which can cause errors for built-up area classification. The threshold applied after pixel aggregation increased the accuracy of built-up area detection. MODIS data can get comparative performance for cities but shows the lowest values for villages. This indicates that the MODIS urban layer can represent built-up areas of cities although it barely represents built-up areas of villages and towns.

Despite its effectiveness, the benchmark experiment can be further improved. Urban area products with optimized spatial and temporal features can be used as reference data. For example, the fact that MODIS data do not include information on infrastructures leads to its low correlation with Ref_Tot. Considering the dramatic urbanization processes between 2000 and 2010 in eastern China, the temporal difference between GHSL and reference data can cause discrepancy. The 30 m resolution global land-cover maps using Landsat TM data in 2009 and 2010 developed by Gong et al. (2013) can be a promising data source. In addition, the selection of different thresholds should be compared since they have a great influence on the assessment results.

11.5   CONCLUSIONS

The GHSL product covering eastern China was produced using CBERS HR data with the JRC I2Q framework. The quality assessment of this product was conducted using visual validation samples and reference datasets. Comparing the urban area products over different settlement patterns, the 10 m GHSL product is a great improvement to the state-of-the-art products. Based on visual validation, the 10 m GHSL had an overall accuracy of 87.04% and the user's accuracy of not built-up areas was 97.81%. From the benchmarking experiment, all GHSL scales ranked better than MOD500 data. In particular, the GHSL at 50 and 200 m resolution showed a high degree of accuracy in both city and village estimation.

As an initial effort of GHSL production for China, the presented experiment proved the effectiveness of HR images for built-up area delineation over a variety of settlement patterns. For further improvement of the GHSL products, images from other Chinese HR satellites such as ZY-3, ZY-2C, and GF-1 can also be integrated into the procedure. The GHSL products will be helpful for population estimation and exposure mapping in disaster risk assessment in China. Information on built-up areas at different times can be applied to depict the evolution of urban extents and analyze the urbanization process in China.

ACKNOWLEDGMENTS

The authors appreciate their colleagues in the JRC who contributed to the production of the Chinese GHSL and the anonymous reviewers for their constructive comments. The Landuse2000 dataset was obtained from the Data Center of Resources and Environment at the Chinese Academy of Sciences. This research was supported by the Major International Cooperation and Exchange Project of the National Natural Science Foundation of China "Comparative study on global environmental change using remote sensing technology" under grant No. 41120114001.

REFERENCES

Arino, O., Ramos, J., Kalogirou, V., Defourny, P., and Achard, F., 2010. GlobCover 2009, *Proceedings of the Living Planet Symposium*, SP-686, Bergen, Norway, June 2010.

Bal, I.E., Bommer, J., Stafford, P., Crowley, H., and Pinho, R., 2010. The influence of geographical resolution of urban exposure data in an earthquake loss model for Istanbul. *Earthquake Spectra*, 26L, 619–634.

Deng, X., Huang, J., Rozelle, S., and Uchida, E., 2008. Growth population and industrialization and land expansion of China. *Journal of Urban Economics*, 63, 96–115.

EEA (European Environment Agency), 2010. *The European Environment—State and Outlook 2010: Land Use*, Publications Office of the European Union, Luxembourg, p. 52.

Ehrlich, D. and Tenerelli, P., 2013. Optical satellite imagery for quantifying spatio-temporal dimension of physical exposure in disaster risk assessments. *Natural Hazards*, 68(3), 1271–1289. DOI: 10.1007/s11069-012-0372-5.

Ehrlich, D., Zeug, G., Gallego, J., Gerhardinger, A., Caravaggi, I., and Pesaresi, M., 2010. Quantifying the building stock from optical high-resolution satellite imagery for assessing disaster risk. *Geocarto International*, 25(4), 281–293.

Esch, T., Taubenböck, H., Roth, A., Heldens, W., Felbier, A., Thiel, M., Schmidt, M., Müller, A., and Stefan, D., 2012. Tandem-x mission—New perspectives for the inventory and monitoring of global settlement patterns. *Journal of Applied Remote Sensing*, 6(1), 1–21.

Esch, T., Thiel, M., Schenk, A., Roth, A., Mehl, H., and Dech, S., 2010. Delineation of urban footprints from TerraSAR-X data by analyzing speckle characteristics and intensity information. *IEEE Transactions on Geoscience and Remote Sensing*, 48(2), 905–916.

Gong, P., Wang, J., Yu, L. et al., 2013. Finer resolution observation and monitoring of global land cover: First mapping results with Landsat TM and ETM+ data. *International Journal of Remote Sensing*, 34(7), 2607–2654.

Guo, H., 2010. Understanding global natural disasters and the role of earth observation. *International Journal of Digital Earth*, 3(3), 221–230.

ISDR (International Strategy for Disaster Reduction), 2009. United Nations international strategy for disaster risk reduction. Terminology on Disaster Risk Reduction. http://www.unisdr.org/we/inform/terminology. Accessed on February 28, 2013.

Liu, J., Liu, M., Zhuang, D., Zhang, Z., and Deng, X., 2003. Study on spatial pattern of land-use change in China during 1995–2000. *Science in China D—Earth Sciences*, 46(4), 373–384.

Lu, D., Tian, H., Zhou, G., and Ge, H., 2008. Regional mapping of human settlements in southeastern China with multisensory remotely sensed data. *Remote Sensing of Environment*, 112, 3668–3679.

Lu, L., Guo, H., Pesaresi, M., Soille, P., and Ferri, S., 2013. Automatic recognition of built-up areas in China using CBERS-2B HR data. *Proceedings of the 2013 Joint Urban Remote Sensing Event (JURSE)*, 65–68.

Ma, T., Zhou, C., Pei, T., Haynie, S., and Fan, J., 2012. Quantitative estimation of urbanization dynamics using time series of DMSP/OLS nighttime light data: A comparative case study from China's cities. *Remote Sensing of Environment*, 124, 99–107.

Mao, Y., Ye, A., and Xu, J., 2012. Using land use data to estimate the population distribution of China in 2000. *GIScience & Remote Sensing*, 49(2), 228–250.

Ministry of Civil Affairs of the People's Republic of China, 2011. *China Civil Affairs Statistic Year Book 2011*, China Statistic Press, Beijing, China.

Miyazaki, H., Shao, X., Iwao, K., and Shibasaki, R., 2013. An automated method for global urban area mapping by integrating ASTER satellite images and GIS data. *IEEE Journal of Selected Topics in Applied Earth Observations and Remote Sensing*, 6(2), 1004–1019.

National Bureau of Statistics of China, 2011. 2010 Sixth national population census data gazette (No. 1). http://www.stats.gov.cn/tjfx/jdfx/t20110428_402722253.htm. Accessed on February 28, 2013.

Pesaresi, M. and Ehrlich, D., 2009. A methodology to quantify built-up structures from optical VHR imagery. In *Global Mapping of Human Settlement Experiences, Datasets, and Prospects*, P. Gamba and M. Herold, eds. CRC Press, Boca Raton, FL, Chapter 3, pp. 27–58, DOI: 10.1201/9781420083408-c3.

Pesaresi, M., Ehrlich, D., Caravaggi, I., Kauffmann, M., and Louvrier, C., 2011. Toward global automatic built-up area recognition using optical VHR imagery. *IEEE Journal of Selected Topics in Applied Earth Observations and Remote Sensing*, 4, 4923–4934.

Pesaresi, M., Guo, H., Blaes, X. et al., 2013. A global human settlement layer from optical HR/VHR RS data: Concept and first results. *IEEE Journal of Selected Topics in Applied Earth Observations and Remote Sensing*, 6(5), 2102–2131.

Schneider, A., Friedl, M.A., and Potere, D., 2009. A new map of global urban extent from MODIS satellite data. *Environment Research Letter*, 4, 044003. DOI: 10.1088/1748-9326/4/4/044003.

Schneider, A., Friedl, M.A., and Potere, D., 2010. Monitoring urban areas globally using MODIS 500 m data: New methods based on urban ecoregions. *Remote Sensing of Environment*, 114(8), 1733–1746.

Tang, X. and Xie, J., 2012. Overview of the key technologies for high resolution satellite mapping. *International Journal of Digital Earth*, 5(3), 228–240.

Taubenböck, H., Esch, T., Felbier, A., Wiesner, M., Roth, A., and Dech, S., 2012. Monitoring urbanization in mega cities from space. *Remote Sensing of Environment*, 117, 162–176.

Wang, L., Li, C.C., Ying, Q. et al., 2012. China's urban expansion from 1990 to 2010 determined with satellite remote sensing. *Chinese Science Bulletin*, 57, 2802–2812. DOI: 10.1007/s11434-012-5235-7.

Yang, X., Jiang, G., Luo, X., and Zheng, Z., 2011. Preliminary mapping of high-resolution rural population distribution based on imagery from Google Earth: A case study in the Lake Tai basin, eastern China. *Applied Geography*, 32, 221–227.

12 Climatological and Geographical Impacts on the Global Pandemic of Influenza A (H1N1) 2009

Bing Xu, Zhenyu Jin, Zhiben Jiang, Jianping Guo,
Michael Timberlake, and Xiulian Ma

CONTENTS

12.1 INTRODUCTION

A new subtype of influenza A virus, H1N1 of swine, human, and avian origin, emerged in the United States and Mexico in April of 2009, and quickly and extensively spread around the globe through human-to-human transmission. As of August 1, 2010, worldwide more than 214 countries have reported laboratory-confirmed cases, including over 18,449 deaths (WHO 2010). Billions of dollars have been spent on preventing and controlling the global influenza pandemic of this virus. However, the mechanisms of its spread remain poorly understood. A better understanding of the relationships between the transmission of this novel H1N1 virus and the seasonal environmental changes, global patterns of human travel, and other physical and social factors will provide useful information for developing effective control strategies and improvements in public health preparedness and emergency responses.

Previous studies indicate that humidity and temperature affect both the survival and transmission of influenza viruses. Investigations of the effect of relative humidity and temperature on influenza virus transmission among guinea pigs indicated that both cold and dry conditions favor transmission (Lowen et al. 2007). However, some other studies have indicated that peak periods of influenza A and B occur during the rainy and hot seasons in Western India, Dakar, Senegal, and Northeast Brazil (Weber and Stilianakis 2008). A recent study found that absolute humidity constrains both transmission efficiency and influenza virus survival to a much greater extent than relative humidity (Shaman and Kohn 2009).

Undoubtedly, patterns of transmission of the A(H1N1) influenza virus are related not only to the physical environment but also to human social activities and connectivity on various spatial and temporal scales (Jiang et al. 2012). Driven by increased productivity and the advancement of science and technology, particularly technological advancement in the information, communication, and transportation sectors, information and material exchanges have been dramatically intensified, leading to well-connected international networks of economies, science, technology, and culture (Xu et al. 2013). Globalization appears to be causing profound, sometimes unpredictable, changes in the ecological, biological, and social conditions that shape exposure to infectious diseases in certain populations (Saker et al. 2004; Wu et al. 2013). Globalization, manifest in the increasing frequency and velocity in the circulation of commodities, people, and ideas, has fundamentally altered the patterns of virus spread and intensified the level of transmission. Human migration and travel have been sources of epidemics throughout history. It is well recognized that human and cargo traffic facilitates the movement of pathogens from place to place across the world (Aron and Patz 2001). Population mobility produces health outcomes that can affect the migrant and host populations either positively or negatively (Xu et al. 2006).

We wanted to estimate the surviving ability of the virus by examining its living environment as we did for other disease vectors (Seto et al. 2002; Xu et al. 2003, 2004). In this study, we first explored the relationship between climatological conditions and the number of A(H1N1) cases worldwide. Then we used the latest available tourism flow data across countries and air passenger data across cities worldwide to describe the level of flows of the human population over the earth's surface and to infer the potential for virus transmission.

12.2 DATA AND METHOD

12.2.1 GEOCODED INFLUENZA A (H1N1) OUTBREAK DATA

We cross-checked and compiled geocoded outbreak data downloaded from http://www.mapcruzin.com*. This global geographic A(H1N1) dataset dating from April 20 to May 24, 2009, registered each outbreak event as located at the centroid of its nearest city or county. We estimated most of the positional registration errors in densely populated areas to be within 50 km.

---

* http://www.mapcruzin.com. H1N1 novel swine flu ArcGIS shapefile and data. The data behind were compiled by Dr. Henry Niman, a biomedical researcher in Pittsburgh, Pennsylvania, using technology provided by Rhiza Labs and Google. Accessed July 7, 2013.

12.2.2 Climatological Data Processing and Modeling

The climatological data are extracted from the MERRA* data. The MERRA (Modern-Era Retrospective Analysis for Research and Applications) dataset is reanalyzed data from the NASA Goddard Earth Observing System Data Assimilation System Version 5 (GEOS-5). Retrospective analyses integrate a variety of observing systems with numerical models to produce a temporally and spatially consistent synthesis of observations and analyses of variables not easily observed. MERRA data were derived by combining all available global surface observations every 3 h and interpolating the surface. Daily climate variables, including air temperature, relative humidity, atmospheric pressure, and precipitation at a spatial resolution of half-degree latitude by two-third-degree longitude, composed the input dataset. Using the date and location of the outbreak, we extracted the corresponding temperature, relative humidity, atmospheric pressure, and precipitation at that location.

Absolute humidity is defined as vapor density or vapor concentration. In a system of moist air, it is the ratio of water vapor present to the volume occupied by the mixture, namely, the density of the water vapor component. Relative humidity is defined as the ratio of the vapor pressure to the saturation vapor pressure with respect to water (Dai 2006). Absolute humidity is described by the following equations:

$$\rho_\omega = \frac{e}{4.615 \cdot T} \tag{12.1}$$

$$e = r \cdot e_s \tag{12.2}$$

$$e_s = \begin{cases} 6.1078 \ \exp\left[\dfrac{17.2693882\,(T - 273.16)}{T - 35.86}\right], & \text{if } T \geq 273.16 \\[4mm] 6.1078 \ \exp\left[\dfrac{21.8745564\,(T - 273.16)}{T - 7.66}\right], & \text{otherwise} \end{cases} \tag{12.3}$$

where
  ρ_ω is the absolute humidity (kg/m^3)
  e is the vapor pressure (hPa)
  e_s is the saturation vapor pressure (hPa)
  T is the temperature (K)
  r is the relative humidity (hPa/hPa)

We calculated the absolute humidity and found that its relationship with the number of confirmed cases occurring on each particular day and location was more significant than that of relative humidity. The frequency curves of temperature, absolute

---

* MERRA, Modern-Era Retrospective Analysis for Research and Applications, were provided by the Global Modeling and Assimilation Office (GMAO) at NASA Goddard Space Flight Center through the NASA GES DISC online archive. http://gmao.gsfc.nasa.gov/research/merra/intro.php. Accessed July 7, 2013.

humidity, and precipitation all conformed to normal distributions. Thus, we applied a moving average smoother to each frequency plot, and fitted the data with nonlinear Gaussian functions. We first processed the daily temperature, absolute humidity, and precipitation data on a month-by-month basis from January 1 to December 31, 2008. Their corresponding normal distribution curves were then applied, and finally, predicative risk maps for each month were obtained by calculating the average of the three density function values. Analysis of the global outbreaks of the influenza has given rise to a new understanding of the characteristic physical conditions that favor the survival and transmission of the virus.

12.2.3 COUNTRY-LEVEL TOURISM FLOW DATA SUMMARY

The World Tourism Organization (UNWTO) publishes Tourism Satellite Account (TSA) data that are compiled by countries around the world (UNWTO 2009). We compiled nonresident tourist data at national borders, including data on the number of tourists arriving by air, land, and sea. This tourist category includes travelers taking trips to major destinations outside their usual environment other than to be employed by resident entities in the countries visited. We averaged the total number of departures and arrivals of tourists at the country level and normalized the data, defining this as the tourism index.

12.2.4 SPATIAL POINT PATTERN ANALYSIS

On a global scale, outbreaks can be considered as point events in space. A point pattern analysis of A(H1N1) outbreaks was used here to determine where the cases were spread as well as to determine which spatial scales are optimal for disease clustering. Choosing the right scale is critical for subsequent analyses. We used exploratory spatial statistical techniques to examine the patterns of A(H1N1) outbreaks. Ripley's K function describes how the expected value of a point process changes over different spatial lags (Gatrell et al. 1996). A peak value in Ripley's K function indicates a clustering at the scale of the corresponding lag. Because our analysis here was on a global scale, the spatial lag is in spherical distance.

An estimate of the spatial K function can be calculated by (Ripley 1981):

$$K(d) = \frac{R}{n^2} \sum_{i=1}^{n} \sum_{j \neq i} I_d(i, j) \tag{12.4}$$

where
  R is the total area of the study
  n is the number of observed events
  $I_d(i, j)$ is an indicator function that takes the value of 1 when the spherical distance between point i and point j is less than d

In clustering, $K(d)$ would be greater than $A(d)$, which is the area of the spherical circle of arc-radius d, and less than $A(d)$ under regularity. We then apply a transformation to $K(d)$ to have $L(d) = R * \arccos(1 - (K(d)/2\pi R^2)) - d$ and plot $L(d)$ against d.

The upper and lower bounds are determined by undertaking Monte Carlo simulations 20 times. For each simulation, we generate the same number of random points as the cases, and then calculate their K functions. To each lag, the upper and lower bounds are the minimum and maximum K values among the set of 20 simulations.

12.2.5 CITY-LEVEL AIR PASSENGER FLOW NETWORK ANALYSIS

Global intercity air passenger flow data were obtained from the International Civil Aviation Organization (ICAO), which provides estimates of the number of air passengers traveling between pairs of cities on commercial airlines (ICAO 1990). We first compiled a city-level matrix-based network dataset that provided information on the origin–destination passenger travel. The network consisted of the 338 cities and 2284 edges that connect major cities worldwide. The topology of the network can be symbolized by a 338 × 338 connectivity matrix C, whose entry C_{ij} is 1 if there is a link pointing from node i to node j. The in-degree and out-degree of a given node in the network is defined as the number of flowing-in and flowing-out connections of the city. The passenger flow intensity was modeled by another 338 × 338 matrix, where each row–column combination represented the volume of passenger travel or the amount of connectivity between cities. These two matrices allowed us to measure the networks and analyze their structural properties. We were able to obtain the degree and the intensity of connectivity among cities, and to measure the significance of city and network linkages.

The clustering coefficient indicates the position of a node in its neighborhood. The neighborhood Γ_υ of a given city υ is a graph, which includes all nodes that have flights to and from υ. The clustering coefficient $C\left(\Gamma_\upsilon\right)$ of neighborhood Γ_υ of city υ characterizes the extent to which cities in Γ_υ are connected to every other city (Watts and Strogatz 1998; Li and Cai 2004).

$$C\left(\Gamma_\upsilon\right) = \frac{E\left(\Gamma_\upsilon\right)}{C_m^2} \tag{12.5}$$

where

$E\left(\Gamma_\upsilon\right)$ is the number of real connections in Γ_υ consisting of m cities

C_m^2 is the total number of all possible connections in Γ_υ

It tells us how well connected the neighborhood of the node is. If the neighborhood is fully connected, the clustering coefficient is 1, and a value close to 0 means that there are far fewer connections in the neighborhood compared to the node itself.

Air traffic data are known to be symmetrical with a nearly perfect correlation between in-degree and out-degree (Li and Cai 2004; Ma and Timberlake 2008). In our case, the bivariate correlation was as high as 0.996. The mean values of out-degree and out-flow passengers were 6.8 and 882,434, respectively. This indicates that each city, on average, is connected to around seven other cities and that the average in/out flow of passengers for each city was somewhere close to 1 million annually.

12.3   RESULTS AND DISCUSSION

We have characterized the circulation of the current A(H1N1) virus from climatological and geographical perspectives. We have used both physical and social data to understand the roles that these factors have played in the recent global pandemic of A(H1N1) influenza.

We first explored the climatological impact on the H1N1 pandemic. We derived the absolute humidity from the relative humidity, atmospheric pressure, and temperature and constructed a frequency plot for each climatological variable (Figure 12.1). Then we applied a moving average smoother to each frequency curve and fitted the data with the nonlinear Gaussian functions *Ntemperature*, *Nabs_humidity*, *Nprecipitation*, and *Nrel_humidity*, giving means and standard deviations of 18.28 and 6.07, 9.8 and 4, 20.2 and 10.52, and 75.76 and 23.36, respectively.

We found that absolute humidity was a more significant predictor than relative humidity. The correlation coefficient and the root mean squared error using absolute humidity to predict the number of cases are 0.7 and 119.4, in comparison with those using relative humidity, 0.6 and 122.42, respectively. We processed the daily temperature, absolute humidity, and precipitation data from January 1 to December 31, 2009. The predicative risk maps for each month were obtained and visualized (Figure 12.2).

Therefore, seasonal patterns of human exposure to the A(H1N1) virus at a particular location can be inferred by seasonal variation in climatological conditions, such as the absolute humidity, precipitation, and temperature, at that location. These variables can be used as physical predictors for assessing the spatial and temporal distribution of disease exposure risk. Here, the disease exposure risk was defined as the average frequency score from the frequency values found on the fitted curves in Figure 12.1 using these predictors (Figure 12.2).

From Figure 12.2, we can clearly see seasonal trends in the predicted risks of influenza outbreaks. High-risk areas in the northern hemisphere gradually move northward between April and August from the subtropics to the cold temperate regions, such as Canada, Europe, areas in the Far East, and the west coast of the United States, and then gradually move southward between November and February to the subtropics, such as Mexico, Florida, India, and parts of South Asia. In the southern hemisphere, in regions such as the west coast and eastern part of South America, the northern part of Australia, and the mid-south of Africa, we see a similar pattern of northerly movement between April and August, except that the amplitude of the movement is reduced. A pattern of southerly movement between November and February is observed in regions such as the southwest coastal areas of South America, Africa, the southern part of Australia, and New Zealand. Once again, the amplitude of movement is smaller than that observed in the northern hemisphere. It is noteworthy that tropical regions around the equator such as most of Southeast Asia including Indonesia, Malaysia, Singapore, the Philippines, and Thailand were not exposed to high risks throughout the year. We hypothesize that this may be due to the fact that absolute humidity is greater than 20 g/m^3 and precipitation is greater than 44 mm per day all year round and that the corresponding risk for these high values was low according to the fitted normal distribution curves. Climatological conditions in these regions are predicted to be unfavorable for harboring the virus. However, we are still uncertain

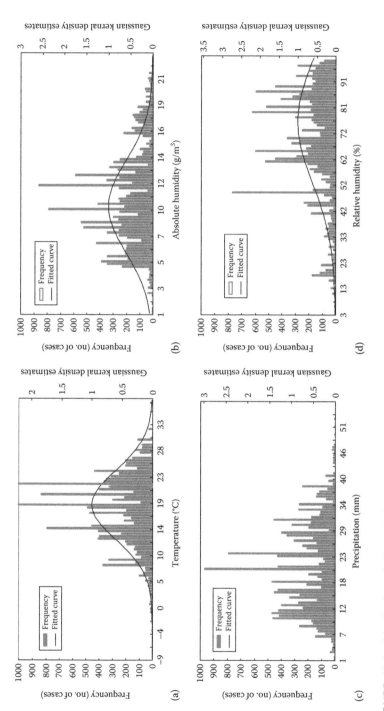

FIGURE 12.1 Distribution of climatological factors during the observation period. Frequency plots for (a) temperature, (b) absolute humidity, (c) precipitation, and (d) relative humidity. Superimposed on the frequency plots are the kernel density estimates using Gaussian distribution curves with (a) *Ntemperature* (18.28, 6.07), (b) *Nabs_humidity* (9.8, 4), (c) *Nprecipitation* (20.2, 10.52), and (d) *Nrel_humidity* (75.76, 23.36). Figures in parentheses represent the means and standard deviations for each function. Climatological data were obtained from MERRA data provided by the NASA Goddard Earth Observing System Data Assimilation System Version 5 (GEOS-5).

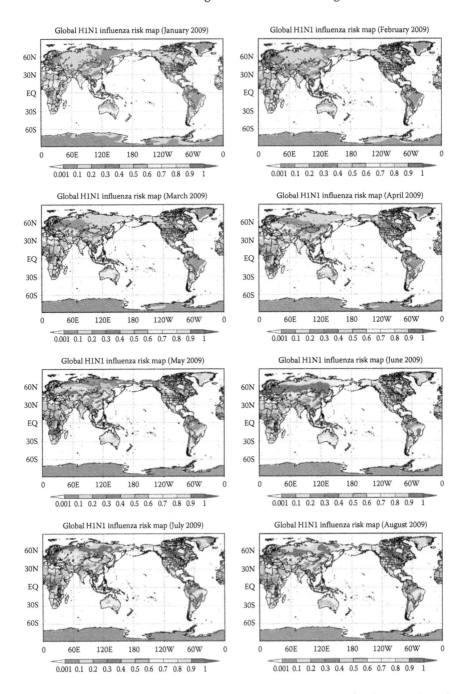

FIGURE 12.2 (See color insert.) Global risk prediction maps for January through December. The warmer the color, the more suitable conditions were for virus survival, and the higher the risk predicted.

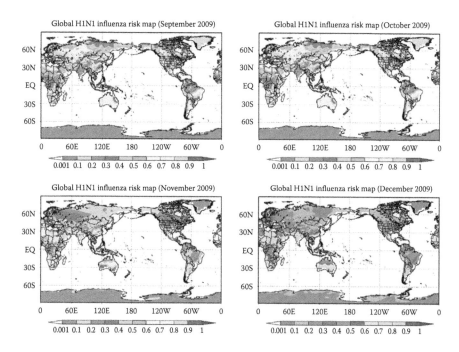

FIGURE 12.2 (continued)   (See color insert.) Global risk prediction maps for January through December. The warmer the color, the more suitable conditions were for virus survival, and the higher the risk predicted.

about the role that social factors such as human travel network play in the transmission of the pandemic H1N1 virus.

We then explored the geographical impact on the H1N1 pandemic. Figure 12.3 shows the tourism index for each country superimposed on the predicted disease risks for June and indicates that the tourism index corresponds well with the total number of confirmed cases at the country level.

The 20 countries with the highest tourism indices were chosen for further analysis and are listed in Table 12.1. The total number of confirmed influenza A(H1N1) cases in these 20 countries accounts for 98.2% of all cases worldwide, with the top 10 countries accounting for 95.5% of all cases. We investigated whether most of the cases were dispersed by the large volume of tourists traveling between the most frequently visited countries, particularly among the top 10 countries. It is interesting to note that Singapore, China, Malaysia, and Thailand have not had a large number of cases so far in spite of their relatively large volume of tourists. An examination of the transmission risk maps for the virus shows that these countries are low-risk areas for the entire year, with the exception of the northern part of Thailand that is predicted to be suitable for virus survival in December and the southern part of China that is predicted to be suitable for virus survival in the spring and fall of each year. The outbreak cases are also dependent on the strength of the containment strategies that each country implements. It is notable that China implemented prevention measures more strictly than some other countries. These prevention measures include body

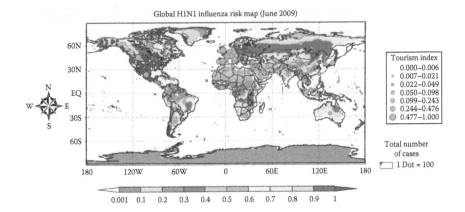

FIGURE 12.3 (**See color insert.**) Tourism index—a useful indicator of the size of influenza outbreaks. The tourism index for each country (0–1) is superimposed on the predicted disease risks (0–1) for June. Pink circles depict the total number of confirmed cases worldwide reported until May 24, 2009.

TABLE 12.1
Total Number of Confirmed Infections in the 20 Countries with the Highest Tourism Indices

Country Name	Tourism Index	Number of Cases
United States	1.00	7796
Germany	0.87	17
United Kingdom	0.75	165
Spain	0.63	148
France	0.63	18
Italy	0.48	23
Netherlands	0.39	3
Canada	0.36	773
Mexico	0.32	5165
Singapore	0.24	0
China	0.21	9
Malaysia	0.17	2
Turkey	0.17	1
Switzerland	0.17	1
Japan	0.16	367
Portugal	0.14	1
Thailand	0.13	4
Austria	0.10	1
South Korea	0.10	10
Saudi Arabia	0.09	0

temperature checks for all passengers in cities, compulsory isolation of adjacent passengers once an infection is detected, and voluntary self-isolation of passengers if they have come from countries that have a high rate of infections. Both unfavorable physical conditions and isolation measures may have delayed the outbreaks in these countries by several weeks to several months. Japan and South Korea have higher population densities in their major cities, thus increasing the risk of infection due to close interaction in public places. In addition, these two countries are predicted to have their highest virus activity period during May and October.

We studied the spatial pattern of the outbreak locations. Figure 12.4 shows values of the K function calculated from all the global outbreaks. There are two peaks located at approximately 600 and 3000 km. We also calculated the K function of major global cities, whose peak value at 600 km coincides with that of disease outbreaks. The distance of 600 km corresponds to the average travel distance from state to state in the United States, and the average travel distance within country or across countries in Europe. Therefore, the spatial pattern of the outbreaks is strikingly similar to that of global air travel patterns and average state-to-state travel. We hypothesize that the outbreaks are mainly due to air passenger flow worldwide and that the virus may have been dispersed from state to state in the United States and from country to country in Europe in this manner. From Figure 12.4 we can see that there is a continuous trend of increasing clustering distances up to 600 km and can thus infer that within states or between points separated by shorter distances, the virus may spread by other means of transportation, such as automobiles or trains. We also calculated the average distance between clustered outbreak centers and found that it peaked around 3000 and 7000 km, corresponding to cross-country travel distance

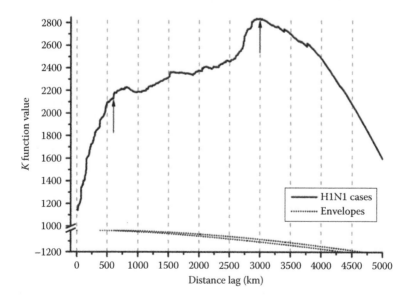

FIGURE 12.4 Ripley's K function values at different spatial lags. The curve peaks at around 600 and 3000 km. The dashed lines are upper and lower bounds of 20 times Monte Carlo simulation indicating randomness.

in North America and the cross-continent travel distance on the global scale, respectively, indicating virus dispersal patterns at various spatial scales.

Using air passenger data at the city level, we then examined how A(H1N1) cases are distributed in relation to global cities, in other words, the impact of major cities on the spread of the influenza virus. We compiled a city-level matrix-based network dataset that provided information on the origin–destination passenger travel. The network consisted of the 338 cities and 2284 edges that connect major cities worldwide. These matrices allowed us to measure the networks and analyze their structural properties. We were able to obtain the degree and the intensity of connectivity among cities and to measure the significance of city and network linkages. The clustering coefficient was calculated for each city. The closer the value is to 0, the more dominant is its position in its neighborhood. For each city, we gathered information on all confirmed influenza A(H1N1) cases within a 300 km radius buffer zone, to examine if the degree and the intensity of the network connectivity could explain the disease spread, namely, the impact of cities on the disease spread. We chose a 300 km radius buffer zone around each city based on the clustering peak of outbreaks starting at around 600 km determined earlier, and the average distance between global cities. The number of cases attributed to any two cities will not overlap if they are more than 600 km apart. If one outbreak fell within the buffer zones of more than two cities, it would be counted with the city closest to it. We chose the first 30 cities whose clustering coefficients are closest to 0, and the total number of confirmed cases, out-degree, and out-passengers for these cities is given in Table 12.2.

From Table 12.2, it is noteworthy that Phoenix, accounting for 517 cases, ranks second only to London, although it has the least out-passenger flow (292,052) and out-degree connections (5) among the top 30 cities. Phoenix is more connected with the outside world in comparison with its neighboring cities. In other words, Phoenix has the most important position in terms of connectivity among its neighbors. This may explain in part why a city with relatively small passenger flows has a large number of infections.

We found that cases occurring in the buffer zones of these 30 cities accounted for 46% (6,284/13,659) of the cases that occurred around all 338 cities in the dataset and 43% (6,284/14,783) of all cases worldwide. Choosing the clustering coefficient as the criteria for ranking cities better explained the data variation by 10% more than out-degree or out-passenger. To examine which cities contributed the remaining 54% of cases, we examined the situation in North America. Cases in North America alone account for 88% of those that occurred around the 338 cities in the dataset and 81% of all cases worldwide. When we consider all 70 cities in North America, rather than the first 13 cities in Table 12.2, the power of the explanation for the data variation increases significantly to 51% (6,992/13,659) for the 338 cities in the dataset and to 47% (6,992/14,783) of all cases. Of these, the United States accounts for 30% (4,131/13,659) of cases in the 338 cities and 28% (4,132/14,783) of all cases, whereas Mexico accounts for 19% (2,637/13,659) of cases in the 338 cities and 18% (2,637/14,783) of all cases. In this way, 97% (13,276/13,659) of global cases that fall within a 300 km buffer zone of the 338 cities and 90% (13,276/14,783) of all cases worldwide are explained by the 87 cities that include the top 30 global cities and the extra 57 cities in North America.

The international human mobility and global air traffic patterns that define human connectivity at various spatial scales will help to make recommendations on targeted

TABLE 12.2
Thirty Cities in Order of Increasing Value of Clustering Coefficient, along with Their Out-Degree, Out-Passenger Numbers and the Total Number of Confirmed Cases That Fall within a 300 km Buffer Zone of the City

City Name	Confirmed Cases	Clustering Coefficient	Out Degree	Out Passenger
London	120	0.09	100	27,941,442
Phoenix, AZ	517	0.10	5	292,053
Paris	14	0.13	80	16,839,504
Toronto	263	0.14	41	4,769,019
Seoul	3	0.15	49	8,930,291
New York, NY	687	0.15	58	9,031,816
Frankfurt	0	0.15	67	12,347,765
Miami, FL	14	0.15	32	4,392,462
Mexico City	2327	0.16	37	3,291,871
Dubai	0	0.18	41	6,140,128
Los Angeles, CA	159	0.19	39	6,092,931
Amsterdam	3	0.19	53	10,425,018
Houston, TX	173	0.20	11	1,376,558
Singapore	0	0.21	46	11,068,494
Boston, MA	424	0.22	13	1,113,498
Montreal	66	0.22	12	1,495,484
Lisbon	1	0.22	20	2,563,461
Bangkok	4	0.23	43	8,355,832
Calgary	61	0.24	7	483,906
Las Vegas, NV	93	0.24	5	424,020
Tokyo	16	0.24	41	11,179,695
Cologne	0	0.24	7	588,086
Chicago, IL	920	0.25	22	3,312,411
Munich	9	0.25	30	3,756,055
Moscow	1	0.25	42	3,551,047
Madrid	31	0.26	40	6,467,379
Buenos Aires	0	0.28	28	2,665,830
Hong Kong	6	0.28	37	10,188,085
Orlando, FL	21	0.29	5	593,971
Osaka	351	0.29	25	3,528,752

intervention strategies and facilitate implementation of pandemic preparedness and emergency responses at various levels of agencies.

However, the study has its limitations. The risk prediction based on climate is limited by the fact that the H1N1 outbreak data in the study mostly reflect the early outbreaks of H1N1 and do not necessarily give a full picture of the climatological suitability for viral survival. The relatively coarse spatial resolution of climate data used, relatively low positioning accuracy of the outbreak data, and the quality of the outbreak data itself also limit the modeling results.

12.4 CONCLUSIONS

A time series of global risk maps were developed to predict environmental exposure based on a model of gridded reanalysis data including daily temperature, precipitation, and absolute humidity. These maps reveal clear seasonal changes and the consequent environmental risks over various parts of the world and provide information for developing early warning signs. Seasonal trends in the predicted risks of influenza outbreaks are clearly captured. High-risk areas in the northern and southern hemispheres gradually move northward between April and August, and then gradually move southward between November and February. The amplitude of movement in the southern hemisphere is smaller than that observed in the northern hemisphere. It is noteworthy that tropical regions around the equator were not exposed to high risks throughout the year. However, social factors such as global human travel played an important role in the introduction and transmission of the pandemic H1N1 virus.

The K function of major global cities peaked at approximately 600 and 3000 km, which coincides with disease outbreaks. The spatial pattern of the outbreaks is strikingly similar to that of global air travel patterns and average state-to-state travel. We hypothesize that the outbreaks are mainly due to air passenger flow worldwide and that the virus may have been dispersed from state to state in the United States and from country to country in Europe in this manner. We can also infer that within states or between points separated by shorter distances, the virus may have been spread by other means of passenger traffic such as by automobiles or trains. The 3000 and 7000 km peaks, corresponding to cross-country travel distance in North America, and the cross-continent travel distance on the global scale, respectively, indicate virus dispersal patterns at various spatial scales.

The tourism index developed in the study corresponds well with the total number of confirmed cases at the country level. The total number of confirmed influenza A (H1N1) cases in the selected 20 countries with the highest tourism indices accounts for 98.2% of all cases worldwide, with the top 10 countries accounting for 95.5% of all cases.

The impact of major global cities on the spread of the influenza virus was also examined through air passenger data at the city level. The clustering coefficient indicates the dominant position of a city in terms of connectivity among its neighbors. This actually explains in part why a city with relatively small passenger flows has a large number of infections. A total of 97% (13,276/13,659) of global cases that fall within a 300 km buffer zone of the 338 cities and 90% (13,276/14,783) of all cases worldwide are explained by the 87 cities that include the top 30 global cities and the extra 57 cities in North America.

In summary, from the indicators that define human travel flow and clustering dominance of connectivity in a global travel network, we found that a relatively small number of countries or cities could account for most of the outbreak cases worldwide. Together with previous findings, all these confirmed that the climatological and geographical factors pose significant impacts on global transmission of the virus and could be incorporated into the development of targeted control and prevention strategies.

ACKNOWLEDGMENTS

This research was supported by the Ministry of Science and Technology, China, National Research Program (2010CB530300, 2012CB955501, 2012AA12A407, 2013AA122003), and the National Natural Science Foundation of China (41271099).

REFERENCES

Aron, J. L. and J. Patz. 2001. *Ecosystem Change and Public Health: A Global Perspective.* Baltimore, MD: Johns Hopkins University Press.

Dai, A. 2006. Recent climatology, variability, and trends in global surface humidity. *Journal of Climate* **19**(15), 3589–3606.

Gatrell, A. C., T. C. Bailey, P. J. Diggle, and B. S. Rowlingson. 1996. Spatial point pattern analysis and its application in geographical epidemiology. *Transactions of the Institute of British Geographers* **21**, 256–274.

ICAO. 1990. On-flight origins and destination. Montreal, Quebec, Canada: International Civil Aviation Organization (ICAO).

Jiang, Z., J. Bai, J. Cai, R. Li, Z. Jin, and B. Xu. 2012. Characterization of the global spatio-temporal transmission of the 2009 pandemic H1N1 influenza. *Journal of Geo-Information Science* **V14**(6), 794–799. doi:10.3724/SP.J.1047.2012.00794.

Li, W. and X. Cai. 2004. Statistical analysis of airport network of China. *Physical Review E* **69**, 046106.

Lowen, A. C., S. Mubareka, J. Steel, and P. Palese. 2007. Influenza virus transmission is dependent on relative humidity and temperature. *PLoS Pathogens* **3**, 1470–1476.

Ma, X. L. and M. F. Timberlake. 2008. Identifying China's leading world city: A network approach. *GeoJournal* **71**, 19–35.

Ripley, B. D. 1981. *Spatial Statistics.* New York: John Wiley.

Saker, L., K. Lee, B. Cattito, A. Gilmore, and D. C. Lendrum. 2004. Globalization and infectious diseases: A review of the linkages. *Special Topics No. 3. Social, Economic and Behavioural (SEB) Research*, 1–62.

Seto, E., B. Xu, S. Liang, R. Spear, P. Gong, W. Wu, G. Davis, D. Qiu, and X. Gu. 2002. The use of remote sensing for predictive modeling of Schistosomiasis in China. *Photogrammetric Engineering & Remote Sensing* **68**(2), 167–174.

Shaman, J. and M. Kohn. 2009. Absolute humidity modulates influenza survival transmission, and seasonality. *Proceedings of the National Academy of Sciences of the United States of America* **106**, 3243–3248.

UNWTO. 2009. *TSA Data around the World.* Madrid, Spain. www.unwto.org.

Watts, D. J. and S. H. Strogatz. 1998. Collective dynamics of 'small world' networks. *Nature* **393**, 440–442.

Weber, P. T. and N. I. Stilianakis. 2008. A note on the inactivation of influenza A viruses by solar radiation, relative humidity and temperature. *Photochemistry and Photobiology* **84**, 1601–1604.

World Health Organization. 2010. Pandemic (H1N1) 2009—Update 112. http://www.who.int/csr/don/2010_08_06/en/index.html. Accessed December 1, 2010.

Wu, X., H. Tian, S. Zhou, L. Chen, and B. Xu. 2013. Impact of global change on transmission of human infectious diseases. *Science China Earth Sciences.* doi:10.1007/s11430-013-4635-0.

Xu, B., P. Gong, G. Biging, S. Liang, E. Seto, and R. Spear. 2004. Snail density prediction for schistosomiasis control using IKONOS and ASTER images. *Photogrammetric Engineering & Remote Sensing* **70**(11), 1285–1294.

Xu, B., P. Gong, S. Liang, E. Seto, and R. Spear. 2003. Snail density estimation for schistosomiasis control by integrating field survey and multiscale satellite images. *Geographic Information Sciences* **9**(1–2), 97–100.

Xu, B., P. Gong, E. Seto, S. Liang, C. H. Yang, S. Wen, D. C. Qiu, X. G. Gu, and R. Spear. 2006. A spatial temporal model for assessing the effects of inter-village connectivity in schistosomiasis transmission. *Annals of the Association of American Geographers* **96**, 31–46.

Xu, G., Q. Ge, P. Gong, X. Fang, B. Cheng, B. He, Y. Luo, and B. Xu. 2013. Societal response to challenges of global change and human sustainable development. *Science China Earth Sciences* **58**(25), 3161–3168.

13 Investigations of the Diurnal Thermal Behavior of Athens, Greece, by Statistical Downscaling of Land Surface Temperature Images and Pattern Analysis

Iphigenia Keramitsoglou

CONTENTS

13.1 INTRODUCTION

Urbanization is tied in with the replacement of natural surfaces by artificial construction materials whose thermal properties are notably different. Urban areas generally have higher solar radiation absorption and greater thermal capacity and conductivity. These differences between the urban and rural areas contribute to the development of the urban heat island (UHI) phenomenon (Oke 1982; Voogt and Oke 2003; Weng and Lu 2008), which is mostly profound during nighttime. Sometimes, in the morning and at midday, an urban heat sink—also called a negative heat island—may be observed, which is considered a brief stage in the development of the UHI that occurs due to differences in the urban–rural warming rates (Oke 1987).

In-depth analysis of this diurnal phenomenon is significant to a range of issues in urban climatology, global environmental changes, human–environment interactions, energy demand, and health-related issues, and it is also important for planning and management practices (Chrysoulakis 2002; Chrysoulakis et al. 2013). The estimated three billion people living in the urban areas in the world are directly exposed to UHI, which will increase in the future as, according to United Nations, projections, urban populations will continue to grow over the next decades (UN 2012). It has been shown (e.g., Keramitsoglou et al. 2013a) that the spatial distribution of heat wave hazard is higher in urban areas due to a series of factors including the presence of UHI. As a result, the urban population is especially vulnerable to heat waves in terms of increased morbidity and mortality (Dousset et al. 2011). Furthermore, the rise in external ambient temperatures in urban environments is also inevitably associated with energy increase to meet raised comfort requirements. There is a distinction between the differential heating of the air above the city known as the canopy layer UHI (Arnfield 2003; Stewart 2011) and the heating of the surface skin called surface UHI (SUHI; Voogt and Oke 2003).

To quantify UHI, the existing permanent meteorological stations offer adequate temporal resolution and long-term archives but are usually located at areas not relevant to monitoring UHI, such as airports. Therefore, they are not suitable to capture the spatial variability of the air and surface temperature fields in and around the city. Remotely sensed thermal infrared (TIR) data are a unique source of information to estimate SUHI, which are also related to canopy layer heat islands. Currently, several space missions have onboard TIR sensors. Meteosat Second-Generation–Spinning-Enhanced Visible and Infrared Imager (MSG-SEVIRI), Geostationary Operational Environmental Satellite (GOES), National Oceanic and Atmospheric Administration–Advanced Very High Resolution Radiometer (NOAA-AVHRR), Terra and Aqua Moderate-Resolution Imaging Spectroradiometer (MODIS), Landsat-7 Enhanced Thematic Mapper (ETM+), Landsat Data Continuity Mission (LDCM, Landsat-8), Thermal Infrared Sensor (TIRS), Terra Advanced Spaceborne Thermal Emission and Reflection Radiometer (ASTER), and Fengyun series have been providing continuous monitoring of land surface temperature (LST) distribution at a spatial resolution ranging from 5 km to 60 m. The temporal resolutions vary from quarter-hour to 16 days, respectively. Voogt and Oke (2003) reviewed studies that had used remote sensing to examine UHI, while Tomlinson et al. (2011) reviewed satellites and sensors relevant to LST measurements in the context of meteorology and climatology. Nevertheless, the trade-off between spatial and temporal resolution of satellite sensors is the limiting factor for the utilization of TIR data for different applications: the TIR datasets have been used extensively in urban thermal applications but are limited by the choices they offer, that is, to use data with either high spatial and low temporal resolution or high temporal and low spatial resolution.

In particular, geostationary satellites (e.g., MSG-SEVIRI viewing Europe and Africa, GOES viewing America, Kalpana viewing India, Fengyun viewing China, and MTSAT observing East Asia) are the only remote sensing platforms that can offer continuous monitoring of LST distribution at quarter-hourly to hourly basis, which is only adequate for the diurnal study of the SUHI phenomenon. Their coarse spatial resolution, however, has prohibited their extensive use for urban studies; yet, recently, scientific interest in these sensors has been revived as computational methods for sharpening them to 1 km (Keramitsoglou et al. 2013b; Zakšek and Oštir 2012) or closer (Bechtel et al. 2012) have become available. An exhaustive review of LST sharpening has been recently published (Zhan et al. 2013).

The challenging option of high spatial and high temporal resolution is demonstrated in this chapter for the assessment of the thermal morphological patterns of cities. This is achieved by downscaling geostationary imagery to quarter-hour LST images of 1 km spatial resolution from May to September 2009 and then extracting the SUHI thermal patterns to study the spatial and temporal (diurnal) variability of the phenomenon. In the results, a special paragraph is included to illustrate the potential of further exploitation of such products for the estimation of city energy demand. The analysis was applied to the Mediterranean coastal city of Athens, Greece, a city that exhibits a strong SUHI phenomenon.

13.2 DATA AND AREA OF INTEREST

13.2.1 DATA

The input datasets were LST products from the SEVIRI on board the MSG satellite. MSG is in a geostationary orbit at 36,000 km above the geographical point of 0° latitude and 0° longitude. SEVIRI has 8 spectral bands in the TIR range out of 12 in total. LST maps are produced operationally by EUMETSAT using the generalized split-window (GSW) algorithm (Wan and Dozier 1996). The MSG LST product is computed within the area covered by the MSG disk, over four specific geographical regions (Europe, North Africa, South Africa, and South America), every 15 min. For each time slot and geographical region (Europe in the case presented here), the LST field and respective quality control data are generated and disseminated by the Land Surface Analysis Satellite Applications Facility (LSA SAF; Trigo et al. 2011). For the present analysis, the quarter-hour LST products from May 1 to September 30, 2009, were used.

Finer spatial resolution input datasets (referred to as "components" hereafter) were used for downscaling the MSG LST images. These can be grouped into dynamic and static components, the former changing within season. Dynamic components included (i) the Normalized Difference Vegetation Index (NDVI) and the Enhanced Vegetation Index (EVI) from the MODIS global MOD13A2 16-day average product, and (ii) emissivity in bands 31 and 32 from the MOD11A2 8-day average product, composed from the daily 1 km clear-sky LST product. Both dynamic components are provided on a 1 km sinusoidal grid and are available free online (e.g., from http://reverb.echo.nasa.gov/reverb/).

The static components included (i) a digital terrain model as of the Shuttle Radar Topography Mission (SRTM; http://srtm.usgs.gov/) from which elevation, aspect, and slope were extracted, and (ii) a European CORINE land cover vector dataset (www.eea.europa.eu). The land cover dataset (thematic information) was decomposed so as to provide percentages (numeric information) of certain land cover classes within a pixel of 1 km. We considered urban, agricultural land, vegetation, and water land cover classes (Keramitsoglou et al. 2013b; Zakšek and Oštir 2012).

13.2.2 ATHENS GREATER AREA

On the southeastern edge of the Greek mainland lies the Athens Greater Area as shown in Figure 13.1. The urban area shown on the top right map is confined by high mountains interrupted by small openings, while it is open to the sea from the south (Saronikos Gulf). Athens is the capital and largest city of Greece. The Athens larger urban zone (LUZ) is the eighth most populated LUZ in the European Union with a population of about 4,000,000, according to Eurostat (http://epp.eurostat.ec.europa.eu/). Mild and relatively wet winters as well as warm dry summers are the characteristics of Athens' warm thermo-mediterranean climate. The city of Athens

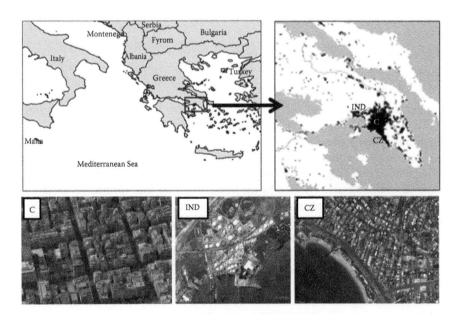

FIGURE 13.1 Location map and urban land cover of Athens Greater Area on the top row. Three areas of the urban fabric are highlighted and representative subareas are presented in the bottom row: C, Athens city center; IND, industrial area of Elefsina-Aspropyrgos; and CZ, coastal urban zone. (Photo from Microsoft Bing Maps Platform data.)

is characterized by a strong UHI effect, mainly caused by the accelerated industrialization and urbanization during recent years (Livada et al. 2002; Santamouris et al. 2007). This has been well documented using in situ meteorological measurements (e.g., Giannopoulou et al. 2011; Santamouris et al. 2007) as well as satellite imagery (e.g., Keramitsoglou et al. 2011; Stathopoulou et al. 2009). The acknowledgment of the problem by the community was shown in an interesting note on heat mitigation, which was a crucial part of the winning proposal submitted to a prestigious competition named "Re-Think Athens" sponsored by the Onassis Foundation for the transformation of the Athens city center (the winner was OKRA www.okra.nl, announced in March 2013).

Apart from the central area of the Attica basin ("C" in Figure 13.1), two more areas are of interest here: one is the industrial area of Elefsina-Aspropyrgos ("IND" in Figure 13.1; see also Keramitsoglou et al. 2011) and the other is the coastal urban zone ("CZ"). The small settlements around the dense urban fabric will not be discussed in this study as their thermal signature is negligible in the scale under consideration.

13.3 METHODOLOGY

The methodology adopted is illustrated in Figure 13.2. Geostationary MSG-SEVIRI LST data of 3–5 km pixel size acquired every 15 min provided the temporal variability required for the diurnal analysis, yet at a coarse spatial resolution. The first critical step was the downscaling of cloud-free geostationary LST images to 1 km/15 min resolutions. At a finer scale of 1 km, dynamic components (emissivity and vegetation indices) were selected appropriately according to their date to match the LST images. Static components (topography and land cover) were also integrated into the downscaling procedure. Downscaling was performed using support vector regression (SVR) machines and gradient boosting (GB; Keramitsoglou et al. 2013b). The output resembled a low Earth orbit LST product—such as the one from MODIS—with a revisiting capability of 15 min, resulting in 96 images per day. SUHI patterns were then identified on the images and extracted. The time series of SUHI patterns as well as the ones of urban and rural LSTs were subsequently processed to reveal the diurnal behavior of the thermal urban morphological patterns.

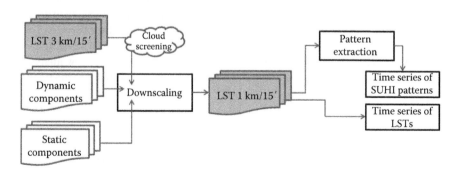

FIGURE 13.2 Conceptual flow of data and procedures to analyze the diurnal variability of SUHI.

13.3.1   Preprocessing

LST datasets were coded in Hierarchical Data Format (HDF5) and contained the following data fields: (i) LST, (ii) quality control information, and (iii) LST error estimate (Frietas et al. 2010; Trigo et al. 2008). The LST SEVIRI-based fields were generated pixel by pixel, maintaining the original resolution of SEVIRI level 1.5 data. These correspond to rectified images to 0° longitude, which present a typical geo-reference uncertainty of about 1/3 of a pixel. Data were kept in the native geostationary projection. We extracted the area of interest around the Athens center (~150 km × 150 km) and generated a binary LST image containing only valid pixels according to the corresponding quality control flags. The identification of cloudy pixels is provided as part of the LST product. To cloud-screen the images, we set a threshold of 30% cloudiness over land and a more stringent one of 10% over Athens city. These two thresholds ensured that only top-quality images were maintained for further analysis: 8240 images were automatically selected and downscaled. A procedure was also developed to select the corresponding file for the finer-scale dynamic components (see Section 13.2.1) to be integrated into downscaling.

13.3.2   Downscaling of MSG

Even though 1 km might not seem an appropriate spatial resolution for monitoring the thermal urban environment, it has been shown, for example, by Rajasekar and Weng (2009) and Keramitsoglou et al. (2011), that this resolution (from MODIS images for instance) is indeed appropriate for the estimation of basic parameters of SUHI: intensity, spatial extent, orientation, and central location. Downscaling of geostationary LST maps (~3–5 km) to a finer resolution (1 km) is one field worth investigating, as the results will combine the 1 km spatial with quarter-hour temporal resolution and thus open the prospect of numerous applications. To improve the spatial resolution of the geostationary imagery, one has to use information layers of better resolution correlated to LST. Specifically for statistical downscaling, the correlation between LST and auxiliary data, such as vegetation cover, topography, and other factors that are offered in a finer scale, is employed (Zakšek and Oštir 2012). The approach developed by Yang et al. (2010) combined many auxiliary datasets: leaf area index, NDVI, soil water content index, and reflectance of visible and near-infrared bands. To find the optimal downscaling solution, they employed an artificial neural network. Zakšek and Oštir (2012) used the correlation existing between LST and the land cover and microrelief parameters to enhance the spatial resolution of the SEVIRI LST over central Europe to 1000 m using the moving window analysis. Keramitsoglou et al. (2013b) used the same underlying principle with sophisticated regression methodologies (such as SVR and neural networks improved by gradient boosting) that were applied globally to the entire image, circumventing limitations of previous studies.

In the present analysis, we used SVR light support vector machines (SVM; Joachims 2002) coupled with gradient boosting (SVR/GB; Friedman 2001), which proved to be the most robust, high-performance methodology reaching correlation coefficients from 0.69 to 0.81 when compared to the other LST maps derived from

MODIS and AVHRR. Keramitsoglou et al. (2013b) provide an assessment of the downscaling procedure against coincident 1 km LST images. In brief, first upscaling was performed to finer spatial resolution components (see Section 13.2.1) and a new set was derived at the geostationary LST image geometry (coarse resolution, ~3 km). The second step was the development of the SVR/GB model: a unique regression model was defined for each input LST coarse-resolution image. The third step was the application of the regression model to the fine-scale components for the generation of the 1 km LST map. By applying this three-step procedure to the geostationary LST imagery, we produced a time series of 1 km spatial resolution LST maps every quarter-hour from May 1 to September 30, 2009.

13.3.3 PATTERNS EXTRACTION

Despite the large number of publications on urban LST and UHIs using satellite and airborne sensors, Voogt and Oke (2003) criticized that thermal remote sensing of urban areas has progressed slowly largely due to qualitative descriptions of thermal patterns and simple correlations between LST and land use/land cover types. There are a number of published works on the characterization of thermal patterns; Rajasekar and Weng (2009), for example, applied a nonparametric model by using fast Fourier transformation (FFT) to MODIS imagery for characterization of SUHI over space, in order to derive SUHI magnitude and other parameters, and Hung et al. (2006) adopted the Gaussian method proposed by Streutker (2002) to measure the spatial extents and magnitudes of SUHIs for eight megacities in Asia. In spite of these advances, new methods for the estimation of SUHI parameters from multitemporal and multilocation TIR imagery are still needed given the increased interest in the urban climate community to use remote sensing data (Weng 2009). Keramitsoglou et al. (2011) extracted the thermal hot spots from more than 3000 MODIS LST maps of Athens, retaining their original values and thus circumventing modeling.

The methodology of Keramitsoglou et al. (2011) has been used in the present analysis to process thousands of derived LST downscaled images, to identify and extract the SUHI patterns, to characterize them in terms of spatial extent and intensity, to study their diurnal behavior, and to investigate any correlations between spatial and thermal attributes. The main computational procedure for thermal pattern analysis was the extraction of pixels whose LST was higher than the suburban reference LST plus a predefined threshold value (let us say 3°C). These were first registered as potential SUHI hot pixels. Separation of hot pixel groups was automatically performed through appropriate segmenting of the initial image by partitioning the potential hot pixels. The grouped SUHI pixels were subsequently treated as different regions (objects), allowing several features related to these objects to be extracted. Once the objects of interest were extracted, a number of parameters per "SUHI object" were calculated and appropriately stored in a database. These parameters included temporal, spatial (weighted centroid location, extent in square kilometers), and thermal information (e.g., minimum, maximum LST). It is worth noting that the methodology is appropriate for both coastal cities, where the discontinuity of sea–land interface would cause complications in a modeling approach (as in Rajasekar and Weng 2009), and inland cities.

13.4   RESULTS AND DISCUSSION

13.4.1   Diurnal Analysis of Heat Island and Sink

A comparison between the LSTs of the dense urban area and a reference suburban area was performed by calculating the average LST value of each area and subsequently calculating the differences between the average urban minus the average rural LST. The evolution of these differences can provide a characterization of the daily evolution of the SUHI phenomenon during the period under consideration (May–September 2009). Based on this, two areas were delineated as urban areas in the city center (located in "C"), including only dense urban land use with an altitude not exceeding 200 m, and a rural area to serve as reference selected at the vicinity of, but not including, the Athens airport, covering nondense urban fabric and other rural land cover. For consistency with the urban area, the reference area was also below 200 m altitude. A total of 8240 downscaled LST images were completely cloud-free over the Athens center and therefore analyzed. In the absence of clouds, a measurement was available every quarter-hour.

The diurnal analysis of the results is shown in Figure 13.3. SUHI takes positive (heat island) and negative (heat sink) values during the day. This finding for Athens is also in agreement with Stathopoulou et al. (2009) at a much finer scale using a single image from Landsat-TM. The figure shows that, on average, SUHI was positive for 18 h and negative for 6 h during the day. Analysis of monthly averages showed identical behavior, and therefore it is not presented (approximate sunrise and sunset times are shown in the figure). Further, it is of interest to note that no shift was observed due to different sunrise and sunset times. The maximum SUHI values occurred from 20:00 to 03:00 (all times are coordinated universal time [UTC],

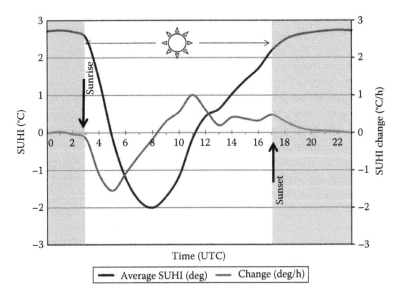

FIGURE 13.3   Average intensity of the diurnal SUHI and sink for Athens. The rate of SUHI change is also shown in gray line.

local time was 3 h ahead) when the phenomenon was well developed and stable (this will be shown by means of more figures during the analysis). A sharp drop occurred from 03:00 to 08:00 with a maximum change rate of 1.55°C/h at 05:00. Subsequently, the difference between urban and rural LSTs increased obtusely reaching 0°C and then became positive while the SUHI phenomenon and pattern (see also Section 13.4.2) was formed. The strong diurnal variability of the SUHI phenomenon can be attributed primarily to the different properties of materials found in rural and urban areas, indicatively (adapted from US EPA 2008):

- Reflectance of solar radiation—urban areas reflect less solar radiation and absorb more than rural areas.
- Thermal emissivity, which is related to the ability to emit TIR radiation—emissivity of construction materials found in Athens is higher than that found in rural areas.
- Heat capacity, which refers to the ability to store heat—cities absorb and store twice the amount of heat compared to rural areas.
- Thermal admittance of materials, which is a measure of the ability of a material to transfer heat in the presence of a temperature difference on opposite sides of the material and which affects the material's thermal response to variation in outside temperature—materials with a low thermal admittance, such as asphalt and brick, found largely in urbanized areas, exhibit greater diurnal temperature variability than materials with high thermal admittance of rural areas.

13.4.2 DIURNAL ANALYSIS OF SUHI PATTERNS

The object-based image analysis procedure of Keramitsoglou et al. (2011) was used to extract the SUHI patterns from the 1 km/15 min downscaled LST maps. This allowed for the analysis of the diurnal behavior of the phenomenon. The spatial and thermal attributes associated with the SUHI objects were calculated and used for the analyses of their intensity, position, and spatial extent. In total, 4507 patterns were extracted from the images between May 1 and September 30, 2009, with a threshold of 3° above the rural LST (R + 3). One of the most informative graphs of this analysis is presented in Figure 13.4. On the top diagram, one can observe the number of occurrences of the SUHI pattern every quarter-hour from 00:00 until 23:45 UTC. For example, 90 (out of the 4507) patterns were observed at 0:00 UTC in the 5 months considered in the analysis, regardless of the date. The stability of the phenomenon during nighttime is apparent. The drop of the number of patterns is steep between 03:00 and 05:00, to a negligible number and subsequently the occurrences gradually rise after 11:00. This behavior is similar to the one observed in Figure 13.3; however, it is worth noting that in this case, we observe patterns rather than LST differences. The bottom panel in Figure 13.4 shows the corresponding average area extent diurnally with the same characteristics. The two panels together imply that the average well-developed SUHI pattern is ~250 km² (for a threshold LST of R + 3°); during the transition phases (steep decay and gradual growth), the area extent varies considerably by an order of magnitude (e.g., from 20 km² at 06:00 to 200 km² at 20:00 UTC).

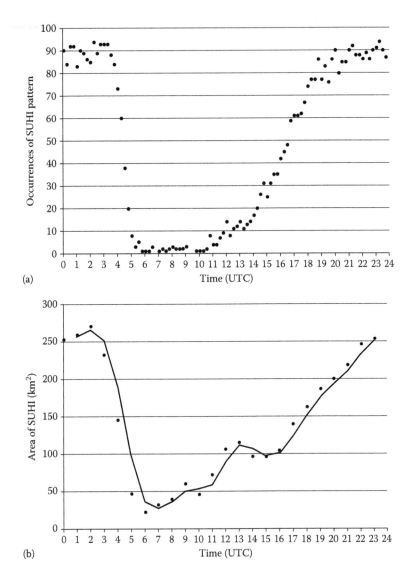

(a)

(b)

FIGURE 13.4 Diurnal variation of SUHI patterns: (a) the occurrence (presence) of SUHI patterns per quarter-hour from May until September 2009 (in total 4507 patterns); (b) the average area of the SUHI patterns diurnally for the same period with the 2 h average line overlaid.

Having realized the diurnal development of the SUHI patterns, two figures elucidate their spatial distribution in the area under study. Figure 13.5 presents hourly snapshots of the pattern and its decay for the night of July 16/17, 2009. The pattern under consideration is shown in orange. The time series from 21:00 to 04:00 on the next day shows the well-developed SUHI of Athens and its decay. The pattern centroid is within area "C" (see also Figure 13.1). The pattern also spreads along the urban coastal zone of Athens (area "CZ") and to a lesser extent to the industrial area of

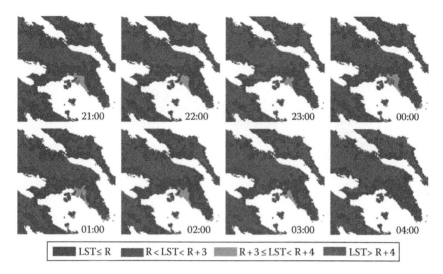

LST ≤ R   R < LST < R + 3   R + 3 ≤ LST < R + 4   LST > R + 4

FIGURE 13.5 (See color insert.) Hourly development and decay of SUHI patterns for a summer day (July 16/17, 2009); time is in UTC. The reader is advised to focus on the orange and red patterns that represent the SUHI patterns at different times.

FIGURE 13.6 Average SUHI pattern for different LST thresholds (R + 3, R + 4, R + 5). The level of gray depicts the "participation" of each pixel in the pattern (from 0% to 100%). Time of the day is indifferent.

Elefsina-Aspropyrgos ("IND" area), for that particular day. The red color on the maps denotes SUHI intensity more than 4°C. Blue and green pixels show LST less than that of the reference rural area (R) and between R and the threshold of R + 3, respectively.

Regarding the spatial extent of SUHI, it is of interest to investigate the "participation" of each pixel in the pattern, regardless of time. This is presented in Figure 13.6. The three panels show the SUHI pattern for three different thresholds of 3°C, 4°C, and 5°C above the LST of the reference rural area. The darker areas show high "participation" of these pixels in the pattern. When increasing the threshold, the spatial extent is smaller. The left panel shows that the 3°C SUHI pattern covers all the three areas of Athens that we considered earlier, that is, the city center, the industrial area ("IND"), and the urban coastal zone ("CZ"). The high intensities (4°C and 5°C threshold) are limited to the city and a persistent feature is present at Piraeus port, the chief port in Greece and the largest passenger port in Europe. This distribution is in agreement with single-image studies at a finer scale (Stathopoulou et al. 2009).

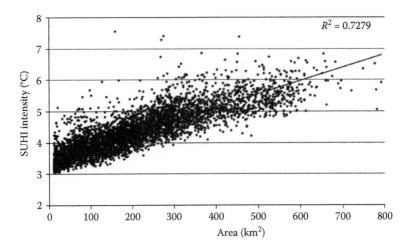

FIGURE 13.7 SUHI pattern analysis shows a strong correlation between the area extent of the pattern and its intensity (4507 points). Time of the day is indifferent.

The analysis showed that the SUHI extent was not correlated directly with any thermal feature such as the maximum or minimum LST; instead, the spatial extent of a hot spot was highly correlated with its intensity. By thermal intensity, we refer to the discrepancy between the maximum LST of the SUHI pattern and the rural reference LST (R). In other words, thermal intensity shows the LST difference between the urban hot spot and the rural area.

Similar to the analysis of Keramitsoglou et al. (2011) for the daytime thermal hot spots, the correlation between the SUHI pattern extent and intensity is presented in Figure 13.7. This finding is new and could not be observed for the SUHI pattern with a correlation coefficient high enough to be trusted in the previous study (the study was carried out using only MODIS LST images).

The spatial extent of a hot spot is defined as the area covered by the aggregated cluster of pixels whose LST is higher than the rural reference plus a predefined threshold value. The construction materials of urban areas exhibit a high thermal inertia (i.e., a low response to temperature changes), and, consequently, they warm up later in the course of the day and continue releasing heat slowly after sunset and during nighttime, when most of the rural surfaces have cooled down. Therefore, as the difference between urban and rural areas increases, the thermal intensity of SUHI increases, and the cluster of warm pixels (SUHI pattern) becomes larger, explaining the strong correlation between them.

13.4.3 COOLING DEGREE HOURS

The rise in external ambient temperatures in urban environments, compared to rural environments, is associated with a series of interconnected impacts, namely, comfort, energy increase to meet raised comfort requirements, and health. Energy demand for cooling of buildings is an indicator of the impact of climate on the energy sector. Conventionally, cooling degree days (CDD) and hours (CDH) are the most common practical methods for assessing the effect of air temperature on the energy performance of a building, and they are used as a reasonable approximation of the cooling energy needs of a city with respect to it. In recent years, the majority of houses and work environments in Athens have air-conditioning systems. In the event of a heat wave, power demand is much increased, often causing significant region-wide blackouts. In such circumstances, the health of the population is at further risk. By definition, a CDH is recorded if the average air temperature for the hour rises above a base temperature that differs between cities. Until recently, CDH could be calculated only at certain points where meteorological stations were installed (e.g., Kolokotroni et al. 2010; Papakostas and Kyriakis 2005; Tselepidaki et al. 1994). The lack of an adequate number of ground stations prohibits any meaningful spatial analysis at and around the urban web.

Here we attempt to estimate CDH by using the downscaled 1 km/1 h LST maps rather than air temperatures. The advantage is the dense space-time grid, which is not available by other measurements. The drawback is that the switch from air temperatures to LST may not be as straightforward as presented here. The motivation for this step originated from two previously published works: one of Stathopoulou et al. (2006) that related midday LST from NOAA-AVHRR data and mean daily air temperature observations recorded at standard meteorological stations in Athens and subsequently estimated CDD from the satellite data, and a recent study by Clinton and Gong (2013) that defined CDD in 2010 globally as data in which the urban *pixel* temperature exceeds 20°C at the time of measurement.

Assuming, thus, that the magnitude of the positive deviation of a pixel with temperature LST at hour h from the base temperature is related to the energy demand, the total number of cooling degree hours for a day based on LST rather than air temperature can be expressed as

$$\text{CDH}(T_b) = \sum_{h=0}^{23} (\text{LST}_h - T_b)^+.$$

The "+" sign indicates that only positive values are summed up. The spatial distribution of CDH in a city is related to the demand for air conditioning. This potential of using downscaled LST images at 1 km on an hourly basis is illustrated in Figure 13.8. Two days have been selected here: one is a typical summer day in July in Athens with maximum air temperature at 33°C (shown in the top row) and another day characterized as hot, with air temperature being above 37°C for more than 3 h (Keramitsoglou et al. 2013a; shown in the bottom row). For the hot day, the spatial distribution of CDH is also of relevance for a higher base LST (above 23°C). In the case of a typical summer day, the figures show that the urban areas of Athens have at least double CDH compared to the rural areas (see also Figure 13.1 for the exact location of the urban fabric in Athens), the patterns of high CDHs reflecting the patterns of SUHI (as in Figure 13.5). The selection of base LST definitely plays an important role, and further analysis should be conducted to correlate air temperature and LST base temperatures (similar to Stathopoulou et al. 2006). In any case, it is common practice (e.g., Papakostas and Kyriakis 2005) to present CDD and CDH results for various base temperatures.

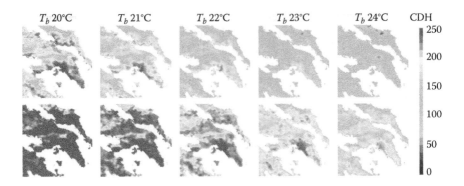

T_b 20°C T_b 21°C T_b 22°C T_b 23°C T_b 24°C   CDH

FIGURE 13.8   (See color insert.) Cooling degree hours in degrees for a single day for various base LSTs (T_b). Top row represents a typical day in July and the bottom row a hot day.

Elevated urban LSTs lead to increased cooling energy consumption and peak electricity demand. The maps of Figure 13.8 are in agreement with previous studies calculating the cooling load of buildings for the city of Athens (Hassid et al. 2000; Santamouris et al. 2001). They reported that during summer noon hours, the cooling load of urban buildings at the city center is about double compared to the respective load in the surrounding Athens area; the peak electricity load for cooling purposes may be tripled, especially for base air temperatures higher than 26°C. However, it is anticipated that local or global climate change will amplify these effects: Assimakopoulos et al. (2012) projected that the energy consumption in buildings in Athens will further increase due to climate change.

It needs to be stressed that the methodology described is a first approximation to relate the spatial distribution of LST degree hours to energy demand and serves more like a demonstration of the potential uses of the downscaled LST images. It needs to be further validated before it can be trusted and used operationally, and this will serve as motivation for future research.

13.5   CONCLUSIONS

In this chapter, the results from the synergistic use of two innovative algorithms, namely, downscaling of geostationary LST images (Keramitsoglou et al. 2013b) and extraction of thermal patterns (Keramitsoglou et al. 2011), have been presented for the study of the diurnal behavior of SUHI in Athens. Five months from May 1 to September 30, 2009, of quarter-hour LST products over Europe have been acquired (13,709 products in total), 60% of which passed all the stringent cloudiness and quality criteria and were further processed.

The SUHI of Athens is stable and well developed from 20:00 to 03:00 UTC. In turn, a sharp drop in LST difference and pattern extent occurs from 03:00 to 08:00, with a maximum SUHI change rate of 1.55°C/h at 05:00. The phenomenon starts developing again after 11:00 UTC. The average well-developed SUHI pattern is ~250 km² (for a threshold LST of R + 3°C), and the pattern extent is highly correlated with its intensity at all times. The SUHI pattern covers the city center, the industrial area of Elefsina-Aspropyrgos, and the urban coastal zone; however, higher intensities are limited to the city and a persistent feature is present at Piraeus port. The downscaled LST images allow for a first approximation of the spatial distribution of CDH, a feature related to energy demand. It was found that during a typical summer day, the urban areas of Athens have at least double CDHs compared to the rural areas and the patterns of high CDHs reflect the patterns of SUHI. Another feature of interest is that between 5:00 and 11:00 UTC in the morning, urban LSTs are lower than rural LSTs, the difference reaching a maximum at 8:00 (urban heat sink).

The results presented here are complementary to the ones previously published for Athens Greater Area, which use either in situ meteorological data (e.g., Giannopoulou et al. 2011; Livada et al. 2002), single-date high-resolution satellite imagery (e.g., Stathopoulou et al. 2009), or a limited number of satellite overpasses at 1 km spatial resolution (Stathopoulou et al. 2006). However, it is the first time that such a long dataset with a spatial resolution of 1 km and the quarter-hour temporal resolution has been processed for the city of Athens. The present approach provides a new insight into the diurnal variations of the spatial distribution of SUHI in Athens.

Apart from the advancement in understanding the particular case study of Athens, the proposed methodologies are fast, use open-source libraries, integrate freely available global datasets, and can therefore be considered for further exploitation of geostationary LST datasets. At present, LST products are archived from different TIR sensors on different geostationary missions (MSG-SEVIRI, US GOES, Chinese FY3). These can be exploited and reanalyzed to generate high-level innovative products designed specifically for cities. The analysis presented here demonstrates that, when appropriately downscaled, the geostationary orbit imagery can reveal a wealth of information at high temporal resolution of 15 min, currently not available by any Earth Observation mission, opening a range of applications. Optimized exploitation can be tailored for different purposes, with several different end users such as urban climate modelers, health responders, and energy demand suppliers.

With global temperatures rising (the warmest September being in 2012), Earth experiencing warmer temperatures than several decades ago, and future climate projections showing that the extreme events of the present might be common phenomena in the future, it is envisaged that such exploitation of satellite-derived products addressing the urban thermal environment will be essential means for supporting urban planning and decision making for the mitigation of SUHI.

ACKNOWLEDGMENTS

I would like to sincerely thank Professor Chris T. Kiranoudis (National Technical University of Athens) for his collaboration in the development of the computer code for the massive processing of the LST images. I would also like to thank Themistocles Herekakis (NOA) for his assistance in the postprocessing analysis of the results. The original LST field and respective quality control data were generated and disseminated by the Land Surface Analysis Satellite Applications Facility (LSA SAF; https://landsaf.ipma.pt/). The comments of three anonymous reviewers on an earlier draft are much appreciated.

REFERENCES

Arnfield, J. 2003. Two decades of urban climate research: A review of turbulence, exchanges of energy and water, and the urban heat island. *International Journal of Climatology* 23:1–26.

Asimakopoulos, D.A., M. Santamouris, I. Farrou, M. Laskari, M. Saliari, G. Zanis, G. Giannakidis et al. 2012. Modelling the energy demand projection of the building sector in Greece in the 21st century. *Energy and Buildings* 49:488–498.

Bechtel, B., K. Zakšek, and G. Hoshyaripour. 2012. Downscaling land surface temperature in an urban area: A case study for Hamburg, German. *Remote Sensing* 4:3184–3200.

Chrysoulakis, N. 2002. Energy in the urban environment: use of Terra/ASTER imagery as a tool in urban planning. *Journal of the Indian Society of Remote Sensing* 30:245–254.

Chrysoulakis, N., M. Lopes, R. San José, C.S.B. Grimmond, M.B. Jones, V. Magliulo, J.E.M. Klostermann et al. 2013. Sustainable urban metabolism as a link between bio-physical sciences and urban planning: The BRIDGE project. *Landscape and Urban Planning* 112:100–117.

Clinton, N. and P. Gong. 2013. MODIS detected surface urban heat islands and sinks: Global locations and controls. *Remote Sensing of Environment* 134:294–304.

Dousset, B., F. Gourmelon, K. Laaidi, A. Zeghnoun, E. Giraudet, P. Bretin, E. Mauri, and S. Vandentorren. 2011. Satellite monitoring of summer heat waves in the Paris metropolitan area. *International Journal of Climatology* 31:313–323.

Freitas, S.C., I.F. Trigo, J.M. Bioucas-Dias, and F.-M. Goettsche. 2010. Quantifying the uncertainty of land surface temperature retrievals from SEVIRI/Meteosat. *IEEE Transactions on Geoscience and Remote Sensing* 48:523–534.

Friedman, J.H. 2001. Greedy function approximation: A gradient boosting machine. *Annals of Statistics* 29:1189–1232.

Giannopoulou, K., I. Livada, M. Santamouris, M. Saliari, M. Assimakopoulos, and Y.G. Caouris. 2011. On the characteristics of the summer urban heat island in Athens, Greece. *Sustainable Cities and Society* 1:16–28.

Hassid, S., M. Santamouris, N. Papanikolaou, A. Linardi, N. Klitsikas, C. Georgakis, and D.N. Assimakopoulos. 2000. Effect of the Athens heat island on air conditioning load. *Energy and Buildings* 32:131–141.

Hung, T., D. Uchihama, S. Ochi, and Y. Yasuoka. 2006. Assessment with satellite data of the urban heat island effects in Asian mega cities. *International Journal of Applied Earth Observation and Geoinformation* 8:34–48.

Joachims, T. 2002. *Learning to Classify Text Using Support Vector Machines: Methods, Theory and Algorithms*. The Springer International Series in Engineering and Computer Science. Kluwer Academic Publishers, Dordrecht, the Netherlands.

Keramitsoglou, I., C.T. Kiranoudis, G. Ceriola, Q. Weng, and U. Rajasekar. 2011. Identification and analysis of urban surface temperature patterns in greater Athens, Greece, using MODIS imagery. *Remote Sensing of Environment* 115:3080–3090.

Keramitsoglou, I., C.T. Kiranoudis, B. Maiheu, K. De Ridder, I.A. Daglis, P. Manunta, and M. Paganini. 2013a. Heat wave hazard classification and risk assessment using artificial intelligence fuzzy logic. *Environmental Monitoring and Assessment* 185:8239–8258.

Keramitsoglou, I., C.T. Kiranoudis, and Q. Weng. 2013b. Downscaling geostationary land surface temperature imagery for urban analysis. *IEEE Geoscience and Remote Sensing Letters* 10:1253–1257.

Kolokotroni, M., M. Davies, B. Croxford, S. Bhuiyan, and A. Mavrogianni. 2010. A validated methodology for the prediction of heating and cooling, energy demand for buildings within the Urban Heat Island: Case-study of London. *Solar Energy* 84:2246–2255.

Livada, I., M. Santamouris, K. Niachou, N. Papanikolaou, and G. Mihalakakou. 2002. Determination of places in the great Athens area where the heat island effect is observed. *Theoretical and Applied Climatology* 71:219–230.

Oke, T.R. 1982. The energetic basis of the urban heat island. *Quarterly Journal of the Royal Meteorological Society* 108:1–24.

Oke, T.R. 1987. *Boundary Layer Climates*. Routledge, Taylor & Francis Group, London, U.K.

Papakostas, K. and N. Kyriakis. 2005. Heating and cooling degree-hours for Athens and Thessaloniki—Greece. *Renewable Energy* 30:1873–1880.

Rajasekar, U. and Q. Weng. 2009. Urban heat island monitoring and analysis by data mining of MODIS imageries. *ISPRS Journal of Photogrammetry and Remote Sensing* 64:86–96.

Santamouris, M., N. Papanikolaou, I. Livada, I. Koronakis, C. Georgakis, A. Argiriou, and D.N. Asimakopoulos. 2001. On the impact of urban climate on the energy consumption of buildings. *Solar Energy* 70:201–216.

Santamouris, M., K. Paraponiaris, and G. Mihalakakou. 2007. Estimating the ecological footprint of the heat island effect over Athens, Greece. *Climatic Change* 80:265–276.

Stathopoulou, M., C. Cartalis, and N. Chrysoulakis. 2006. Using midday surface temperature to estimate cooling degree-days from NOAA-AVHRR thermal infrared data: An application for Athens, Greece. *Solar Energy* 80:414–422.

Stathopoulou, M., A. Synnefa, C. Cartalis, M. Santamouris, T. Karlessi, and H. Akbari. 2009. A surface heat island study of Athens using high-resolution satellite imagery and measurements of the optical and thermal properties of commonly used building and paving materials. *International Journal of Sustainable Energy* 28:59–76.

Stewart, I.D. 2011. A systematic review and scientific critique of methodology in modern urban heat island literature. *International Journal of Climatology* 31:200–217.

Streutker, D.R. 2002. A remote sensing study of the urban heat island of Houston, Texas. *International Journal of Remote Sensing* 23:2595–2608.

Tomlinson, C.J., Chapman, L., Thornes, J.E., and Baker, C. 2011. Remote sensing land surface temperature for meteorology and climatology: A review. *Meteorological Applications* 18(3):296–306. doi:10.1002/met.287.

Trigo, I.F., C.C. DaCamara, P. Viterbo, J.-L. Roujean, F. Olesen, C. Barroso, F. Camacho-de Coca et al. 2011. The satellite application facility on land surface analysis. *International Journal of Remote Sensing* 32:2725–2744.

Trigo, I.F., I.T. Monteiro, F. Olesen, and E. Kabsch. 2008. An assessment of remotely sensed land surface temperature. *Journal of Geophysical Research* 113:D17108.

Tselepidaki, I., M. Santamouris, D.N. Asimakopoulos, and S. Kontoyiannidis. 1994. On the variability of cooling degree days in an urban environment: Application to Athens, Greece. *Energy and Buildings* 21:93–99.

United Nations (UN) Department of Economic and Social Affairs, Population Division. 2012. *World Urbanization Prospects, the 2011 Revision: Highlights.* http://esa.un.org/unup/pdf/WUP2011_Highlights.pdf (accessed March 31, 2013).

U.S. EPA Environmental Protection Agency. 2008. Reducing urban heat islands: Compendium of strategies. U.S. EPA, Climate Protection Partnerships Division. Available online http://www.epa.gov/heatisld/resources/compendium.htm (accessed May 2013).

Voogt, J.A. and T.R. Oke. 2003. Thermal remote sensing of urban climate. *Remote Sensing of Environment* 86:370–384.

Wan, Z. and J. Dozier. 1996. A generalised split-window algorithm for retrieving land-surface temperature from space. *IEEE Transactions on Geoscience and Remote Sensing* 34:892–905.

Weng, Q. 2009. Thermal infrared remote sensing for urban climate and environmental studies: Methods, applications, and trends. *ISPRS Journal of Photogrammetry and Remote Sensing* 64:335–344.

Weng, Q. and Lu, D. 2008. A sub-pixel analysis of urbanization effect on land surface temperature and its interplay with impervious surface and vegetation coverage in Indianapolis, United States. *International Journal of Applied Earth Observation and Geoinformation* 10:68–83.

Yang, G., R. Pu, W. Huang, J. Wang, and C. Zhao. 2010. A novel method to estimate sub-pixel temperature by fusing solar-reflective and thermal-infrared remote sensing data with an artificial neural network. *IEEE Transactions on Geoscience and Remote Sensing* 48:2170–2178.

Zakšek, K. and K. Oštir. 2012. Downscaling land surface temperature for urban heat island diurnal cycle analysis. *Remote Sensing of Environment* 117:114–124.

Zhan, W., Y. Chen, J. Zhou, J. Wang, W. Liu, J. Voogt, X. Zhu, J. Quan, and J. Li. 2013. Disaggregation of remotely sensed land surface temperature: Literature survey, taxonomy, issues, and caveats. *Remote Sensing of Environment* 131:119–139.

Section IV

Innovative Concepts and Techniques in Urban Remote Sensing

14 Integrated Urban Sensing in the Twenty-First Century

Günther Sagl and Thomas Blaschke

CONTENTS

Today, vast volumes of highly diverse sensor data are generated, and this amount is growing exponentially. As highlighted in several chapters of this book (e.g., Chapters 2, 5, 8, or 10), high-resolution remotely sensed data serve day-to-day applications. Virtual Globes such as Google Earth have brought such images to everybody's fingertips. Lesser known to the wider public are two other fields of data generation: real-time in situ sensing of environmental parameters and sensing of human behavior in space and time. Environmental data are mainly sensor-generated. Examples include weather stations or intelligent mobile sensor pods. We call these "machine-generated" data. On the contrary, direct measurements of humans in space and time are predominantly restricted for privacy reasons. Information about persons or groups and their behavior in space and time is either derived from so-called volunteered geographic information (VGI) or it may be derived from proxy data, for example, from mobile communication networks or social media. In this chapter, we argue that multiple coordinated views of spatiotemporal data provide unprecedented opportunities for geographic analysis in times of "big data." Together, these different types of data generation enable an integrated sensing. We focus on urban areas where the density of relevant information is already high. We claim that integrated urban sensing opens new vistas to physical and social dynamics at the environment–human interface. We analyze the intersection of machine-generated (satellite imagery, weather stations) and user-generated (social media, mobile phone data) data and we contend that geographic information systems (GIS) as a tool and geographic information science (GIScience) principles are together the lynchpin of integrated urban sensing. In particular, GIS plays a major role in urban monitoring studies. We demonstrate that GIS-based integrated urban sensing enables analyses, forecasts, and visualizations of a variety of spatial components of socioeconomic phenomena. This includes people, urban commodities, and their respective changes, but also information flows and human interaction with urban commodities as well as the relationships among networks of human interaction and natural environments.

14.1   INTRODUCTION

At the beginning of the twenty-first century, a wide range of technologies are able to sense, directly or indirectly, a variety of environmental, human, and social phenomena—thereby facilitating the "Digital Earth" concept introduced by Gore (1998). Such sensing technologies generate vast and rapidly increasing volumes of digital sensor data. It is claimed that these data may at least partially reflect the dynamics of both environmental and social phenomena in remarkable spatial and temporal detail, and thus open novel research opportunities for the GIScience domain as well (cf. Annoni et al., 2011, Goodchild et al., 2012, Hey et al., 2009).

The focus of this chapter is on urban areas. Conceptually, the methods described would work everywhere the information content is dense enough. We avoid a discussion on what "dense enough" means when targeting cities. The term "city" comprises not only a geographical area characterized by a dense accumulation of people or buildings, but implicitly includes a multilayered construct containing multiple dimensions of social, technological, and physical interconnections and services (Blaschke et al., 2011).

In this chapter, we will discuss concepts which in synopsis may support our vision of integrated urban sensing. In the empirical part, we will concentrate on the research question of how the spatial and temporal nature of the acquired data might be characterized. The hypothesis is that rather than utilizing relatively small samples of individuals, as social sciences may have to, we can gain insights into a "collective behavior" which may characterize some aspects of urban life. The aim is to abstract beyond individual characteristics of probes—moving objects in general including humans and commodities. We will demonstrate methods to analyze, visualize, and explain some of the patterns we identified. The results are not immediately scalable to larger studies but we could prove the appropriateness of the methods in several earlier studies (Sagl et al., 2011, 2012a–c). Indubitably, we build on the ideas of Resch and coworkers (Resch et al., 2011, 2012a,b).

Resch et al. (2012b) even suggest a concept of a "live city," in which the city is regarded as an actuated near real-time control system creating a feedback loop between the citizens, environmental monitoring systems, the city management, and ubiquitous information services, thereby facilitating the "smart city" concept.

In this chapter, we will utilize the clarification of the term "live"—as opposed to common understanding of "near real time"—of Resch et al. (2012b) but we will stay at the analytical level. The aim is to report on the methods of integrated urban sensing, and we will leave out a discussion of a "live city." Such a discussion would need to comprehensively cover privacy and legislative issues. Nevertheless, a visionary outlook can be provided. On purpose, such an outlook is limited to the analytical capabilities and excludes anticipated societal developments. Some analytical capabilities will be demonstrated and it will be briefly discussed as to whether this potential may lead to new vistas of space-time analysis..

Within the context of "live city" and "smart city," the need of an advanced understanding of environmental and social dynamics such as the weather or human behavior is obvious. This also refers to the broad spectrum of sustainable resource management and its various application domains such as electricity, heat, water, transportation, and urban planning as well as safety, security, health care, etc. The cross-integration of multiscale ubiquitous sensor data into spatial, temporal, and spatiotemporal analysis can potentially enhance our understanding of resources' demand and, thus, their efficient allocation (cf. Hancke et al., 2013). Hence, integrated urban sensing enables a more holistic view on urban phenomena and processes, thereby facilitating the concepts of live and smart cities. In fact, integrated urban sensing might be a promising way for quantifying urban performance with respect to both the physical as well as the social and human capital (Ho Van et al., 2009).

The research in the context of integrated urban sensing is diverse. An increasing amount of scholars aim to explore the possibilities of statistical methods of analysis that are better suited for the peculiarities of space-time data. GIS methods are at the core of such options. GIS-based spatial analysis techniques can help unlock and visualize the substantial spatial and temporal components of the geographic phenomena of interest. In addition to the scientific value of such techniques, GIS enables researchers to generate sophisticated visualizations and computer animations that are useful for education. Ultimately, such visualizations serve to convey the results of research on urban systems to a wider public (Blaschke et al., 2012).

In this chapter, we address the need for multiscale integrated sensing for cross-scale integrated monitoring of urban spaces. We do so by linking together three dimensions involved in sensing: machine- or user-generated data, the underlying geographic phenomena, and the type of sensing (Figure 14.1). "Scale" herein refers to both temporal and spatial scales and can be seen as the overarching meta-dimension in the context of sensing, analyzing, and monitoring geographic phenomena. We rely on the concepts of GIScience (Goodchild, 2010) but we do not discuss the role of GIScience as such. There is a significant body of literature about what exactly makes spatial special (e.g., Goodchild, 1991) and we refer to Blaschke and Strobl (2010), who orchestrate various trends and developments in this field under 10 themes.

Specifically, we focus on environmental data, human data including their mobility, and social data. In the following sections, we present several case studies of how "sensors" and "sensor networks" in combination with GIScience concepts can be employed to investigate spatial and temporal characteristics of physical and social phenomena across multiple spatial and temporal scales.

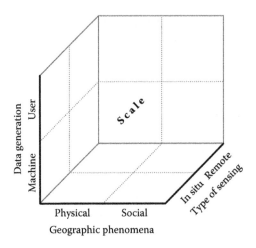

FIGURE 14.1 Dimensions involved in sensing.

14.2 TRIPLE D: DIMENSIONS, DOMAINS, AND DATA OF INTEGRATED URBAN SENSING

In this section, we illustrate and interlink dimensions, domains, and data in the context of integrated urban sensing (Figure 14.2). For instance, social sensor data (representing some social phenomena) are predominately user-generated and sensed in situ while environmental sensor data (representing some physical phenomena) are typically machine-generated (views 2 and 3); there are hardly any sensor data that are sensed remotely and generated by users (view 1). Although such links between the dimensions of sensing might be obvious, we want to make their

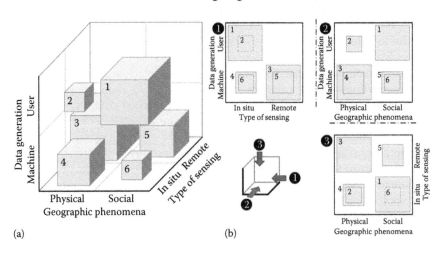

FIGURE 14.2 Blocks of sensor data assigned to the dimensions of sensing: (a) in 3D and (b) from a top/front/side perspective.

contributions to integrated urban sensing more explicit, specifically with respect to different spatial and temporal scales.

Some examples of data types corresponding to Figure 14.2 are as follows:

1. VGI and mobile network traffic
2. VGI in the context of environmental status updates
3. Satellite imagery
4. Measurements from sensors and sensor networks
5. Human settlements extracted from satellite imagery
6. Counter data at entrances and exits of shopping malls, public transport, etc.

14.2.1 CHARACTERIZING DOMAINS OF URBAN SENSING

Sensors and sensor networks generate digital representations of the Earth's surface, or measure the healthiness of vegetation, or the pressure of a snow pack which could potentially destroy a rooftop. Most prominent examples of sensors are weather stations. They measure air temperature, rain fall, solar radiation, particulate matter, etc. They are therefore multisensor stations. Likewise, satellites carry multiple sensors which measure reflectance values of atmospheric parameters such as ozone. However, many sensors, as well as the generation of sensor data, can be characterized in a binary manner (Table 14.1). For instance, air temperature is typically measured periodically and on purpose using a single calibrated in situ sensor that generates accurate measurements. On the other hand, the user-generated traffic in mobile phone networks is managed by the network's backend in order to enable mobile users to communicate wirelessly almost everywhere at any time. As a by-product, the log files from such a backend can reflect the human behavior patterns of millions of mobile users in remarkable spatial and temporal detail. Therefore, the mobile network, which is not intended for sensing, can, however, be used indirectly as a large-scale sensor for human behavior.

TABLE 14.1
Binary Characterization of Sensor
Devices and Sensor Data: Some Examples

In situ	Remote
Single sensor	Multiple sensor
Terrestrial	Aquatic
Machine-generated	User-generated
Singularly	Periodical
Direct	Indirect
On purpose	Not intent for sensing
Demanded	By-product
Voluntarily provided	Involuntarily provided

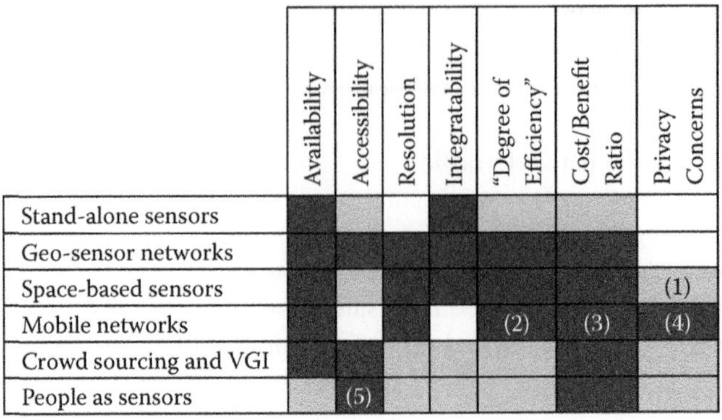

	Availability	Accessibility	Resolution	Integratability	"Degree of Efficiency"	Cost/Benefit Ratio	Privacy Concerns
Stand-alone sensors							
Geo-sensor networks							
Space-based sensors								(1)
Mobile networks					(2)	(3)	(4)
Crowd sourcing and VGI							
People as sensors		(5)					

FIGURE 14.3 Examples of urban sensing techniques and weighting of other characteristics regarding sensor data (white = low/bad, light gray = moderate/medium, dark gray = high/good); (1) specifically high-resolution imagery; (2) well-established infrastructure; (3) requires no additional investment; (4) depending on the aggregation level; and (5) if available, then accessible.

In Figure 14.3, we make an attempt to contrast the following additional characteristics of urban sensing techniques by using three levels: low/bad, moderate/medium, and high/good.

- *Availability* refers to existence and quantity of sensor data.
- *Accessibility* depends on availability and refers to the easiness of data access.
- *Resolution* refers to the spatial, temporal, as well as thematic (i.e., qualitative) granularity.
- *Integratability* refers to the easiness of including sensor data in analysis workflows.
- *Degree of efficiency* is the ratio between information achieved and technical complexity.
- *Cost/benefit ratio* shows the data's added value considering the monetary investment.
- *Privacy concerns* refer to the impingement upon individual or collective privacy rights.

In addition to machine-generated sensor data, the data generated and shared via the Internet voluntarily by individuals (summarized by the term VGI; Goodchild, 2007) and the data generated but shared involuntarily by users of digital systems (e.g., using a mobile phone within a mobile network and thereby generating network traffic) represent an increasingly large and broad sample of the society's behavior (cf. Shoval, 2007).

Different sensors and sensing technologies generate sensor data that represent a geographic phenomenon of interest at different spatial and temporal granularity: remotely sensed data typically have a lower spatial and a lower temporal resolution than in situ sensed data but comprise wider coverage. On the other hand, the granularity of VGI or data from social media is even far from being constant, as is their spatial accuracy, their semantics, and many other data quality parameters.

However, the volume and sample size of data are exploding (Hey et al., 2009), not least due to the use of "thick" mobile devices such as smart phones, which are typically constantly connected to the Internet, social media, and other services of the Web 2.0.

Therefore, we will distinguish two major groups of data. Environmental data are mainly generated by "real" sensors, for instance, weather stations or intelligent mobile sensor pods. We call these "machine-generated" data. On the contrary, direct measurements of humans in space and time are predominantly restricted for privacy reasons. Information about persons or groups and their behavior in space and time is derived either from VGI or from proxy data. Proxy data can stem from, for instance, mobile communication networks or social media. For simplicity, we call this second group "user-generated" data (refer to Figure 14.2).

14.2.2 Why Are Remote Sensing Data Left Out Here?

Although remote sensing is at the core of this book, we will refrain from covering remote sensing concepts herein. Rather, we will focus on other aspects of "Earth observation." First, all the other 17 chapters in this book describe in detail remote sensing platforms and sensors, methods of data acquisition and analysis, and interpretation. Second, remotely sensed data are often seen as the process of generating thematic interpretations from digital signals that model parts of the Earth's surface. Following this definition, we would use the results as thematic layers in integrated sensing applications as categorical data, usually from interval to nominal levels of measurement. Third, Blaschke et al. (2011) have already focused on the integration of remote sensing and other forms of sensing for urban applications. Unclassified image data may be more "objective" and have a greater range of measurement but many applications require classified thematic interpretations. This way, we hope to be able to contribute to more holistic and integrative urban observation systems.

In particular, we postulate non-remote sensing data to be crucially important for the following functions:

1. Characterizing urban ecosystems, built environment, air quality, and carbon emission
2. Developing indicators of population density, environmental quality, and quality of life
3. Characterizing patterns of human, environmental, and infectious diseases

These points reflect goals of the Group of Earth Observation (GEO) task SB-04-C1—Global Urban Observation and Information, for which this book is planned to be a major contribution. We strongly believe that remote sensing data are undoubtedly at the core of these tasks but non-remote sensing (in situ, social network, etc.) data are indispensible, too, and we therefore concentrate on the latter. For example, remote sensing data together with climatological station data are the starting point for modeling urban climate, microclimatological parameters, and phenomena like urban heat islands. Nevertheless, in order to better understand the impacts of global climate change on urban areas, validations are needed and the "people's view" needs to be incorporated—which is a ground view. We postulate that in most cases up to date, people-centered information on the status and development of urban environments and personal spaces will be needed. They may not only be used to fill gaps in global urban observations but they will also in the future be indispensible when characterizing urban ecosystems, population density in built-up environment, air quality, environmental quality, and carbon emission.

For the sake of completeness, it must be stated that remote sensing also delivers data that are used as proxies for environmental parameters without classification. Well-known examples include the Normalized Differenced Vegetation Index (NDVI), the Leaf Area Index (LAI), and land surface temperature (LST), which can be used directly in integrated urban sensing and analysis.

We conclude that remote sensing is a mature technology, particularly for larger-scale observations, that has been significantly utilized in a world increasingly employing geospatial data. However, fine-grained urban remote sensing data are—aside from numerous case studies—hardly available across large areas. In the remainder of this chapter, we will therefore focus on additional and emerging sensing methods that are supposedly less familiar to the target audience of this book.

14.2.3   ENVIRONMENTAL SENSING: "MACHINE-GENERATED" DATA

Environmental sensing, environmental analysis, and environmental monitoring are all well-established fields. The fields may overlap, the terms may be used ambiguously by different communities and their methodologies, and paradigms may undergo changes. They are, nevertheless, not new and do not need to be discussed in detail herein.

What is relatively new is the information technology (IT) framework—typically referred to as the Sensor Web (cf. Delin and Jackson, 2001, Resch, 2012, Zyl et al., 2009)—which enables complex combinations of sensing methods and arrangements of different sensing devices to assess a variety of environmental phenomena. Furthermore, information had to often be constructed out of data only retrospectively, that is, the data acquisition was totally decoupled from the data analysis. Sagl et al. (2012d) describe an exercise where mobile radioactive radiation sensor measurements were spatially interpolated in near real time for supporting rescue forces in time-critical decision making. Although not a typical urban sensing scenario, this can convincingly illustrate the advantage of creating timely information: in a classic workflow, experts would have gathered radiation information in the field and would have created maps containing isolines of certain radiation concentrations afterward. Such a classic mapping exercise could take hours, which could be critical in this example. Sagl et al. (2012d) could show that near–real time and fully automated analysis workflows based on standardized services speed up this process significantly while hiding the heterogeneity of underlying sensors and sensor networks. Purposely, we refer to the term "near real time" as it does not impose rigid deadlines but suggests the dynamic adaptation of a time period according to different usage contexts (Resch, 2012). The terms "live" and "near real time" are seemingly appropriate and used synonymously herein.

14.2.4 Human Sensing: "User-Generated" Data

The digital traces that people continuously leave behind—voluntarily or not—while using communication devices such as mobile phones or interacting with social media platforms reflect their behavior in great detail. These traces can be seen as social sensor data (Sagl et al., 2012c) and can serve as proxy for human activity and mobility. A spatial and temporal analysis of such proxy data can thus provide insights into the social dimension and, moreover, the functional configuration of complex urban systems.

Herein, we focus on the potential of user-generated data from mobile networks and social media. For instance, Sagl et al. (2012b) show that both characteristic and exceptional urban mobility patterns can be derived from handovers (i.e., movements between pairs of radio cells) within a mobile network. Such insights can help to better understand the daily "pulse" of urban movements in the city (Sevtsuk and Ratti, 2010), thereby providing additional information for, for example, public transportation management strategies. In the context of urban mobility, online social networks such as Foursquare can also be used to examine differences and similarities, and derive even universal laws, in human movements across several metropolitan areas around the globe (Noulas et al., 2012). In summary, such studies clearly demonstrate the significance of different user-generated "sensor" data for multiscale integrated urban analysis.

However, the different nature of user-generated data samples results in differences in terms of representativeness and semantic expressiveness: from a user's perspective, "involuntarily" provided mobile network traffic naturally represents a relatively large proportion of the population across social classes; however, it is typically lacking in content. For instance, the number of text messages sent/received is known but not the text itself, or the number and duration of calls is known but not the topic of the talk itself. This is in contrast to social media data, which, first, typically represent a rather specific subgroup of the population, and, second, contain content of some semantic value (e.g., the number and the text of Twitter messages, so-called tweets, are known). In addition, "when data collection is situated 'outside' the thing being studied, observation remains arguably neutral. But when data collection is embedded among the actors within a setting, as in participant observation, a cycle of interactivity is launched in which observation changes behavior that changes observation and so on" (Cuff et al., 2008, p. 28). These aspects shall be taken into account when analyzing user-generated data—or social sensor data in the context of integrated urban sensing.

14.2.5 Combining Environmental with Social Sensing

Several approaches exist that aim to combine environmental (or physical) sensing with social sensing. Typically, such approaches are driven by different contexts. For instance, take the concept of "people as sensors": individual and context-dependent information directly complement sensor measurements of physical phenomena from well-calibrated hardware (Resch, 2013). Following this concept of complementing "real" sensor measurements, Hayes and Stephenson (2011) describe "online sensing" and use blogs, wikis, Twitter, Forums, etc., instead of "people as sensors" directly (although one may say that people are acting as sensors and putting data on these social media sources). In fact, they show how online sensors such as geo-coded images on Flickr or tags from Twitter messages can complement the temporal and spatial coverage of physical sensor measurements, even for cross-correlation. Kamel Boulos et al. (2011) provide a comprehensive review on overlapping domains of sensing including the Sensor Web and citizen sensing in the broad context of environmental and public health surveillance and crisis/disaster management. They argue that crowd sourcing allows for both horizontal and vertical sharing of environmental and social-related information, that is, between and among people using, for example, Twitter, Facebook, etc. (horizontally) or between people and other "machines" (vertically), such as comparing in-house prices with Amazon prices. They claim that "crowd reaching," which is supposed to be the counterpart of crowd sourcing, should be more established to reach the masses with useful and individualized information such as health tips. However, a clear distinction between crowd reaching and location-based services (LBS) remains.

Rather than supporting or complementing data and information from and among different sources, Blaschke et al. (2011) argue for the integration of several geospatial technologies—including remote sensing—in order to gain a more holistic view on urban systems on different spatial and temporal scales. On a rather local or regional scale, Sagl et al. (2011) introduced an approach to bridge the gap between large-scale social sensing and environmental monitoring in order to potentially disclose insights into some instantaneous interactions between people and their environmental context factors. First, they derived basic weather conditions such as "normal" or "adverse" from time series of several meteorological variables (air temperature, rainfall, solar radiation, etc.). Second, these conditions were then linked to aggregated mobile phone usage, which served as a proxy for the collective human behavior, using frequency domain analysis methods. In order to take into account the spatial and temporal domain, this approach was developed further and resulted in the "context-aware analysis approach," which allows for investigating one geographic phenomenon in the context of another; moreover, it allows for quantifying environment–human relationship aspects (Sagl et al., 2012a).

Thus, in the context of integrated urban sensing, diverse technologies can be seen as sensors or sensor networks that are able to generate sensor data reflecting the underlying geographic phenomenon in great detail, thereby contributing to a more holistic understanding of urban phenomena.

14.3   CASE STUDIES

We present three case studies which demonstrate how sensor data from different sensing technologies are combined. Additionally, we present a conceptual framework for fully integrating both environmental and social dynamics. GIScience concepts are implicitly or explicitly used to investigate physical and social phenomena in both time and space.

14.3.1 COLLECTIVE URBAN DYNAMICS

The first example is adopted from Sagl et al. (2012c) for this book chapter. It demonstrates user-generated mobile network traffic and geo-tagged photos from Flickr (a social media platform) can be used to provide additional insights into how collective social activity shapes urban systems. We used different geo-visualization techniques to effectively communicate such insights.

Figure 14.4 shows the overall activity in a mobile network within the course of a typical working day in the city of Udine, northern Italy. While the city center is

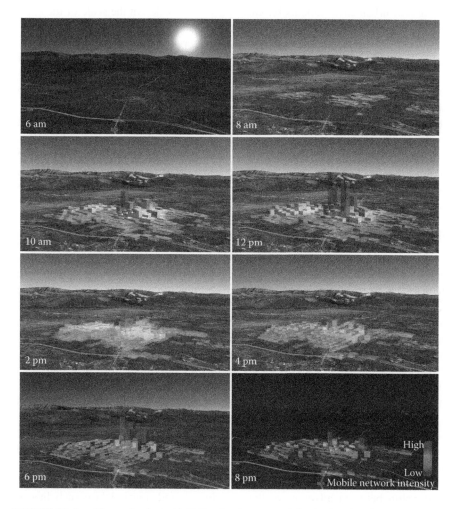

FIGURE 14.4 **(See color insert.)** Collective human activity in the city (Udine, northern Italy)—spatiotemporal mobile communication activity on a typical working day as seen from a mobile network operator's perspective. (From Sagl, G. et al., From social sensor data to collective human behaviour patterns: Analysing and visualising spatio-temporal dynamics in urban environments, in Jekel, T., Car, A., Strobl, J., and Griesebner, G., eds., *GI-Forum 2012: Geovisualization, Society and Learning*, Wichmann Verlag, Berlin, Germany, 2012c.)

clearly identifiable, the urban periphery behaves differently in the morning as compared to the evening. The actual interactive Google Earth application behind these screenshots enables an advanced understanding of where and when people actively use the mobile network.

The second example is adopted from Sagl et al. (2012b). It shows that a purely visual analytics approach can be used to extract characteristic and exceptional urban mobility information from mobile network traffic, more specifically from the number of handovers (i.e., the number of movements between pairs of radio cells). We show, among other things, the symmetry and similarity of the normalized mobility among four urban environments for each day of the week. On the scale of the main administrative urban unit, the overall urban mobility patterns show a surprisingly high degree of similarity and symmetry. All patterns show that the maximal total mobility is reached on Tuesday, closely followed by Wednesday; the minimal total mobility is clearly on Sunday. However, the absolute net migration flows start to diverge on Wednesday and converge again on Sunday. Gorizia, the smallest of the four cities, shows the comparably highest mobility activity on Friday and Saturday, which is confirmed as an asymmetric mobility behavior. In addition, we identified several exceptional patterns in the data and associated them to real-world events such as soccer matches or concerts. This enables an automated identification and classification of exceptional urban mobility behavior and thus potentially facilitates incident management.

14.3.2 CONTEXT-AWARE URBAN SPACES

The consolidation of environmental and social sensor data on a common space-time basis enables a context-aware analysis, that is, the analysis of one geographic phenomenon in the context of another (e.g., human mobility in the context of the weather), thereby facilitating the identification and characterization of relationships, correlations, and possibly even causalities. In a first step, we focused on the evaluation of potential relations between phenomena of interest (e.g., between specific or even extreme weather conditions and the collective human mobility). This includes the use of established as well as the development of novel analysis methods and the evaluation of both. As described in detail in Sagl et al. (2012a), analysis methods from the time, space, and frequency domains have been applied in order to reveal relationships between weather and telecom data. In fact, using the maximal information coefficient (MIC) (Reshef et al., 2011), which is a novel statistic to measure the dependence for two-variable associations, we mapped the strength of that relationship back to the geographic space (Figure 14.5). The locations marked in white (Figure 14.5, L1–L4) indicate that the strength of the relationship between adverse weather conditions and unusual human behavior correlates with the functional configuration of the city—in this case, locations with an obviously high degree of human dynamics: L1 covers a bus hub with a large parking lot; L2 covers the "Centro Studi Volta," a school for multidisciplinary activities; L3 is within a main residential area; and L4 is an official living place for nomadic people and gypsies.

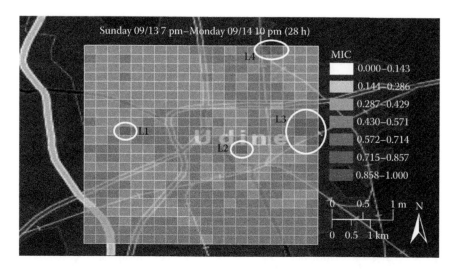

FIGURE 14.5 (See color insert.) Measuring the strength of the relationship between adverse weather conditions and unusual human activity using the MIC. 0, no relationship; 1, functional relationship. (Modified from Sagl, G. et al., *Sensors*, 12, 9835, 2012a.)

14.3.3 Integrated Sensing for a More Holistic Geo-Process Understanding

A variety of "sensors" and "sensor networks" can be used to systematically assess and monitor dynamic geographic phenomena at different spatial and temporal scales. However, the monitoring is typically done for each phenomenon individually (e.g., for air temperature or mobility). In order to enhance—or at least verify—our understanding of both environmental and social processes for multidisciplinary studies, a more holistic monitoring framework is needed. One way to fully integrate the spatiotemporal dynamics of both environmental and social phenomena is the "adaptive geo-monitoring framework" (Figure 14.6). It extends the "adaptive monitoring approach" (Lindenmayer and Likens, 2009) by adding the spatial dimension and the mutual context-awareness of dynamic geographic phenomena.

As described in more detail in Sagl (2012), the adaptive geo-monitoring framework enables further context-aware analysis approaches considering integrated (statistical) analysis methods from the space, time, and frequency domains. Sagl (2012) has demonstrated this concept for a small subset of potential applications. He explored spatial, temporal, and periodic relationships between basic weather conditions and some collective human behavior aspects.

The framework is designed to enable a more holistic process understanding of environmental and social phenomena across spatial and temporal scales. The term "adaptive" refers to, for instance, two or three spatial dimensions, zero or one temporal dimension, and n attribute dimensions; near–real time or "live" as well as postprocessing workflows; aggregation and decomposition of sensor data depending on the thematic focus (e.g., air quality as a composition of particulate matter, CO_2, NO_x, etc.); interpolation and extrapolation of the phenomenon of interest respecting

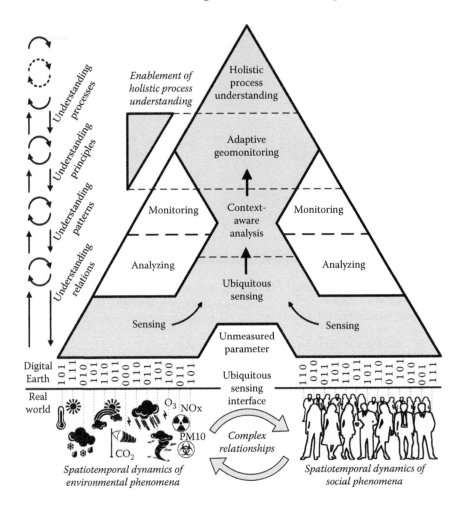

FIGURE 14.6 Adaptive geo-monitoring, a conceptual framework for fully integrating environmental and social dynamics for enabling a more holistic process understanding.

different operational scales (e.g., diurnal and local versus seasonal and regional variability of air temperature); and hybrid reasoning methods by iteratively applying the inductive and deductive research approach.

Hence, the adaptive geo-monitoring approach potentially provides novel capabilities for examining the spatiotemporal behavior of physical and social phenomena through ubiquitous sensing and context-aware analysis.

14.4   DISCUSSION AND CONCLUSION

Within the overall realm of Global Urban Monitoring and Assessment—which is part of the title of this book—this chapter focused on integrated urban sensing. In contrast to the majority of chapters, remote sensing issues were widely excluded here although remote sensing plays a pivotal role for integrated sensing strategies of almost all kinds of information on the Earth's surface. The reason is that Blaschke et al. (2011) explicitly focused on the integration of remote sensing and other forms of sensing for urban applications and we aimed not to repeat ourselves.

In a recent editorial, Tsou and Leitner (2013) postulated an emerging paradigm which aims at mapping cyberspace and social media. A special issue of the journal *Cartography and Geographic Information Science* provides discussions of social theories, innovative mapping methods, sentiment analysis, spatial modeling and statistics, space-time analysis, and geo-visualization examples within this field. We want to particularly highlight the contribution of Li et al. (2013) in this special issue: these authors provide an excellent overview of the big data study in social media by analyzing the spatial pattern of some 20 million Twitter messages and 4.2 million pictures from Flickr. These spatial patterns of big data prove a strong linkage between the uneven distribution of social media messages and the characteristics of local residents (messengers) cross-referenced from census data. Again, we may only be at a starting point of such developments—while remote sensing is believed to be a relatively mature field.

Our chapter highlighted only very few and limited examples within the realm of such trends, or even a paradigm shift as claimed by Tsou and Leitner. The urban social dynamics derived from user-generated sensor data demonstrate that "social sensor networks," specifically mobile networks and social media, significantly support our view on dynamic urban systems. From the novel context-aware analysis approach, we conclude that it is a promising way to provide insights into environment–human interface aspects, thereby potentially enabling a holistic process understanding of environmental and social dynamics in urban spaces.

However, we want to stress that we are aware that we had only explored isolated aspects of human–environment interaction. Such an analysis alone would be too reductionistic. Without venturing into a debate of environmental determinism, it is stated that many factors influence the collective human behavior. In this respect, weather is certainly only one factor that can influence human behavior. The respective case study in Section 14.3 therefore demonstrated mainly that correlations can be investigated—in principle. To what degree the results explain causalities will depend on the application. In fact, we need to be careful since many of the data sources and methods described in this chapter are used as proxies for processes that are much more complex. Such processes can only partially be explained by the datasets and methods used. For social behavior, one needs to critically examine the extent to which mobile phone data and particularly social media data may help us to better understand social communication behavior. Communication is not bound to mobile phone calls and texts. We cannot use these alone to learn about social communication although an increasing number of people use electronic means to communicate. Nevertheless, we strongly believe that our research outcomes can be the basis for further research on environment–human interface aspects and may stimulate interdisciplinary research activities toward the development of an adaptive framework for real-time monitoring and modeling of environment–human feedback loops.

After all these new technological opportunity potentials, the reader finally needs to be reminded of the limitations of all technology: any technology is at its best as good as a wise user has planned for it to be. We have stressed the increasing ability to add location to almost all existing information. In a way, we may claim that this will unlock the wealth of existing knowledge about social, economic, and environmental matters. Furthermore, it could play a vital role in understanding and addressing many of the challenges we face in an increasingly complex and interconnected world. Nevertheless, we are still at the beginning of an era of data affluence in mankind and we will have to guide day-to-day users in this field. There are many remaining and even some new issues of data privacy and new educational duties. In this respect, we do not believe that some space-time behavior analysis methods are now sold as new approaches to old problems only (Timmermans et al., 2002). Rather, new problems and new research questions arise.

Integrated urban sensing is a novel attempt to identify another dimension of cities as a "living space." The approach is clearly in its infancy, but we believe that we will soon see applications where decisions makers will use such information. Maybe a major in the future will better know where her or his people are at what time of the day. While integrated urban sensing will help us understand the spatiotemporal pattern of humans and of groups—even with anonymous and aggregated data—it will not tell us "why." In the times of "big data," we will sophisticate our reasoning methods but we need to keep in mind that only part of social interaction has a spatial component.

REFERENCES

Annoni, A., Craglia, M., Ehlers, M., Georgiadou, Y., Giacomelli, A., Konecny, M., Ostlaender, N. et al. 2011. A European perspective on Digital Earth. *International Journal of Digital Earth*, 4, 271–284.

Blaschke, T., Donert, K., Gossette, F., Kienberger, S., Marani, M., Qureshi, S., and Tiede, D. 2012. Virtual globes: Serving science and society. *Information*, 3, 372–390.

Blaschke, T., Hay, G. J., Weng, Q., and Resch, B. 2011. Collective sensing: Integrating geospatial technologies to understand urban systems—An overview. *Remote Sensing*, 3, 1743–1776.

Blaschke, T. and Strobl, J. 2010. Geographic information science developments. *GIS Science. Zeitschrift für Geoinformatik*, 23(1), 9–15.

Cuff, D., Hansen, M., and Kang, J. 2008. Urban sensing: Out of the woods. *Communications of the ACM*, 51, 24–33.

Delin, K. A. and Jackson, S. P. 2001. The Sensor Web: A new instrument concept. *Proccedings of the SPIE International of Optical Engineering*, 4284, 1–9.

Goodchild, M. F. 1991. Geographic information systems. *Progress in Human Geography*, 15(2), 194–200.

Goodchild, M. F. 2007. Citizens as sensors: The world of volunteered geography. *GeoJournal*, 69, 211–221.

Goodchild, M. F. 2010. Twenty years of progress: GIScience in 2010. *Journal of Spatial Information Science*, 1, 3–20.

Goodchild, M. F., Guo, H., Annoni, A., Bian, L., de Bie, K., Campbell, F., Craglia, M. et al. 2012. Next-generation Digital Earth. *Proceedings of the National Academy of Sciences*, 109, 11088–11094.

Gore, A. 1998. *The Digital Earth: Understanding Our Planet in the 21st Century* [Online]. Available at: http://portal.opengeospatial.org/files/?artifact_id=6210 (accessed 2011/10/08).

Hancke, G., Silva, B., and Hancke, G. P. 2013. The role of advanced sensing in smart cities. *Sensors*, 13, 393–425.

Hayes, J. and Stephenson, M. 2011. Bridging the social and physical sensing worlds: Detecting coverage gaps and improving sensor networks. *The First Workshop on Pervasive Urban Applications (PURBA)* in conjunction with the *Ninth International Conference on Pervasive Computing*, San Francisco, CA.

Hey, T., Tansley, S., and Tolle, K. 2009. *The Fourth Paradigm: Data-Intensive Scientific Discovery*. Microsoft Corporation, Redmond, Washington.

Ho Van, Q., Astrom, T., and Jern, M. 2009. Geovisual analytics for self-organizing network data. *IEEE Symposium on Visual Analytics Science and Technology* (*VAST*), Atlantic City, NJ, October 12–13, 2009, pp. 43–50.

Kamel Boulos, M., Resch, B., Crowley, D., Breslin, J., Sohn, G., Burtner, R., Pike, W., Jezierski, E., and Chuang, K.-Y. S. 2011. Crowdsourcing, citizen sensing and Sensor Web technologies for public and environmental health surveillance and crisis management: Trends, OGC standards and application examples. *International Journal of Health Geographics*, 10, 67.

Li, L., Goodchild, M. F., and Xu, B. 2013. Spatial, temporal, and socioeconomic patterns in the use of Twitter and Flickr. *Cartography and Geographic Information Science*, 40(2), 61–77.

Lindenmayer, D. B. and Likens, G. E. 2009. Adaptive monitoring: A new paradigm for long-term research and monitoring. *Trends in Ecology & Evolution*, 24, 482–486.

Noulas, A., Scellato, S., Lambiotte, R., Pontil, M., and Mascolo, C. 2012. A tale of many cities: Universal patterns in human urban mobility. *PLoS One*, 7, e37027.

Resch, B. 2012. Live geography—Standardised geo-sensor webs for real-time monitoring in urban environments, Dissertations in Geographic Information Science, Heidelberg, Germany, Akademische Verlagsgesellschaft AKA GmbH.

Resch, B. 2013. People as sensors and collective sensing-contextual observations complementing geo-sensor network measurements. In: Krisp, J. M. (ed.) *Progress in Location-Based Services*. Springer, Berlin, Germany.

Resch, B., Britter, R., Outram, C., Chen, X., and Ratti, C. 2011. Standardised geo-sensor webs for integrated urban air quality monitoring. In: Ekundayo, E. O. (ed.) *Environmental Monitoring*. InTech. Available at: http://www.intechopen.com/books/environmental-monitoring/standardised-geo-sensor-webs-for-integrated-urban-air-quality-monitoring.

Resch, B., Britter, R., and Ratti, C. 2012a. Live urbanism—Towards the senseable city and beyond. In: Pardalos, P. A. R. S. (ed.) *Sustainable Architectural Design: Impacts on Health*. Springer, New York.

Resch, B., Zipf, A., Beinat, E., Breuss-Schneeweis, P., and Boher, M. 2012b. Towards the live city–paving the way to real-time urbanism. *International Journal on Advances in Intelligent Systems*, 5, 470–482.

Reshef, D. N., Reshef, Y. A., Finucane, H. K., Grossman, S. R., McVean, G., Turnbaugh, P. J., Lander, E. S., Mitzenmacher, M., and Sabeti, P. C. 2011. Detecting novel associations in large data sets. *Science*, 334, 1518–1524.

Sagl, G. 2012. Towards adaptive geo-monitoring: Examining environmental and social dynamics and their relationships for holistic process understanding. *Seventh International Conference on Geographic Information Science* (*GIScience*), Columbus, OH, September 18–21, 2012.

Sagl, G., Beinat, E., Resch, B., and Blaschke, T. 2011. Integrated geo-sensing: A case study on the relationships between weather conditions and mobile phone usage in northern Italy. *First IEEE International Conference on Spatial Data Mining and Geographical Knowledge Services* (*ICSDM*), Fuzhou, China, June 29–July 1, 2011, pp. 208–213.

Sagl, G., Blaschke, T., Beinat, E., and Resch, B. 2012a. Ubiquitous geo-sensing for context-aware analysis: Exploring relationships between environmental and human dynamics. *Sensors*, 12, 9835–9857.

Sagl, G., Loidl, M., and Beinat, E. 2012b. A visual analytics approach for extracting spatio-temporal urban mobility information from mobile network traffic. *ISPRS International Journal of Geo-Information*, 1, 256–271.

Sagl, G., Resch, B., Hawelka, B., and Beinat, E. 2012c. From social sensor data to collective human behaviour patterns: Analysing and visualising spatio-temporal dynamics in urban environments. In: Jekel, T., Car, A., Strobl, J., and Griesebner, G. (eds.), *GI-Forum 2012: Geovisualization, Society and Learning*. Wichmann Verlag, Berlin, Germany.

Sagl, G., Resch, B., Mittlboeck, M., Hochwimmer, B., Lippautz, M., and Roth, C. 2012d. Standardised geo-sensor webs and web-based geo-processing for near real-time situational awareness in emergency management. *International Journal of Business Continuity and Risk Management*, 3, 339–358.

Sevtsuk, A. and Ratti, C. 2010. Does urban mobility have a daily routine? Learning from the aggregate data of mobile networks. *Journal of Urban Technology*, 17, 41–60.

Shoval, N. 2007. Sensing human society. *Environment and Planning B: Planning and Design*, 34, 191–195.

Timmermans, H., Arentze, T., and Joh, C.-H. 2002. Analysing space-time behaviour: New approaches to old problems. *Progress in Human Geography*, 26, 175–190.

Tsou, M.-H. and Leitner, M. 2013. Visualization of social media: Seeing a mirage or a message? *Cartography and Geographic Information Science*, 40(2), 55–60.

Zyl, T. L. v., Simonis, I., and McFerren, G. 2009. The Sensor Web: Systems of sensor systems. *International Journal of Digital Earth*, 2, 16–30.

15 Object-Based Image Analysis for Urban Studies

Vivek Dey, Bahram Salehi, Yun Zhang,
and Ming Zhong

CONTENTS

15.1 INTRODUCTION

Remote sensing (RS) has proven to be an indispensable technology for urban studies because of its ability to frequently update (in a few weeks) the information of urban areas. Although RS includes airborne imagery, light detection and ranging (LIDAR), radio detection and ranging (RADAR) imagery, etc., this work limits its analysis to space-borne satellite imagery. This is because space-borne RS is a cost-effective alternative for urban land-use mapping and urban feature change detection, as compared to airborne imagery and LIDAR data (Weng and Quattrochi 2006, Netzband et al. 2007, Yang 2011).

Although RS technologies have existed for nearly four decades, RS has achieved its grounds in urban studies only in the last decade. This is attributed to the tremendous growth in the sensor technologies for RS. As a result, the spatial resolution of remotely-sensed Earth observation satellite images has been refined continuously from 80 m of Landsat-1 (launched in 1972) to 0.46 m of WorldView-2 (launched in 2009). Similarly, spectral resolution has been refined from 4 multispectral bands to 220 hyperspectral bands of Hyperion satellite (launched in 2000) sensor, but they have low spatial resolution (30 m). While spatial resolution of hyperspectral bands is good for a wide range of applications in urban studies (such as urban forestry, urban heat island, and impervious surface mapping), very high resolution (VHR) imagery (having <1 m spatial resolution) is required for the identification of individual urban features (such as roads, buildings, and trees) at local scales of urban areas (Blaschke 2010, Blaschke et al. 2011). The products of VHR images are more crucial for urban studies (such as urban planning) because of the level of detail they can provide.

Behind the development of RS for urban studies is the paradigm shift of technology from pixel-based image analysis to object-based image analysis (OBIA) for analyzing VHR satellite images (Castilla and Hay 2008). This shift is the result of the change in the US government laws which relaxed the restriction on the spatial resolution of commercially available satellite images and the availability of commercially available object-based image processing software, eCognition™ since 2000 (Aplin et al. 1997, Blaschke and Strobl 2001, Castilla and Hay 2008). The reason behind the failure of the pixel-based image analysis was its inability to match or even come closer to human perception of images and to achieve a better classification result with a high-spatial resolution image as compared to a low-spatial resolution image (Blaschke and Strobl 2001, Flanders et al. 2003, Blaschke et al. 2006). Humans perceive an image feature as a group of pixels, and this perception is much more pronounced in a VHR image (<1 m) as compared to a Landsat-7 TM image of 30 m resolution. Figure 15.1 shows the difference in visual perception (e.g., identification of individual houses and trees) with an increase in spatial resolution. Fortunately, OBIA has been widely researched in other disciplines, such as computer vision and medical image processing. Consequently, OBIA was easily able to support increasing amounts of research

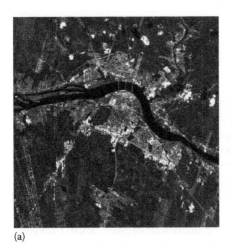 

(a)   (b)

FIGURE 15.1 **(See color insert.)** Depiction of the visual details in low- and high-resolution satellite images of same dimension, 512 × 512 pixels of downtown Fredericton, Canada: (a) Landsat TM 30 m MS and (b) QuickBird 2.44 m MS. (From Wuest, B.A., Towards improving segmentation of very high resolution satellite imagery, MScE thesis, Department of Geodesy and Geomatics Engineering Technical Report No. 261, University of New Brunswick, Fredericton, New Brunswick, Canada, 2008.)

in the last decade on the processing of RS images (Blaschke 2010). However, the context of application of OBIA for RS images and in other disciplines is different. Hence, OBIA techniques are required to be adapted for VHR RS images (Castilla and Hay 2008).

Because of this increasing interest in OBIA in RS and urban studies, this chapter reviews existing OBIA techniques, provides overall information on OBIA techniques, and draws conclusions on the advantages and limitations of current OBIA techniques. The overall information on OBIA techniques includes the concept of OBIA, its renaming to geographic object-based image analysis (GEOBIA), its major technical developments and applications for urban studies (mainly segmentation), commercially available software for OBIA, accuracy assessment for OBIA, and future prospects of OBIA.

15.2 CONCEPT OF OBIA AND RENAMING TO GEOBIA

The fundamental and most problematic step of OBIA is image segmentation. The next step of OBIA is to classify these homogeneous objects into a meaningful geographic representation, such as residential houses and roads, using internal features of the objects and their mutual relationships with other objects (Hay and Castilla 2008). For example, one can identify the boundaries of the residential houses, small residential buildings, large commercial buildings, and possibly some trees in Figure 15.1b of urban downtown of Fredericton, Canada. These boundaries correspond to boundaries of human-perceived homogeneous geo-objects. This example reveals that the application of OBIA on high–spatial resolution images has two major benefits, which are specified as follows:

1. It can provide geographic information systems (GIS) ready information, which consists of vector boundaries of the geo-objects and their features as the associated attributes, for example, building feature extraction (Dey et al. 2011).
2. It enables urban mapping at small scales. For example, OBIA enables mapping at a small scale of a community area in a city. Such mapping is an essential aspect of urban studies, for example, urban planning, transportation planning, and land-use development.

These benefits assume that the results from computer-based OBIA and human-perceived reference results are comparable (Zhang et al. 2008). For accurate urban studies, fully automated OBIA is still not used due to visually pleasing aspect of the results of image segmentation (Zhang et al. 2008, Marpu et al. 2010).

Before delving into image segmentation, it is important to understand the context of OBIA for RS images and why the new acronym, GEOBIA, is closer to this context than conventional OBIA. Hay and Castilla (2008) stated that the shift of RS image analysis from pixel-based to OBIA is a paradigm shift, and this shift has been widely embraced by the RS community that deals with geographic objects. In order to have a distinct identity pertinent to challenges and problems of GIScience, the term GEOBIA was coined during the first international conference on OBIA for RS images held at the University of Salzburg on July 4–5, 2006 (Hay and Castilla 2008). Further, Hay and Castilla (2008) defined the primary objective of GEOBIA as "a discipline is to develop theory, methods and tools sufficient to replicate (and/or exceed experienced) human interpretation of RS images in automated/semi-automated ways." As stated by Hay and Castilla (2008), the emphasis of GEOBIA should be to produce geographic information from RS images (i.e., classification of segmented polygons to geographic objects) using OBIA techniques rather than developing new OBIA techniques, which should be adapted from rich existing and continuous research on image segmentation from other disciplines, such as computer vision and medical image processing.

Nevertheless, it is important to understand the existing techniques of image segmentation to analyze which one is best suited for GEOBIA. Therefore, the next section describes in more detail image segmentation and its techniques, which are suitable for the analysis of VHR RS images.

15.2.1 IMAGE SEGMENTATION FOR URBAN VHR RS IMAGES

In an image segmentation process, the image is partitioned into segments (also known as regions or objects) such that each segment is distinct from its surrounding segments based on the homogeneity criteria (Pal and Pal 1993, Blaschke 2010). Homogeneity criteria are employed to determine discontinuity (i.e., edges), similarity (i.e., regions), or both (Gonzalez and Woods 2002). This constitutes two distinct approaches of image segmentation, namely, region-based and edge-based segmentation. Edge-based segmentation methods mainly involve edge detection followed by a contour-generating algorithm (Schiewe 2002). Edge-based methods often identify noise as edges, which result in oversegmentation. Due to these problems, edge-based methods are not discussed further. The second approach is a region-based process, which is further subcategorized as a top-down approach, that is, breaking an image into grids to achieve segmentation, and a bottom-up approach, that is, growing a region/segment from a single pixel. Both approaches are followed by merging and refinement of regions (Guindon 1997, Wuest and Zhang 2009, Corcoran et al. 2010). These approaches state how pixels can be grouped to achieve image segmentation. In addition, they also give an idea about the possible techniques that can be employed for image segmentation.

15.2.2 ADVANTAGES OF IMAGE SEGMENTATION

Though image segmentation is often categorized as one of the most critical tasks in image processing, its benefits supersede its drawbacks (Pal and Pal 1993, Blaschke et al. 2006). The major benefits of image segmentation–based object formation in the RS image analysis are as follows: (a) it identifies image objects (regions) as perceived by the human eye; (b) it enables the use of shape, size, and contextual information for analysis; (c) it allows the use of topological relationships for vector-based GIS operations; (d) it decreases the execution time of classification and increases its accuracy; (e) it minimizes the modifiable areal unit problem (MAUP), caused by the dependency of statistical results (e.g., mean and standard deviation) on the spatial units (the chosen spatial resolution of study); and (f) it reduces the fuzzy boundary problems (Blaschke et al. 2006, Blaschke 2010). However, these benefits are outcomes of using an appropriate OBIA technique of RS domain for a certain application. The appropriateness of an OBIA technique is determined based on the homogeneity criteria used in that technique. For example, Song and Civco (2004) used the shape index and density features, which are useful for linear objects, to identify road objects after segmentation of a preclassified IKONOS satellite image using eCognition™.

15.2.3 Homogeneity Criteria for Image Segmentation

Digital production of geo-objects aims at imitating the visual cognition of humans in analyzing images (Blaschke and Strobl 2001). Hence, it is obvious that homogeneity criteria should be derived from features of the visual cognitive ability of humans, that is, visual cues for image analysis (Estes 1999). These visible cues can also be called image interpretation elements. In the digital world, these image interpretation elements are derived from features of objects (groups of aggregated pixels), such as area, brightness, contrast, and geometry. The major image interpretation elements include (1) spectral; (2) spatial, which includes texture, morphology (shape and size), and context; (3) connectivity; and (4) association. A brief discussion of each of these image interpretation elements is provided next.

Spectral features consist of mainly contrast, brightness, and color and are the most important features in the interpretation of both pixel-based image analysis and OBIA. On the other hand, spatial interpretation is more prominent in OBIA and consists of features related to neighborhoods of a pixel or an object. For example, the density feature of an object gives an idea about the compactness of the object, whereas a high-density object represents a near-square building object and a very-low-density object represents a road object (Benz et al. 2004, Definiens AG 2009). Textures are basically distinct spatial patterns on an image. For example, high residential density areas and commercial areas have distinct textures on a VHR RS image.

Shape and size features determine the geometry of an object and are important for scale- and shape-based segmentation. For example, a segmentation procedure of an urban area image might exploit the fact that all the residential buildings have a rectangular shape at different scales and sizes. Figure 15.2 illustrates the concept

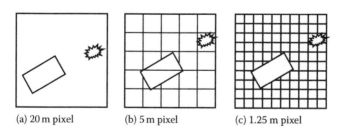

(a) 20 m pixel   (b) 5 m pixel   (c) 1.25 m pixel

FIGURE 15.2 Concept of the appropriate scale of representation of an object. Two objects of rectangular, R, and star, S, shapes have been taken as examples, where (a) at 20 m spatial resolution (SR) shows that both objects are undetectable; (b) at 5 m SR shows that R is detectable but S is not; and (c) at 1.25 m SR shows that both objects are detectable. (Reprinted from *ISPRS J. Photogram. Remote Sensing*, 65, Blaschke, T., Object based image analysis for remote sensing, 2–16. Copyright 2010, from Elsevier.)

of scale in GEOBIA. Among shape and size features, compactness and smoothness features are the most widely used for image segmentation because of their utility in the segmentation technique of eCognition™. A square-shaped building is a compact shape whereas a road feature is less compact. On the other hand, a building feature is a highly smooth object whereas a tree is less smooth.

Context is a spatial property because it is always based on the spatial neighborhood of an object or pixel. Context proves crucial for human cognition in identifying objects having similar spectral and shape properties. For example, a shadow of an object gives a clue that the object has a height component and it might be a building. Connectivity is important in identifying special geo-objects such as roads and rivers. Connectivity utilizes topological properties of adjacency for object analysis. Finally, association features in conjunction with context provide cues for associating an object with its geographical counterpart. For example, a parking lot is associated with multiple cars and thus is differentiable from spectrally similar roads in a VHR image of an urban area. These image interpretation elements are crucial in both segmentation and classification aspects of GEOBIA (Dey et al. 2010, Salehi et al. 2012a).

A plethora of research work has been done on GEOBIA with these interpretation elements, and the major techniques used in these works are as follows: (1) clustering, (2) mathematical model-based techniques, and (3) multiscale models. The following sections deal with the development of these techniques in the context of GEOBIA.

15.2.3.1 Clustering Techniques

Clustering techniques have been traditionally employed for unsupervised image classification to group pixels based on certain homogeneity criteria. Segmentation and clustering are conceptually different with respect to their origins. While traditional clustering techniques rely on aggregation of pixels in the spectral domain, segmentation attempts to aggregate pixels in a spatial domain (Haralick and Shapiro 1985). Therefore, this technique has remained unpopular. However, a hybrid of clustering and other techniques can be used for image segmentation. For example, Wang et al. (2010) proposed a multiscale region-based image segmentation (RISA), which utilizes a hybrid of K-means clustering, and a region-merging approach. They showed that RISA generates results comparable with results of eCognition™ using scale, spectral, compactness, and smoothness criteria of objects for urban land cover mapping.

15.2.3.2 Mathematical Model–Based Techniques

This group of techniques employs widely used mathematical concepts, which can also be utilized for image segmentation purposes. The major techniques of this group employed for image segmentation in GEOBIA are Markov random field (MRF), fuzzy logic, artificial neural network (ANN), and level set model. While MRF uses the concept of spatial neighborhood for image segmentation and is particularly good in determining textured objects, fuzzy logic uses the concept of fuzziness to determine boundaries of objects (Pal and Pal 1993, Shankar 2007, Dey et al. 2010). Similarly, ANN and the level set model use several optimization techniques to outline the objects (Shankar 2007). In most of their implementation, these four techniques require several parameters to be optimized for image segmentation (Dey et al. 2010). A few recent applications of these techniques include the following: (1) Lizarzo and Barros (2010) applied fuzzy logic for a complete GEOBIA, that is, including both segmentation and classification, on a QuickBird image using various spectral, shape, size, and customized spatial features; and (2) Karantzalos and Argialas (2009) applied level set techniques for segmentation-based urban building and road feature detection from a QuickBird image. However, these techniques are relatively less popular for GEOBIA because they inherently lack the usage of the concept of scale in a geographic sense (Castilla and Hay 2008, Dey et al. 2010). Further, there is a lack of commercially available software utilizing these techniques.

15.2.3.3 Multiscale Model–Based Techniques

Multiscale model is a conceptual model that inherits the concept of scale, a core feature of geo-objects, in its analysis (Castilla and Hay 2008). The idea of a multiscale model is to identify different objects at different scales such that they are hierarchically related to each other. For example, a single-family residential house is extracted at a lower scale as compared to a community comprising multiple residential houses. The multiscale model is widely popular for applications of GEOBIA on VHR images of both urban areas and other land covers (e.g., agriculture and glacier) (Blaschke 2010). In addition, commercially available software, eCognition™ (available since 2000), has implemented a segmentation algorithm based on multiscale region growing and merging and named the algorithm Fractal net evaluation approach (FNEA). eCognition™ has complete capability of GEOBIA application, which includes object-based classification as well as creation of refined vector boundaries of geo-objects (Baatz and Schäpe 2000, Definiens AG 2009). Segmentation outputs of FNEA have become a de facto standard for comparison of other developing segmentation techniques in GEOBIA and have been used for all sorts of applications in urban studies such as urban sprawl, urban heat island, land-use/land-cover mapping, environmental mapping, urban disaster management, etc. (Neubert et al. 2008, Blaschke 2010, Blaschke et al. 2011). Owing to the popularity of the multiscale model, particularly FNEA, it is further expatiated in the following sections.

15.2.3.3.1 Analysis of Multiscale Model

Apart from FNEA, none of the other multiscale models gained wide attention for GEOBIA. A few other examples include multiscale object-specific analysis (MOSA) by Hay and Marceau (2004), which was applied for the segmentation of an IKONOS image for agricultural landscape, and RISA by Wang et al. (2010), which was applied for the segmentation of an urban QuickBird image to determine land covers. Although FNEA benefits a lot by being implemented in commercial software, it owes its popularity to its ability of solving/softening a majority of five major problems in a general segmentation technique. These five major problems are as follows:

1. Optimal parameter estimation: Most of the segmentation techniques require parameter estimation before segmentation. For example, FNEA, a multiscale model, requires estimation of three parameters, namely, scale, shape, and compactness, and these parameters impact the segmentation results to a significant level (Benz et al. 2004, Möller et al. 2007, Tian and Chen 2007). The traditional approach of estimation of these parameters is

a trial-and-error approach (Benz et al. 2004). Efficient implementation of FNEA in commercial software and several researches on efficient parameter estimation of FNEA have significantly reduced the effort of parameter estimation (Möller et al. 2007, Tian and Chen 2007, Drăgut et al. 2010, Tong et al. 2012).

2. Reproduction of results: In region-based segmentation approaches, a bottom-up approach of region growing requires random seed pixel generation (Pal and Pal 1993). This randomization produces a different segmentation result for each run of segmentation with the same parameters. Hence, one cannot use a trial-and-error approach for parameter estimation because a reliable result cannot be regenerated with the same set of parameters. eCognition™ has implemented a proprietary algorithm for this random seed pixel generation, which ensures reproduction of the same segmentation result on every run with the same parameters.

3. Inclination on operator: Despite the development of many objective assessment techniques of segmentation accuracy, visual/subjective assessment of segmentation results is widely used (Zhang 1997, Castilla and Hay 2008, Lang et al. 2009, Corcoran et al. 2010, Dey et al. 2010, Marpu et al. 2010). Hence, an optimum segmentation result involves an operator's discretion. Although eCognition™ does not directly provide any objective assessment techniques, it provides predefined and customizable features of an object, which can be used in comparison with the reference results of a vector-based GIS dataset.

4. Undersegmentation and oversegmentation: Each and every segmentation result has to deal with the problem of undersegmentation and oversegmentation (Zhang 1997, Marpu et al. 2010). Figure 15.3 depicts a case of under- and oversegmented results. Therefore, deciding on an optimum segmentation result is the key in GEOBIA. Castilla and Hay (2008) provided an excellent definition of an optimal yet practical segmentation result: "[A] good segmentation is one that shows little over-segmentation and no under-segmentation, and a good segmentation algorithm is one

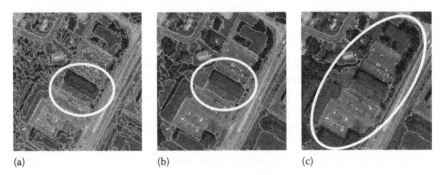

(a)   (b)   (c)

FIGURE 15.3 Illustration of (a) oversegmentation, (b) correct segmentation, and (c) undersegmentation of a building object (in black outline) of a QuickBird image of Fredericton, Canada.

that enables the user to derive a good segmentation without excessive fine tuning of input parameters." FNEA also suffers from this problem. GEOBIA still needs human interpretation for a decision on optimal segmentation (Zhang et al. 2008).

5. Execution time of algorithm: Due to an accelerated increase in the computational power, execution time has become a less significant factor. Nevertheless, a trial-and-error based approach, also required in FNEA, for an optimal segmentation result deprives us from a quicker analysis of terabytes of RS images being generated daily (Möller et al. 2007, Tian and Chen 2007, Drăgut et al. 2010, Tong et al. 2012).

15.2.4 OTHER SEGMENTATION TECHNIQUES

Among other model-based segmentation techniques, watershed-based and hierarchical split and merge region (HSMR) techniques are the latest. Watershed-based techniques utilize a seed pixel–based region-growing method whereas HSMR uses a top-down region-based approach of image splitting and merging, followed by boundary refinement (Beucher 1992, Wuest and Zhang 2009). Wuest and Zhang (2009) applied their HSMR technique for urban land-use mapping of a QuickBird image using scale, shape, and size features. Watershed-based techniques have seen some recent developments for urban VHR image segmentation, such as size-constrained region merging (SCRM) by Castilla et al. (2008) on a QuickBird image of an agricultural area. In an evaluation of available segmentation techniques in GEOBIA, Marpu et al. (2010) found that SCRM produces results comparable to that of eCognition™. However, SCRM lacks the implementation of a multiscale representation of image objects (Castilla et al. 2008).

Among commercially available software, ENVI EX feature extraction (since 2007), ERDAS IMAGINE objective (since 2008), and Feature Analyst (since 2001) are alternatives to eCognition™ for GEOBIA. ENVI EX feature extraction software uses a multiscale segmentation algorithm that uses more than five user-defined parameters for GEOBIA with an ability to show a window-based preview of segmentation. ERDAS IMAGINE objective uses machine-learning algorithms along with morphological, edge-based, and vector operations for GEOBIA (Mehta et al. 2012, Susaki 2012). Feature Analyst also uses machine-learning algorithms for GEOBIA (Optiz and Blundell 2008). A more comprehensive list of commercial segmentation software and their performance comparisons can be found in Marpu et al. (2010). All of these commercial software incorporate crucial image properties (e.g., spectral, spatial, texture, shape, size, and association) used for GEOBIA. In addition to these commercially available software, programming packages for segmentation are also available for quick development of customized algorithms. Two of the examples include the commercially available segmentation package of MATLAB® and the open-source segmentation package from OPENCV. However, these software and programming packages need more research to establish their effectiveness in solving the five major problems of image segmentation stated in Section 15.2.3.3.1 (Blaschke 2010, Marpu et al. 2010).

Based on the last several sections on image segmentation techniques for GEOBIA, it can be concluded that the RS community has widely relied on eCognition™ for image segmentation and classification in GEOBIA. Secondly, there is a lack of an objective measure to assess the accuracy of image segmentation. Finally, visual assessments of segmentation results are considered to be a norm in GEOBIA.

15.3   OTHER PROCESSES IN GEOBIA

Apart from image segmentation, GEOBIA also involves classification of objects into valid geo-objects based on scales of the object. Even with a fixed segmentation process, there are many groups of techniques for classification. A few examples from earlier sections are the FIRME model of Lizarzo and Barros (2010) and the segmentation-based classification of objects by Blaschke et al. (2006). Although the same set of image interpretation elements can be utilized for both segmentation and classification, their conceptual interpretations are different. Unlike segmentation, classification mostly uses ancillary data and height data for the categorization of objects into geo-objects (Salehi et al. 2012a). eCognition™, ENVI EX, Feature Analyst, and ERDAS IMGINE objective also provide a wide range of inbuilt and customized attributes/features of objects for performing a rule-based classification (Benz et al. 2004). Similar to segmentation, classification opens a whole gamut of techniques for GEOBIA. However, in this chapter, the analysis is limited to past progressions of image segmentation in GEOBIA. Authors suggest referring to Salehi et al. (2012b) for a comprehensive review of classification techniques for GEOBIA.

Although image segmentation and subsequent object-based classification provide a vector-based output, the result is not directly GIS-ready. Figure 15.4 shows

FIGURE 15.4   Typical image segmentation result of urban VHR GeoEye-1 image, 0.41 m for panchromatic and 1.65 m for multispectral bands, of Hobart city. (From GeoEye-1 image of Hobart, Tasmania, Australia. Licensed to authors for reproduction in research papers.)

an example of an image segmentation process of an urban VHR pan-sharpened GeoEye-1 RS image of Hobart city in Australia (Zhang 2004). The image reveals that a typical image segmentation process does not result in smooth boundaries as seen in vector GIS data of an urban area comprising boundaries of buildings, roads, parcels, etc. Hence, after association of these segments into geographic objects, further processing is required to obtain the results in a GIS-ready format. These processing steps include mainly boundary smoothening and attribute table generation for vector boundaries. However, these steps are standard procedures and are not listed among bottlenecks of GEOBIA.

15.4 CONCLUSION

This chapter reviewed the developments of OBIA in the context of urban studies. As OBIA in RS is associated with geo-objects, the research community of RS agreed to rename it as GEOBIA to more clearly reflect its adherence to RS (Castilla and Hay 2008). GEOBIA is crucial in urban studies due to its capability of analyzing a VHR RS image similar to human interpretation of images and of identifying urban features and their changes at different spatio-temporal scales. As a bonus, GEOBIA results in the production of vector boundaries, which are close to a GIS-ready format. The application of GEOBIA in urban studies includes urban land-use mapping, urban heat island, transportation mapping, urban forest analysis, urban sprawl, and urban pollution monitoring (Netzband et al. 2007).

This chapter mainly focused on the segmentation aspect of GEOBIA. The chapter categorized the segmentation techniques into seven major categories: (1) clustering, (2) MRF, (3) fuzzy logic, (4) ANN, (5) level set, (6) multiscale, and (7) watershed model. Table 15.1 provides a list of selected review papers having theoretical possibilities for future applications. FNEA, a multiscale model implemented by eCognition™, is the most widely used technique among the GEOBIA research community and has become a standard technique, similar to the Gaussian maximum likelihood technique for pixel-based classification. Hence, for urban studies, the best OBIA technique is FNEA using eCognition™ for all urban studies because it has shown consistent results in all fields of urban applications (Blaschke 2010). Nevertheless, commercially available software such as ENVI EX, ERDAS IMAGINE objective, and Feature Analyst are emerging, which are in the same league as eCognition™.

Despite several advances in the objective accuracy assessment of image segmentation techniques, visual assessment of segmentation results by a human operator is still widely used (Zhang et al. 2008, Marpu et al. 2010). Therefore, it is suggested that both visual and objective assessment techniques should be used for determining optimal image segmentation for detection of over- and undersegmentation. In addition to image segmentation, GEOBIA includes classification steps, which have their own pros and cons and are also a bottleneck in GEOBIA. The use of ancillary data is proving to be effective in image segmentation and is the way to go for successful classification even with eCognition™-based classification (Salehi et al. 2012a).

TABLE 15.1

List of Selected Reviewed Papers on Seven Major Categories of Segmentation Techniques with Possible Potential Future Applications

Authors (Method)	Categorization				
	Interpretation Elements	Approach	Image Used	Evaluation Method	Application
Clustering approach					
Wang et al. (2010) (RISA)	Spectral, spatial, scale, shape, and size	Region growing and merging	SPOT-5 and QuickBird	Classification accuracy	Urban area (implemented as software)
Fuzzy model					
Fan et al. (2009) (SWFCM)	Spectral, spatial, and prior knowledge	Cluster growing	Landsat TM	Classification accuracy and cluster validity indices	Agriculture mixed water land
Lizarazo and Barros (2010) (FIRME model)	Spectral, spatial, and contextual	Region growing	QuickBird	Classification accuracy 83%	Urban
Level set					
Karantzalos and Argialas (2009)	Spectral, shape, size, and scale	Region based	QuickBird	Visual assessment	Urban
Multiscale model					
Baatz and Schäpe (2000)	Spectral, spatial, size, and shape	MR segmentation	Almost all VHR RS imagery	Visual assessment	Implemented as software
MOSA by Hay and Marceau (2004)	Spectral, size, scale, and spatial	Region based	IKONOS Pan	Visual assessment	Forest
Watershed model					
SCRM by Castilla et al. (2008)	Spectral, shape, size, and spatial	Region growing	QuickBird	Visual assessment	Agricultural and urban
HSMR model					
Hu et al. (2005)	Spectral, texture, size, and scale	Region splitting and merging	QuickBird and IKONOS (MS and Pan)	Visual assessment	Urban
Wuest and Zhang (2009)	Spectral, texture, size, and scale	Region splitting and merging	QuickBird MS	Visual assessment	Urban

REFERENCES

Aplin, P., Atkinson, P. M., and P. J. Curran. 1997. Fine spatial resolution satellite sensors for the next decade. *International Journal of Remote Sensing* 18:3873–3881.

Baatz, M. and A. Schäpe. 2000. Multiresolution segmentation—An optimization approach for high quality multi-scale image segmentation. In *AGIT Symposium, Angewandte Geographische Informations-Verarbeitung XII*, ed. J. Strobl et al., Salzburg, Germany. www.ecognition.cc/download/baatz_schaepe.pdf (accessed March 20, 2013).

Benz, U. C., Hofmann, P., Willhauck, G., Lingenfelder, I., and M. Heynen. 2004. Multi-resolution, object-oriented fuzzy analysis of remote sensing data for GIS ready information. *ISPRS Journal of Photogrammetry & Remote Sensing* 58:239–258.

Beucher, S. 1992. The watershed transformation applied to image segmentation. In *10th Pfefferkorn Conference on Signal and Image Processing in Microscopy and Microanalysis*, 16–19 September, Cambridge, U.K., pp. 299–314.

Blaschke, T. 2010. Object based image analysis for remote sensing. *ISPRS Journal of Photogrammetry and Remote Sensing* 65:2–16.

Blaschke, T., Burnett, C., and A. Pekkarinen. 2006. Image segmentation methods for object based analysis and classification. In *Remote Sensing Image Analysis: Including the Spatial Domain,* eds. S. M. de Jong and F. D. van der Meer. Springer-Verlag, Dordrecht, the Netherlands, pp. 211–236.

Blaschke, T., Hay, G. J., Weng, Q., and B. Resch. 2011. Collective sensing: Integrating geospatial technologies to understand urban systems—An overview. *Remote Sensing* 3:1743–1776.

Blaschke, T. and J. Strobl. 2001. What's wrong with pixels? Some recent developments interfacing remote sensing and GIS. *GIS—Zeitschrift für Geoinformationssysteme* 6:12–17.

Castilla, G. and G. J. Hay. 2008. Image objects and geographic objects. In *Object-Based Image Analysis: Spatial Concepts for Knowledge-Driven Remote Sensing Applications*, eds. T. Blaschke, S. Lang, and G. J. Hay. Springer-Verlag, Berlin, Germany, pp. 91–110.

Castilla, G., Hay, G. J., and J. R. Ruiz-Gallardo. 2008. Size-constrained region merging (SCRM): An automated delineation tool for assisted photointerpretation. *Photogrammetric Engineering & Remote Sensing* 74:409–419.

Corcoran, P., Winstanley, A., and P. Mooney. 2010. Segmentation performance evaluation for object-based remotely sensed image analysis. *International Journal of Remote Sensing* 31:617–645.

Definiens AG. 2009. *eCognition Developer 8.0, Reference Book*. Definiens AG, München, Germany.

Dey, V., Zhang, Y., and M. Zhong. 2010. A review on image segmentation techniques with remote sensing perspective. *International Archives in the Photogrammetry, Remote Sensing and Spatial Information Sciences* 38:31–42.

Dey, V., Zhang, Y., and M. Zhong. 2011. Building detection from Pan-Sharpened GeoEye-1 satellite imagery using context based multi-level image Segmentation. In *Proceedings of International Symposium on Image and Data Fusion (ISIDF)*, Tengchong, Yunnan, China, August 9–11, 2011, pp.1–4.

Drăgut, L., Tiede, D., and S. R. Levick. 2010. ESP: A tool to estimate scale parameter for multiresolution image segmentation of remotely sensed data. *International Journal of Geographic Information Science* 24:859–871.

Estes, J. E. 1999. Lecture 2: Elements, aids, techniques and methods of photographic/image interpretation. http://userpages.umbc.edu/~tbenja1/umbc7/santabar/vol1/lec2/2lecture. html (accessed March 20, 2013).

Fan, J., Han, M., and J. Wang. 2009. Single point iterative weighted fuzzy C-means clustering algorithm for remote sensing image segmentation. *Pattern Recognition* 42:2527–2540.

Flanders, D., Hall-Beyer, M., and J. Pereverzoff. 2003. Preliminary evaluation of eCognition object-based software for cut block delineation and feature extraction. *Canadian Journal of Remote Sensing* 29:441–452.

Gonzalez, R.C. and R. E. Woods. 2002. Image segmentation. In *Digital Image Processing*, 2nd edn., Prentice-Hall, Upper Saddle River, NJ, USA, pp. 566–589.

Guindon, B. 1997. Computer-based aerial image understanding: A review and assessment of its application to planimetric information extraction from very high resolution satellite images. *Canadian Journal of Remote Sensing* 23:38–47.

Haralick, R. M. and L. G. Shapiro. 1985. Image segmentation techniques. *Computer Vision, Graphics, and Image Processing* 29:100–132.

Hay, G. J. and G. Castilla. 2008. Geographic object-based image analysis (GEOBIA): A new name for a new discipline. In *Object-Based Image Analysis: Object-Based Image Analysis: Spatial Concepts for Knowledge-Driven Remote Sensing Applications*, eds. T. Blaschke, S. Lang, and G. J. Hay. Springer-Verlag, Berlin, Germany, pp. 75–89.

Hay, G. J. and D. J. Marceau. 2004. Multiscale object-specific analysis (MOSA): An integrative approach for multiscale landscape analysis. In *Remote Sensing and Digital Image Analysis. Including the Spatial Domain*. Book Series: Remote Sensing and Digital Image Processing, eds. S. M. de Jong and F. D. van der Meer. Kluwer Academic Publishers, Dordrecht, the Netherlands, pp. 71–92.

Hu, X., Tao, C. V., and B. Prenzel. 2005. Automatic segmentation of high-resolution satellite imagery by integrating texture, intensity, and color features. *Photogrammetric Engineering & Remote Sensing* 71:1399–1406.

Karantzalos, K. and D. Argialas. 2009. A region-based level set segmentation for automatic detection of man-made objects from aerial and satellite images. *Photogrammetric Engineering & Remote Sensing* 75:667–677.

Lang, S., Schöpfer, E., and T. Langanke. 2009. Combined object-based classification and manual interpretation—Synergies for a quantitative assessment of parcels and biotopes. *Geocarto International* 24:99–114.

Lizarazo, I. and J. Barros. 2010. Fuzzy image segmentation for urban land-cover classification. *Photogrammetric Engineering & Remote Sensing* 76: 151–162.

Marpu, P. R., Neubert, M., Herold, H., and I. Niemeyer. 2010. Enhanced evaluation of image segmentation results. *Journal of Spatial Science* 55:55–68.

Mehta, P., Kanakappan, S., Jayaraj, C., and D. K. Raju. 2012. Urban temperature profiling using Worlview-2 8-band imagery and ERDAS IMAGINE. 2012 geospatial challenge, http://geospatial.intergraph.com/resources/whitepapers.aspx?Page=4 (accessed May 30, 2013).

Möller, M., Lymburner, L., and M. Volk. 2007. The comparison index: A tool for assessing the accuracy of image segmentation. *International Journal of Applied Earth Observation and Geoinformation* 9:311–321.

Netzband, M., Redman, C. L., and W. L. Stefanov. 2007. Remote sensing as a tool for urban planning and sustainability. In *Applied Remote Sensing for Urban Planning, Governance and Sustainability*, ed. C. L. Redman. Springer-Verlag, Dordrecht, the Netherlands, pp. 1–23.

Neubert, M., Herold, H., and G. Meinel. 2008. Assessing image segmentation quality— Concepts, methods and applications. In *Object Based Image Analysis: Object-Based Image Analysis: Spatial Concepts for Knowledge-Driven Remote Sensing Applications*, eds. T. Blaschke, S. Lang, and G. J. Hay. Springer-Verlag, Berlin, Germany, pp. 760–784.

Optiz, D. and S. Blundell. 2008. Object recognition and image segmentation: The Feature Analyst® approach. In *Object Based Image Analysis: Object-Based Image Analysis: Spatial Concepts for Knowledge-Driven Remote Sensing Applications*, eds. T. Blaschke, S. Lang, and G. J. Hay. Springer-Verlag, Berlin, Germany, pp. 153–167.

Pal, N. R. and S. K. Pal. 1993. A review on image segmentation techniques. *Pattern Recognition* 26:1277–1294.

Salehi, B., Zhang, Y., Zhong, M., and V. Dey. 2012a. Object-based classification of urban areas using VHR imagery and height points ancillary data. *Remote Sensing* 4:2256–2276.

Salehi, B., Zhang, Y., Zhong, M., and V. Dey. 2012b. A review of the effectiveness of spatial information used in urban land cover classification of VHR imagery. *International Journal of GeoInformatics* 8:35–51.

Schiewe, J. 2002. Segmentation of high-resolution remotely sensed data-concepts, applications and problems. In *ISPRS Technical Commission IV Symposium: Geospatial Theory, Processing and Application*, vol. XXXIV Part 4, July 9–12, 2002, Ottawa, Ontario, Canada, pp. 358–363.

Shankar, B. U. 2007. Novel classification and segmentation techniques with application to remotely sensed images. In *Lecture Notes on Computer Science: Transactions on Rough Set VII*, ed. J. F. Peters, A. Skowron, V. W. Marek, E. Orlowska, R. Slowinski, and W. Ziarko, vol. 4400. Springer-Verlag, Berlin, Germany, pp. 295–380.

Song, M. and D. Civco. 2004. Road extraction using SVM and image segmentation. *Photogrammetric Engineering & Remote Sensing* 70:1365–1371.

Susaki, J. 2012. Development of building segmentation algorithm for dense urban areas from aerial photograph. In *2012 IEEE International Geoscience and Remote Sensing Symposium (IGARSS)*, Munich, Germany, July 22–27, 2012, pp. 550–553.

Tian, J. and D.-M. Chen. 2007. Optimization in multi-scale segmentation of high-resolution satellite images for artificial feature recognition. *International Journal of Remote Sensing* 28:4625–4644.

Tong, H., Maxwel, T. L., Zhang, Y., and V. Dey. 2012. A supervised and fuzzy-based approach to determine optimal multi-resolution image segmentation parameters. *Photogrammetric Engineering & Remote Sensing* 78:1029–1044.

Wang, Z., Jensen, J. R., and J. Im. 2010. An automatic region-based image segmentation algorithm for remote sensing applications. *Environmental Modelling and Software* 25:1149–1165.

Weng, Q. and D. A. Quattrochi. 2006. *Urban Remote Sensing*. CRC Press, Taylor & Francis, London, U.K., pp. 1–432.

Wuest, B. and Y. Zhang. 2009. Region based segmentation of QuickBird multispectral imagery through band ratios and fuzzy comparison. *ISPRS Journal of Photogrammetry and Remote Sensing* 64:55–64.

Wuest, B. A. 2008. Towards improving segmentation of very high resolution satellite imagery. MScE thesis, Department of Geodesy and Geomatics Engineering Technical Report No. 261, University of New Brunswick, Fredericton, New Brunswick, Canada.

Yang, X. 2011. *Urban Remote Sensing: Monitoring, Synthesis, and Modeling in the Urban Environment*. John Wiley & Sons, Upper Saddle River, NJ, pp. 1–408.

Zhang, H., Fritts, J. E., and S. A. Goldman. 2008. Image segmentation evaluation: A survey of unsupervised methods. *Computer Vision and Image Understanding* 110:260–280.

Zhang, Y. 2004. Highlight article: Understanding image fusion. *Photogrammetric Engineering & Remote Sensing* 70:675–661.

Zhang, Y. J. 1997. Evaluation and comparison of different segmentation algorithms. *Pattern Recognition Letters* 18:963–974.

16 Defining Robustness Measures for OBIA Framework

A Case Study for Detecting Informal Settlements

Peter Hofmann

CONTENTS

16.1 WHY AUTOMATED DETECTION OF INFORMAL SETTLEMENTS FROM REMOTE SENSING DATA?

16.1.1 INFORMAL SETTLEMENTS IN THE CONTEXT OF WORLDWIDE URBANIZATION

According to estimations of UN-HABITAT (2011), the number of people living in urban areas worldwide will increase from 3.5 billion in 2010 to 4.9 billion in 2030. This is equivalent to an urbanization rate of 1.81% per year from 2010 to 2020 and 1.60% per year between 2020 and 2030, respectively. That is, on average, the number of people living in urban areas worldwide is increasing per year by 1.71%. The largest portion of urbanization will take place in developing countries: from 2010 to 2020, the rate will be at 2.21% per year and from 2020 to 2030, it will decrease to 1.92% per year. From the group of developing countries, the African continent will have the highest urbanization rates: 3.21% per year for the period from 2010 to 2020 and 2.91% per year from 2020 to 2030, of which the largest portion belongs to sub-Saharan Africa with 3.51% per year (2010 to 2020) and 3.17% per year (2020 to 2030). Irrespective of the reasons for such enormous migration movements, these numbers indicate that cities in the developing countries, especially the sub-Saharan countries, will face an enormous increase in population pressure. Following UN-HABITAT (2007), among the "nearly one billion people alive today one in every six human beings are slum dwellers, and that number is likely to double in the next thirty years." This number is estimated to be 2 billion by 2030. That is, besides a general increase in urbanization, the portion of the urban population living under slum conditions will increase from 28.57% from now to 40.82% in 2030. This means that by 2030 there will be 2 billion people in urban areas lacking access to safe water, improved sanitation, and secure tenure living in overcrowded and unsecure housing structures with environmental degradation. "Alarmingly, there is currently little or no planning to accommodate these people or provide them with services" (UN-HABITAT 2007). With this background, programs such as the "United Nations Millennium Development goal to improve the lives of at least 100 million slum dwellers by 2020" (UN-HABITAT 2007) were started. However, regarding the increase in rates of slum dwellers, it is obvious that global and local policies and instruments are necessary to improve the living conditions in informal settlements and to fight poverty at all. "Much more political will is needed at all levels of government to confront the huge scale of slum problems that many cities face today, and will no doubt face in the foreseeable future" (Anna Kajumulo Tibaijuka, Executive Director of UN-HABITAT, in UN-HABITAT 2007). The described situation makes it clear that there is an increasing need to detect informal settlements at least for inventory reasons but also for continuous monitoring, mapping, and finally upgrading in terms of providing a minimum standard of housing conditions.

16.1.2 DYNAMICS OF INFORMAL SETTLEMENTS

Although established informal settlements and the cores of informal settlements show rather durable structures, the situation is different in reality at the informal settlements' borders. Due to their informal character, dwellings are relatively easily built, extended, or demolished, which allows the dwellings' residents to react very flexibly on changing living conditions and therefore either to move quickly and easily if necessary or to extend their dwellings if necessary and possible. This has an impact on the location and physiognomy of informal settlements including the footprints and general structure, which can change very rapidly. However, although alternating structures are observable in some cases, in most cases, peripheral growth takes place. That is, informal settlements located at a city's periphery in general grow faster than the centrally located ones. Moreover, informal settlements usually grow faster at their outer border than in the core areas (APHRC 2002; Sartori et al. 2002; Kuffer 2003; Weber and Puissant 2003; Radnaabazar 2004; Davis 2006). This growth has an impact on neighboring land coverage (Sliuzas and Kuffer 2008), be it settlement, agriculture, or uncultivated land with habitats for potentially invaluable species and their ecosystems. In some informal settlements, even vertical growth can be observed (Canham and Wu 2008; Sliuzas et al. 2008; UNESCO 2012). Because of these expeditious changes of informal settlements' shape, conventional methods of mapping fail. Feasible methods are rather implemented by integrating the settlements' inhabitants themselves. They can contribute to an up-to-date spatial data acquisition by volunteered mapping. The most prominent example of such an approach is the Map Kibera Project (MKP) (http://mapkiberaproject.org/). Besides the production of shared geo-information, MKP enforced community and neighborhood building. In this context, Veljanovski et al. (2012) report the supporting role of very-high-resolution (VHR) remote sensing data in conjunction with object-based image analysis (OBIA) methods, especially for population estimation. For ex post change detection, that is, monitoring of already elapsed periods, remote sensing data and appropriate analysis methods proved to be the only reliable data source. Such ex post approaches are the basis for estimating past population sizes and densities together with their development. This information can be used to project respective future population developments by comparing past and recent image patterns with socioeconomic data. This way, hot spots of urbanization within informal settlements are identifiable by means of remote sensing image analysis. But even if there is no accompanying ground mapping, remote sensing is a valuable instrument for monitoring the development of informal settlements (Kuffer 2003; Sliuzas et al. 2008).

16.1.3 COMMON AND DIFFERENT PATTERNS OF INFORMAL SETTLEMENT

As several authors (Sliuzas et al. 2008; Kit et al. 2012; Taubenböck and Kraff 2013) have pointed out, there is an increasing need for mapping and monitoring informal settlements globally. Although several individual approaches for detecting informal settlements from mostly VHR remote sensing data have been introduced so far, there is no unique standard method available to delineate or even analyze them from remote sensing data (Sliuzas et al. 2008). The reasons therefore are manifold and are explainable by the individual characteristics of informal settlements rather than by a lack of understanding of the phenomenon. That is, each informal settlement shows an individual fingerprint and simultaneously fits typical general characteristics of informal settlements (Hofmann 2005; Taubenböck and Kraff 2013). The latter can be defined as general slum ontology (GSO), as introduced by Kohli et al. (2012). The GSO acts as a top-level (Guarino 1997a,b) or canonical (Subieta 2000) ontology for informal settlements. For operational tasks (e.g., the definition of OBIA rule sets), the GSO can be reapplied, adapted, and extended as needed. The resulting ontology then reflects the individual characteristics of the informal settlement of concern. That is, there exists already a general description of what makes an informal settlement, namely, a dense and irregular network of usually unpaved roads and lanes and a dense and irregularly arranged grid of small shacks or small houses, just to name those that are detectable from remote sensing data. However, this description is rather fuzzy and the de facto pattern of an individual informal settlement depends very often on local criteria such as cultural background, available construction material, topography, existing infrastructure, and existing formal settlement structures. The degree of local influence factors on the deviation from the "ideal," that is, top-level or canonical, informal settlement is hardly predictable. However, the defined ontologies can act as input for the creation of an OBIA rule set, which needs to be adapted and extended according to the current situation. Therefore, recent methods of automated detection of informal settlements from remote sensing data still include a relatively high proportion of manual adaptation to local conditions (Hofmann, 2005; Hofmann et al. 2008a,b, Sliuzas et al. 2008).

16.2 AUTOMATION AND ROBUSTNESS IN THE CONTEXT OF OBIA

As stated by several authors (Blaschke and Strobl 2001; Hofmann 2001; Benz et al. 2004; Niebergall et al. 2008; Sliuzas et al. 2008; Veljanovski et al. 2012), OBIA has numerous advantages, especially in the domain of analyzing VHR remote sensing data, since it operates on image objects as aggregates of pixels rather than on single pixels. Thus, on the one hand, effects such as the salt-and-pepper effect (Blaschke and Strobl 2001) are avoided and, on the other hand, a large feature space can be used for further image object analysis (Benz et al. 2004). Sometimes, OBIA has been criticized due to its dependency on the image segmentation used (Hay and Castilla 2006; Smith and Morton 2008). In this context, many discussions have been held about the suitability and performance of different segmentation algorithms (Meinel and Neubert 2004; Van Coillie et al. 2008; Zhang et al. 2008) and how to assess segmentation quality (Neubert and Herold 2008; Neubert et al. 2008). However, OBIA is an iterative process, starting with arbitrary initial image segmentation and continuing with step-by-step, knowledge-based improvement of image segments according to the analysis task (Baatz et al. 2008). The resulting image objects can be considered as the image representatives of the real-world objects that are to be detected. With this background, a very important point when regarding the segmentation quality and representation capabilities of image objects is whether scale is represented reasonably by the image object hierarchy. That is, do sub- and superlevel image objects reflect the interscale relationships of the real-world objects they represent? And can these interscale relationships be expressed with sufficient quality?

For the case of detecting informal settlements, this means that, on a lower scale, typical structural elements, such as small buildings and shacks, shadows of small buildings and shacks, as well as small roads and lanes, need to be outlined well enough in order to indicate on a higher scale a high density of dwellings and an irregularly shaped network of small roads—one of the major properties of informal settlements. Automation of informal settlement detection increases with the robustness of the underlying rule set. That is, for a given rule set, the fewer the manual adaptations and interactions necessary to produce sufficient results in similar images, the more robust the rule set is considered to be. Consequently, a highly robust rule set increases the automation of informal settlement detection.

16.2.1 DIFFERENT IMAGE DATA AND THE NEED FOR ADAPTING INITIAL SEGMENTATION PARAMETERS

When developing OBIA rule sets, this is usually done using one or two reference images reflecting a subset of the image data to be used and depicting the objects of interest to be detected. That is, the rule set to be developed is generated for a relatively clear defined task concerning objects of interest and the type of image data to be used. For reapplying developed classification rules on different images, the initial image objects should be comparable in size and shape. However, the spatial resolution of VHR remote sensing data can vary from approximately 0.25 to 5 m. Different spatial resolutions lead to a varying number of pixels per real-world object to be represented. Thus, in order to produce comparable image objects, the initial segmentation parameters need to be adapted with respect to the different spatial resolutions used (Hofmann et al. 2008a). Besides, the radiometric resolution has an impact on image object generation, as it increases or decreases details of local contrast. For recent remote sensing images, the radiometric resolution can be of 8, 11, 12, or even 16 bits. That is, the radiation at a pixel's location is quantized in $2^8 = 256$, $2^{11} = 2,048$, $2^{12} = 4,096$, or $2^{16} = 65,536$ discrete values. When segmenting images of different radiometric resolution, more or less randomly shifted object borders can arise (see Figure 16.1). Last but not least, the spectral coverage of the sensors' bands lets objects of interest appear differently, and therefore they can have an impact on initial segmentation results as well.

16.2.2 ADAPTING IMAGE SEGMENTATION PARAMETERS

Most segmentation algorithms directly or indirectly take local contrast into account. Changing the radiometric resolution has an impact on local contrast and thus on the generation of comparable image objects. Consider a spectral difference segmentation that agglomerates pixels to image objects if their mutual gray value differences are below a given threshold. Different quantization leads to different gray value gradients in the pixels' neighborhood. Consequently, in the image with higher radiometric resolution, object borders are generated at positions where they would not appear in the lower radiometric resolution. Vice versa,

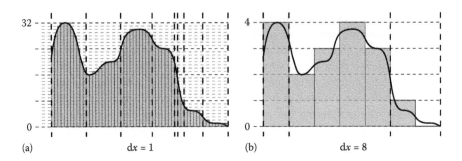

FIGURE 16.1 Different radiometric and spatial resolutions and their impact on object generation. (a) High resolution (32 gray values and 1 spatial unit). (b) Low resolution (4 gray values and 8 spatial units). Object borders are indicated by dotted vertical lines; scanning units are indicated by dotted horizontal lines. The original signal is indicated by the solid line; the scanned signal is given in gray bars.

existing gradients in the image with lower radiometric resolution get smoothed in the image with higher radiometric resolution. Consequently, local contrast is too low for generating an object border and the border disappears (see Figures 16.1 and 16.2). Hence, a generic segmentation adaptation for images with different radiometric resolutions is hardly feasible.

For region-growing algorithms taking the object size into account, the spatial resolution is relevant, too: the smaller the pixel size, the more pixels need to be agglomerated in order to create objects of similar size (Figure 16.1). For multiresolution segmentation (MRS), as introduced by Baatz and Schäpe (2000), Hofmann et al. (2008a) demonstrated a method for compensating different spatial resolutions. Different bandwidths are not compensable at all. But bands with bandwidths only existing in one image can be excluded from segmentation, while redundant bands can be merged into one band and similar bands can be used equally. Especially when working with pan-sharpened data, this issue can have an impact on object generation: for pan-sharpening, usually only those multispectral channels should be used that are covered by the spectrum of the panchromatic channel. But the latter can vary from sensor to sensor. Thus, pan-sharpened data from one sensor are not necessarily equal to that of another sensor.

16.2.3 Robustness of OBIA Rule Sets

The term robustness is applied in a variety of domains (Jen 2003). Constructions, for example, are considered to be robust if they function stable even beyond their specifications. Organisms are called robust if they are able to adapt to changing living conditions in terms of survival and reproduction (Kitano 2007). Societal structures can be seen as robust if they continue to exist under changing socioeconomic conditions (Berman 1997). Computer software is often called robust if it keeps functioning under conditions it was intentionally not made for, such as unexpected user behavior, invalid input data, or other stressful environmental

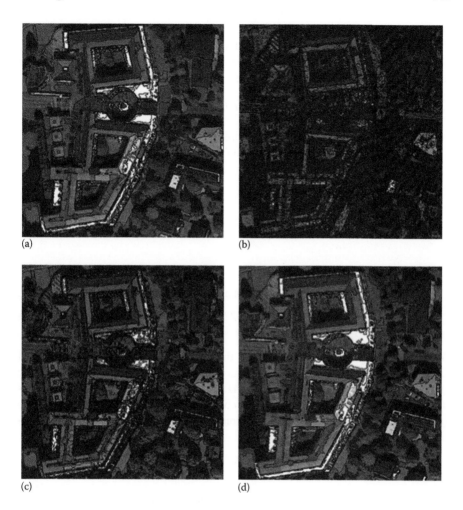

FIGURE 16.2 Different segmentation results (spectral difference) for different radiometric resolutions. (a) 8-bit, spectral difference threshold = 10. (b) 16-bit, spectral difference threshold = 10. (c) 16-bit, spectral difference threshold = 30. (d) 16-bit spectral difference threshold = 90.

conditions, for example, hardware faults (IEEE 1990; Kropp et al. 1998; Fernandez et al. 2005; Shahrokni and Feldt 2013). In the context of OBIA, rule sets can be considered robust if they produce similar results with similar quality on similar images with minimum adaptation effort (Hofmann et al. 2011). As we saw already in Section 16.2.2, different image properties have an impact on the initial segmentation results and therefore on the object quality. Thus, a prerequisite for a sensible rule set evaluation is comparable image objects. After a rule set has been adapted to an image and has produced acceptable classification results, the rule set's deviations in conjunction with the achieved classification accuracy can be investigated. Analyzing these deviations can be considered as the robustness analysis of a rule set. It becomes more reliable the more often it is applied on different varying but similar images. Classification rules are usually of the following form:

If <condition> is fulfilled then assign Object O to Class C

In OBIA classification, rules are used to assign image objects to respective classes. They can be nested, that is, objects which fulfill a variety of (pre)conditions can be selected for class assignment:

If <condition₁> is fulfilled then

    If <condition₂> is fulfilled then

      ...

        If <conditionₙ> is fulfilled then assign Object O to Class C

The conditions 1 to n can be pooled into one condition using a logical *AND* operator:

If <condition₁> AND <condition₂> AND ... <conditionₙ> are fulfilled then assign Object O to Class C

When classifying, for all image objects the conditions are evaluated in terms of *TRUE* and *FALSE*. Thus, nested rules have the advantage of reducing computing time, since for the first condition to be *FALSE*, the evaluation of all following rules is skipped. Consequently, the number of objects to be fully evaluated is reduced to the number of objects fulfilling all conditions. However, conditions can also be combined with logical *OR* operators:

If <condition₁> OR <condition₂> OR ... <conditionₙ> is fulfilled then assign Object O to Class C

In such cases, all conditions need to be evaluated per object, since the object is only not assigned to class C if all of the conditions 1 to n are *FALSE*. Classification rules as described earlier can also be considered as class descriptions. That is, class C is described by the conditions to be fulfilled per object in order to assign the object to class C. Elaborate class descriptions can consist of a variety of combined *AND* and *OR* conditions, as well as of explicit negations (*NOT* or ¬). In order to analyze the robustness of an existing rule set, it is necessary to measure its deviations if it is adapted and applied to similar images. Rule set deviations can be of the following forms:

1. Adding or subtracting classes to or from the rule set
2. Adding or subtracting single rules to or from class descriptions
3. Changing logical operators in rules
4. Changing relational operators in rules
5. Changing thresholds in rules

Enumerating these deviations is a first attempt to quantifying a rule set's robustness: the more the deviations, the less robust the rule set is. In the case that fuzzy classification rules (Benz et al. 2004) are applied, changes in the shape of each fuzzy membership function need to be considered, too (Hofmann et al. 2011), which is somewhat equivalent to points (4) and (5). Taking classification accuracy into account means comparing the accuracy that was achieved in the original image(s) the rule set was developed on with the accuracy achieved in all different images with respective adapted rule sets. For comparison reasons, the accuracy needs to be measured for the original rule set and all adapted rule sets identically, whereas the method used is indifferent but should be chosen adequately. Although there is a variety of accuracy assessment methods available (Van Rijsbergen 1979; Congalton and Green 1999), not all of them are suited for particular cases and not all of them produce equal values; therefore, for quantifying a rule set's robustness, normalizing the classification accuracy is necessary. Thus, measuring a rule set's robustness is always bound to the chosen method of accuracy assessment. For a rule set developed on one image and being adapted and reapplied on a similar image, we can formally describe the rule set's robustness r as follows:

$$r = \frac{q_2/q_1}{d+1} \tag{16.1}$$

with q_1 the accuracy achieved in the original image, q_2 the accuracy achieved in the image the rule set was reapplied on, and $q_1, q_2 \in \{0\ldots1\}$. d is the sum of all deviations of the rule set after adaptation as outlined under points (1) to (5) earlier. After a rule set is adapted and applied on several images, its mean robustness can be calculated easily. To determine the deviation for fuzzy rules (cases 4 and 5), the following points need to be considered: a membership function expresses the degree of membership μ with $\mu \in \{0\ldots1\}$ to a class regarding a value range vr with an upper bound v_u and a lower value bound v_l of a given property. A value of $\mu(v) = 0$ indicates for an object no membership concerning property value v. A value of $\mu(v) = 1$ in contrast means a full membership. The center value a of the membership function is given by $a = v_l + (vr/2)$ or $a = v_u - (vr/2)$. It indicates the crisp property value the membership function represents in terms of a classification rule. Fuzzy membership functions can be roughly categorized as depicted in Table 16.1.

In principle, fuzzy membership functions can have any kind of shape. However, in practice, three types have been established since they are easier to interpret and understand than complex shape functions. Nevertheless, membership functions can also be of linear shape, that is, without soft transitions at the extremes. Combining a fuzzy-lower-than with a fuzzy-greater-than function leads to a t-norm function, whereas v_u of the greater-than function is identical to v_l of the lower-than function. Both are identical to a of the created t-norm function. t-norm functions can also have a value range of $\mu(v) = 1.0$, which gives them a plateau-like shape. In the case where the slope of the membership function is at $\mu'(v) = 1.0$ for value v and the membership at this value is at $\mu(v) = 1.0$ or $\mu(v) = 0.0$, the membership function is called crisp,

TABLE 16.1

Principal Categories of Fuzzy Membership Functions

Category	Symbol	$\mu(v_l)$	$\mu(a)$	$\mu(v_u)$
t-Norm (triangular)		0.0	1.0	0.0
Fuzzy-lower-than		1.0	0.5	0.0
Fuzzy-greater-than		0.0	0.5	1.0

which is equal to threshold setting. If the function has only one property value v with $\mu'(v) = 1.0$ and $\mu(v) = 1.0$, the function is called a singleton, which is the same as the identity: $\mu(v) \equiv 1.0$. The deviation δF of a membership function after adaptation to a similar image is the sum of the membership function's shift δa and its stretch or compression δv:

$$\delta F = \delta a + \delta v \qquad (16.2)$$

where $\delta a = 0$ if the function of concern is not shifted and $\delta v = 0$ if the function is neither stretched nor compressed. In the case where the membership function has been shifted in a positive direction along the v-axis, its deviation is given by $\delta a = 1 - (a_2/a_1)$. For a negative shift, it can be determined by $\delta a = 1 - (a_1/a_2)$. Analogously, the function's stretch or compression can be determined by $\delta v = 1 - (vr_2/vr_1)$ for stretching a function and $\delta v = 1 - (vr_1/vr_2)$ for compressing it. By systematically analyzing all possible rule set deviations as described earlier, critical rules can be determined automatically: rules that deviate in a wide range, that is, have a high value for δF, or classes that often have different descriptions have a high negative impact on the overall robustness of the rule set (see Table 16.3). For a good rule set design in terms of transferability, one should consider skipping or exchanging such rules or classes by other, potentially more robust (i.e., less deviating) rules.

16.3   OBIA RULE SETS FOR DETECTING INFORMAL SETTLEMENTS

16.3.1   Image Ontology for Informal Settlements

Developing an OBIA rule set in principle means to define rules that translate object properties into semantically meaningful real-world classes (Arvor et al. 2013). This is either done implicitly by sample-based classification mechanisms or explicitly by defining respective classification rules (Section 16.2.3). The latter has the advantage of being easily adapted to changing imaging conditions, if necessary, and allows formulating additional expert knowledge, such as spatial relationships. However, for being transferable, an OBIA rule set should reflect at least the underlying top-level ontology (Section 16.1.3). That is, the structure of the rule set, its classes, and

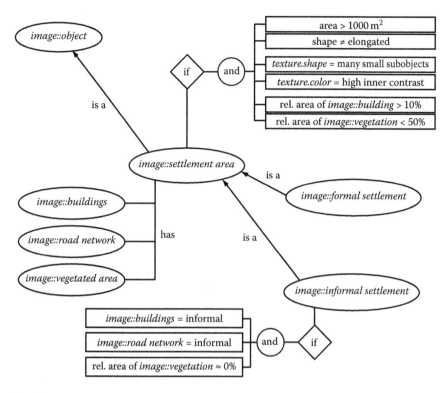

FIGURE 16.3 Top-level ontology for informal settlements, their components, and their appearance in remote sensing data (image::).

the classes' spatial dependencies together with some basic concepts should be similar to the respective ontology (Hofmann 2005; Kholi et al. 2012, Figure 16.3).

The domain description *image::* indicates that the ontology describes object classes as they can be observed in remote sensing data. Each *image::* class refers to a respective real-world class. The class *image::settlement area* is described by two principal characteristics: existence prerequisites of other object classes and physical object conditions. The former are described by the *has*-relations: *image::settlement area has image::buildings, image::road network, image::vegetated area*. The latter describe measurable thresholds in a fuzzy manner, such as *many small subobjects* or *shape ≠ elongated*. Consequently, if an image object cannot refer to building, road, or vegetation subobjects or if it is too elongated, it cannot be a settlement area at all. The class *image::informal settlement* is described as a subclass of *image::settlement area* by the *is_a* relation. Thus, it inherits the properties of *image::settlement area*. That is, the same prerequisites are valid for *image::informal settlement* but it distinguishes itself by its informal characteristics (*image::buildings = informal, image::road network = informal*, and *rel. area of image::vegetation ≈ 0%*). The rules that make *image::buildings = informal* and *image::road network = informal* are to be defined in separate ontologies. The same holds for the fuzzy concepts *elongated* or *high inner contrast*.

16.3.2   Transforming the Ontology into a Rule Set

For detecting informal settlements in VHR remote sensing images, a respective OBIA rule set has been developed, based on a pan-sharpened IKONOS scene depicting the so-called Cape Flats in Cape Town. The scene was captured on March 19, 2000. The rule set was adapted and reapplied on a pan-sharpened QuickBird scene showing the Ilha do Governador in Rio de Janeiro on May 14, 2002. As a development framework, the cognition network language (CNL) has been used, which is implemented in the software eCognition® (Trimble 2013).

16.3.2.1   Initial Segmentation Rules

The rule set starts with a two-level MRS, whereas on the top level, the average object size is at 5923 m² and on the base at 49 m². The segmentation parameters for the IKONOS scene were determined empirically by trial and error. Inspecting the segmentation results visually, the top-level segmentation depicts relatively good settlement structures at block level, including informal settlements. On the base level, small structures such as road segments, shacks, small buildings, and shadows are relatively well outlined. However, in many cases, the shacks and their shades are visually hardly distinguishable and neither is the segmentation. Nevertheless, with the two-level segmentation (see Figure 16.4) approach, the ontological relationships *image::settlement area has image::buildings*, *image::settlement area has image::road network*, and *image::settlement area has image::vegetated area* can be described as spatial-hierarchical sub- and superobject relationships. Similarly, the properties *texture.shape* and *texture.color* can be described by statistical parameters of the subobjects shape and color properties per superobject, such as the mean area of subobjects per superobject or the mean spectral difference of subobjects per superobject (Trimble 2012a). The segmentation of the QuickBird scene has been adapted according to the ratio between the sensors' pixel size (Section 16.2.1; Table 16.2) and applied with comparable results (Figure 16.5).

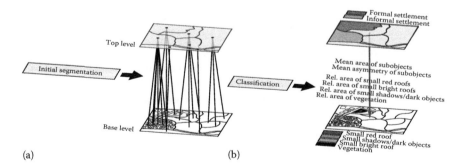

FIGURE 16.4   (See color insert.) Object-hierarchical relationships after segmentation (parameterization, see Table 16.2) and classification between top- and base-level objects. For detailed class descriptions, see Table 16.3. Every top-level object relates to its subobjects in the base level and vice versa (a). Relationships to subobjects can be used for classification of superobjects (b).

TABLE 16.2

MRS Parameters Used for Initial Image Segmentation

	IKONOS, Cape Town		QuickBird, Rio de Janeiro	
	Top Level	Base Level	Top Level	Base Level
Scale parameter	100	10	144	14
w_{color}	0.2	0.1	0.2	0.1
w_{shape}	0.8	0.9	0.8	0.9
Compactness	0.5	0.5	0.5	0.5
Smoothness	0.5	0.5	0.5	0.5

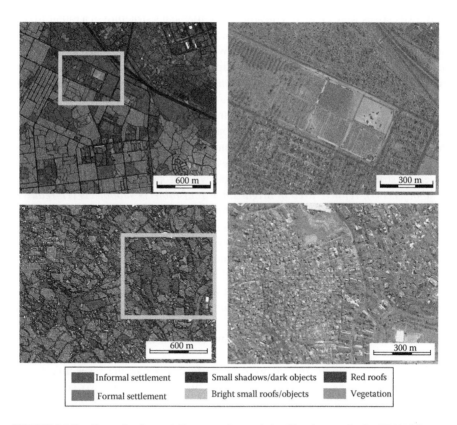

FIGURE 16.5 (**See color insert.**) Segmentation and classification results for IKONOS (top) and QuickBird (bottom). Left: top segmentation level with respective classes. Right: base segmentation level with respective classes. Blue rectangle in the left images indicates the location of the right images.

Since the initial segmentation result reflects informal settlement areas clearly, no further segmentation enhancements were performed.

16.3.2.2 Classification Rules

The classification rules were directly applied on the initially generated image objects, where the class descriptions intended to reflect the underlying ontology as well as possible. Respective classes described by fuzzy membership functions were developed, where the rule set initially consisted of three top-level classes: *settlement, informal settlement,* and *formal settlement. Formal settlement* and *informal settlement* are subclasses of *settlement* and therefore inherit its properties (see Section 16.3.1; Trimble 2012b: 95–111 and Table 16.3). The class *formal settlement* then acts as the inverse of class *informal settlement.* That is, objects fulfilling the criteria of *settlement* in general but not those of *informal settlement* are a *formal settlement* if they are fuzzy-greater than 1850 m². At base level, single shacks or other buildings with informal character (*image::buildings = informal*) are not undoubtedly identifiable. Thus, two classes were created indicating the settlement's structure in a rather fuzzy manner: *bright small roofs/objects* and *small shadows/dark objects.* While the former roughly outlines small square objects with bright roofs (e.g., shacks with roofs made from metal sheets), the latter just outlines dark objects that can be a shack or its shadow or both. The classes *red roofs* (*image::buildings ≠ informal*) and *vegetation* (*image::vegetation*) are rather clear: *red roofs* are simply determined by a high ratio of the red band to the green band in an object, while *vegetation* shows a high fraction of the near-infrared band in an object. At the top segmentation level, the ontological concepts for informal settlement *image::buildings = informal* is realized by pointing to the corresponding classes at base level, that is, evaluating the settlement structure. For this purpose, the following properties have been used: *area of subobjects (1), relative area of bright small roofs/objects subobjects (1), relative area of red roofs subobjects (1),* and *relative area of small shadows/dark objects subobjects (1)* (see Figure 16.4 and Table 16.3). The concept *image::road network = informal* is realized by the rule *asymmetry of subobjects: mean (1).* Asymmetry measures how elongated an object is. The higher the asymmetry, the more elongated the object is. An irregular road network, such as that of informal settlements, leads to a relatively lower mean asymmetry within a settlement area. Accordingly, the *asymmetry of subobjects: mean (1)* of informal settlement areas must be lower than that of formal settlement areas. The concept *relative area of image::vegetation ≈ 0%* was directly implemented as a fuzzy membership function. For detailed class descriptions, refer to Table 16.2.

The class descriptions applied were all performed using the respective fuzzy membership functions (see Section 16.2.3). This approach has two advantages: (1) it allows to better express fuzzy concepts, such as *red* or *rectangular,* and (2) slight property variations in the data can be more easily captured. This is of advantage, especially for the detection of informal settlements, since even on a local level their patterns can be varied. Additionally, transitional forms of settlements are detectable by their membership degree to *informal settlement* and *formal settlement,* respectively. That is, a transitional settlement type has an overall membership to both classes by $\mu \neq 0$ and at least $\mu > 0.3$, which is the threshold set for defuzzification.

TABLE 16.3

Fuzzy Class Descriptions and Deviations for IKONOS (Cape Town) and QuickBird (Rio de Janeiro) Rule Set

Class	Property	Membership Function	Cape Town		Rio de Janeiro		Deviations		
			vr_1	a_1	vr_2	a_2	δv	δa	δF
Top level (level 2)									
Settlement	Area of subobjects (1)		3.00	43.50	3.00	51.50	0.00	0.18	0.18
	Asymmetry		0.01	0.96	0.01	0.96	0.00	0.00	0.00
	Average mean difference to neighbors of subobjects (NIR-channel) (1)		233.00	210.20	230.00	235.00	0.01	0.12	0.13
	Relative area of small shadows/dark objects subobjects (1)		0.00	0.01	0.00	0.01	0.00	0.00	0.00
	Relative area of vegetation subobjects (1)		0.10	0.45	0.10	0.45	0.00	0.00	0.00
Formal settlement	Area [m²]		100.00	1850.00	100.00	1850.00	0.00	0.00	0.00
	Not informal settlement								
Informal settlement	Area of subobjects (1) [m²]		2.00	40.00	2.00	37.00	0.00	0.08	0.08
	Asymmetry of subobjects: mean (1)		0.02	0.56	0.02	0.62	0.00	0.10	0.10
	Relative area of bright small roofs/objects subobjects (1)		0.01	0.04	0.01	0.06	0.00	0.71	0.71
	Relative area of red roofs subobjects (1)		0.0002	0.006	0.01	0.02	49.00	1.79	50.79

(continued)

TABLE 16.3 (continued)

Fuzzy Class Descriptions and Deviations for IKONOS (Cape Town) and QuickBird (Rio de Janeiro) Rule Set

Class	Property	Membership Function	Cape Town		Rio de Janeiro		Deviations		
			vr_1	a_1	vr_2	a_2	δv	δa	δF
	Relative area of small shadows/dark objects subobjects (1)		0.02	0.03	0.02	0.03	0.00	0.00	0.00
	Relative area of vegetation subobjects (1)			—	0.01				
Base level (level 1)									
Small shadows/dark objects	Area		5.00	37.50	5.00	37.50	0.00	0.00	0.00
	Ratio blue channel		0.13	0.26	0.13	0.32	0.00	0.20	0.20
Vegetation	Ratio NIR-channel		0.01	0.30	0.01	0.30	0.00	0.00	0.00
Red roofs	Ratio red channel/ratio green channel		0.01	1.10	0.05	1.28	4.00	0.16	4.16
Bright small roofs/objects	Area		20.00	50.00	20.00	50.00	0.00	0.00	0.00
	Brightness		25.00	762.50	25.00	762.50	0.00	0.00	0.00
	Shape index				0.10				
								$\Sigma\delta F$	56.35

Note: For detailed property descriptions, see Trimble 2012a.

Assuming that the developed rule set reflects the ontology for informal settlements at best, for each classified object it can be expressed to what degree (of membership) it fulfills the criteria of the informal settlement prototype. Vice versa, each image object is a gradual member of the class (concept) informal settlement or formal settlement of the ontology.

16.4 ROBUSTNESS ANALYSIS OF THE DEVELOPED RULE SET

16.4.1 Robustness Measurement

In order to analyze the robustness of the developed rule set, it was reapplied to the segmented QuickBird scene of Rio and single rules were adapted manually until acceptable classification results were obtained. The respective deviations were determined as outlined in Section 16.2.3 and displayed in Table 16.3. In both scenes, the classification accuracy has been generated by comparing each classification with a complete manual reference map. In the IKONOS scene, 215 ha (true positives) of informal settlements were classified correctly, whereas 50 ha were omitted by the classifier (false negatives) and 93 ha were mapped wrongly as informal settlement (false positives). This leads to precision *prec*, recall *rec*, and quality *qual* (Van Rijsbergen 1979; Heipke et al. 1997) as follows:

$$prec = \frac{\text{True positives}}{\text{True positives} + \text{false positives}} = \frac{2,151,138\,\text{m}^2}{2,151,138\,\text{m}^2 + 930,314\,\text{m}^2} = 0.70$$

(16.3)

$$rec = \frac{\text{True positives}}{\text{True positives} + \text{false negatives}} = \frac{2,151,138\,\text{m}^2}{2,151,138\,\text{m}^2 + 499,624\,\text{m}^2} = 0.81$$

(16.4)

$$qual = \frac{\text{True positives}}{\text{True positives} + \text{false positives} + \text{false negatives}}$$

$$= \frac{2,151,138\,\text{m}^2}{2,151,138\,\text{m}^2 + 930,314\,\text{m}^2 + 499,624\,\text{m}^2} = 0.60$$

(16.5)

As *prec* $\in$ {0...1}, *rec* $\in$ {0...1}, and *qual* $\in$ {0...1}, no normalization for robustness analysis is necessary. In the QuickBird scene, we could achieve accuracies of *prec* = 0.52, *rec* = 0.68, and *qual* = 0.31. Regarding the rule set deviation, in the present case, two rules were added to the QuickBird rule set: *relative area of vegetation subobjects* (1) for the description of informal settlements (for the Cape Town scene, this rule was not necessary) and *shape index* for describing the fuzzy class *bright small roofs/objects*. No classes were added or deleted and no logical or relational operators were added, deleted, or changed. That is, the rest of the deviations are changes of the fuzzy membership functions' values δF. According to Table 16.2, they sum up to

$\Sigma\delta F = 56.35$, which leads to an overall deviation of $d = 56.35 + 2 = 58.35$. Together with the achieved accuracies, we obtain a robustness of $r_{prec} = 0.012$ if we use precision, $r_{rec} = 0.014$ if we use recall, and $r_{qual} = 0.009$ if we use quality as the criterion. Considering that for $r > 1$, classification results are improving ($q_2 > q_1$) with little or no deviation ($d \approx 0$) and that for $r < 1$, the rule set was adapted ($d > 0$) but results did not improve ($q_2 \leq q_1$), the rule set must be seen as not very robust. Vice versa, if r was at ~1.0 or higher, the rule set would be very robust.

16.4.2 INTERPRETATION OF ROBUSTNESS MEASUREMENT RESULTS

Regarding the deviations of the fuzzy membership functions δF, there are some rules with no deviation ($\delta F = 0.0$), some with slight deviation ($0.0 < \delta F \leq 0.1$), one rule with higher deviation ($\delta F = 4.16$), and one rule with extreme deviation ($\delta F = 50.79$). While the rules with no and slight deviation can be interpreted as robust, the remaining rules seem to react more sensitively on image variations. In relation to the overall deviation $d = 58.35$, the impact of the extreme deviating rule on the rule set's robustness is very high (~87% of the overall deviation but only 1 rule out of 20). This indicates that the rule *relative area of red roofs subobjects (1)* is not easily transferable and should therefore be skipped or substituted, if possible. The impact of the rule *ratio red channel/ratio green channel* is comparably ~8.2% low although the sum of all other deviations equals 2.8%. Since the rule *relative area of red roofs subobjects (1)* indicates the density of small buildings with red roofs, the following considerations make the rule's high deviation plausible: while in South Africa the shacks' roofs are mainly made of plastic, iron sheets, or wood, in Rio, brick is more common. Thus, the *relative area of red roofs subobjects (1)* per informal settlement object must be higher in Rio. The deviation for the class *red roofs* could be explained by different construction material, too.

Excluding δF for *relative area of red roofs subobjects* (1) from the calculation of d, the robustness core parameters change to $\Sigma\delta F = 5.56$ and $d = 7.56$, leading to a slightly increased robustness of $r_{prec} = 0.09$, $r_{rec} = 0.10$, and $r_{qual} = 0.06$, respectively. If, additionally, the deviation for *ratio red channel/ratio green channel* of the class *red roofs* is excluded, overall deviations of $\Sigma\delta F = 1.4$ and $d = 3.4$ are produced, leading to a robustness of $r_{prec} = 0.17$, $r_{rec} = 0.19$, and $r_{qual} = 0.12$.

16.5 OUTLOOK TOWARD SEMIAUTOMATED TECHNIQUES OF MAPPING INFORMAL SETTLEMENTS

Although OBIA is a reasonable method for analyzing VHR remote sensing data, especially in the context of detecting and monitoring informal settlements, the design of rule sets is of core importance. For transferability and flexibility reasons, it should reflect the underlying ontology of the objects of concern. Simultaneously, it needs to take into account the imaging situation of the data used. This implies that there is no general rule set for the detection of informal settlements possible, but the effort for adaptation can be reduced if the rule set reflects the top-level ontology as well as possible. By measuring the robustness of rule sets, on the one hand, the suitability of a given rule set for a given application can be determined and, on the other hand, critical rules can be identified; that is, rules that might need to be adapted if the rule set is applied on similar data. As long as the ontology is not violated, such rules should be avoided or substituted. In the presented case, the rule set was applied directly on the initially generated image objects with relatively good results. As pointed out, a generic adaptation of initial segmentation parameters to varying image data is hardly feasible. However, classification results could certainly be improved if further dedicated (re)segmentation procedures were applied, generating optimal image objects. Especially the structural elements *buildings* and *road network*, which are key elements in identifying (informal) settlements, were described and detected in a fuzzy manner. Although robustness analysis has been applied only on two images, the results indicate a majority of robust rules in the developed rule set. Recent OBIA technologies allow creating solutions for highly automated image analysis. When reapplied on similar images, the necessary adjustments of a rule set can be performed even by a nonspecialist very easily. An example is given by eCognition Architect (Trimble 2013). For the present case, it would be easily possible to embed the developed rule set in a respective eCognition Architect environment. Necessary adaptations, especially those for critical rules, could be performed using respective slider widgets and/or buttons in a graphical user interface (GUI). This way, a variety of VHR images could be analyzed fast and as automated as possible.

ACKNOWLEDGMENTS

The author would like to thank Heinz Rüther of the University of Cape Town and Thomas Blaschke and Joseph Strobl of Salzburg University for their support, Space Imaging for providing the IKONOS scene, and INTERSAT for providing the QuickBird scene.

REFERENCES

African Population and Health Research Center (APHRC). 2002. *Population and Health Dynamics in Nairobi's Informal Settlements*. Nairobi, Kenya: African Population and Health Research Center.

Arvor, D., Durieux, L., Andrés, S., and M.-A. Laporte. 2013. Advances in geographic object-based image analysis with ontologies: A review of main contributions and limitations from a remote sensing perspective. *ISPRS Journal of Photogrammetry and Remote Sensing*, 82, 125–137.

Baatz, M., Hoffmann, C., and G. Willhauck. 2008. Progressing from object-based to object-oriented image analysis. In *Object Based Image Analysis*, eds. T. Blaschke, S. Lang, and G. Hay. New York: Springer.

Baatz, M. and A. Schäpe. 2000. Multiresolution segmentation an optimisation approach for high quality multi-scale image segmentation. In *Angewandte Geographische Informationsverarbeitung*, vol. XII., eds. Strobl, J. et al., pp. 12–23. Karlsruhhe: Wichman.

Benz, U.C., Hofmann, P., Willhauck, G., Lingenfelder, I., and M. Heynen. 2004. Multi-resolution, object-oriented fuzzy analysis of remote sensing data for GIS-ready information. *ISPRS Journal of Photogrammetry & Remote Sensing*, 58, 239–258.

Berman, S. 1997. Civil society and the collapse of the Weimar Republic. *World Politics*, 49, 401–429.

Blaschke, T. and J. Strobl. 2001. What's wrong with pixels? Some recent developments interfacing remote sensing and GIS. *GIS Zeitschrift für Geoinformationssysteme*, 6, 12–17.

Canham, S. and R. Wu. 2008. *Portrait from above Hong Kong's Informal Rooftop Communities*, 2nd edn. Berlin, Germany: Peperoni Books.

Congalton, R.G. and K. Green. 1999. *Assessing the Accuracy of Remotely Sensed Data: Principles and Practices*. Boca Raton, FL: Lewis Publishers.

Davis, M. 2006. *Planet of Slums*. London, U.K.: Verso.

Fernandez, J., Mounier, L., and C. Pachon. 2005. A model-based approach for robustness testing. In *Testing of Communication Systems, Lecture Notes in Computer Science 3502*, eds. F. Khendek and R. Dssouli, *Proceedings of 17th IFIP TC6/WG 6.1 International Conference TestCom 2005*, Montreal, Canada, May 31–June 2, 2005.

Guarino, N. 1997a. Some organizing principles for a unified top-level ontology. In *Proceedings of AAAI Spring Symposium on Ontological Engineering*. Stanford, CA: AAAI Press.

Guarino, N. 1997b. Semantic matching: Formal ontological distinctions for information organization, extraction, and integration. In *Information Extraction: A Multidisciplinary Approach to an Emerging Information Technology*, ed. M.T. Pazienza. Berlin, Germany: Springer, pp. 139–170.

Hay, G.J. and G. Castilla. 2006. Object-based image analysis: Strengths, weakness, opportunities and threats (SWOT). *International Archives of the Photogrammetry, Remote Sensing and Spatial Information Sciences*, XXXVI-4/C42. CD-ROM.

Heipke, C., Mayer, H., and C. Wiedemann. 1997. Evaluation of automatic road extraction. *International Archives of the Photogrammetry, Remote Sensing and Spatial Information Sciences*, 32(3-4W2), 151–160.

Hofmann, P. 2001. Detecting informal settlements from IKONOS image data using methods of object oriented image analysis—An example from Cape Town (South Africa). In *Proceedings of the Second International Symposium Remote Sensing of Urban Areas*, ed. C. Jürgens, Regensburg, Germany, pp. 107–118.

Hofmann, P. 2005. Übertragbarkeit von Methoden und Verfahren in der objektorientierten Bildanalyse das Beispiel informelle Siedlungen (Transferability of methods and procedures in object-oriented image analysis the example of informal settlements). Dissertation, University of Salzburg, Salzburg, Germany.

Hofmann, P., Blaschke, T., and J. Strobl. 2011. Quantifying the robustness of fuzzy rule sets in object-based image analysis. *International Journal of Remote Sensing*, 32(22), 7359–7381.

Hofmann, P., Strobl, J., and T. Blaschke. 2008a. A method for adapting global image segmentation methods to images of different resolutions. In *Proceedings of the Second International Conference on Geographic Object-Based Image Analysis*, eds. G. Hay, T. Blaschke, and D. Marceau, University of Calgary, Calgary, Alberta, Canada. CD-ROM.

Hofmann, P., Strobl, J., Blaschke, T., and H.J. Kux. 2008b. Detecting informal settlements from QuickBird data in Rio de Janeiro using an object-based approach. In *Object Based Image Analysis*, eds. T. Blaschke, S. Lang, and G.J. Hay. Heidelberg, Germany: Springer, pp. 531–554.

IEEE Std 610.12-1990, IEEE standard glossary of software engineering terminology, 1990.

Jen, E. 2003. Essays & commentaries: Stable or robust? What's the difference? *Complexity*, 8(3), 12–18.

Kohli, D., Sliuzas, R., Kerle, N., and A. Stein. 2012. An ontology of slums for image-based classification. *Computers, Environment and Urban Systems*, 36, 154–163.

Kit, O., Lüdeke, M., and D. Reckien. 2012. Texture-based identification of urban slums in Hyderabad, India using remote sensing data. *Applied Geography*, 32, 660–667.

Kitano, H. 2007. Towards a theory of biological robustness. *Molecular Systems Biology*, 3, 137.

Kropp, N.P., Koopman, P.J., and D.P. Siewiorek. 1998. Automated robustness testing of off-the-shelf software components. In *Proceedings of the Twenty-Eighth Annual International Symposium on Fault-Tolerant Computing*, Munich, Germany, June 23–25, 1998, pp. 230–239. IEEE Computer Society.

Kuffer, M. 2003. Monitoring the dynamics of informal settlements in Dar Es Salaam by remote sensing: Exploring the use of SPOT, ERS and small format aerial photography. In *Proceedings of CORP 2003*, ed. M. Schrenk. Vienna, Austria: Technical University Vienna, pp. 473–483.

Meinel, G. and M. Neubert. 2004. A comparison of segmentation programs for high resolution remote sensing data. *International Archives of the Photogrammetry, Remote Sensing and Spatial Information Sciences*, XXXV-B4, 1097–1102.

Neubert, M. and H. Herold. 2008. Assessment of remote sensing image segmentation quality. *International Archives of the Photogrammetry, Remote Sensing and Spatial Information Sciences*, XXXVIII-4/C1. CD-ROM.

Neubert, M., Herold, H., and G. Meinel. 2008. Assessing image segmentation quality concepts, methods and applications. In *Object Based Image Analysis*, eds. T. Blaschke, S. Lang, and G. Hay. New York: Springer.

Niebergall, S., Loew, A., and W. Mauser. 2008. Integrative assessment of informal settlements using VHR remote sensing data: The Delhi case study. *IEEE Journal of Selected Topics in Applied Earth Observation and Remote Sensing*, 1(3), 193–205.

Radnaabazar, G., Kuffer, M., and P. Hofstee. 2004. Monitoring the development of informal settlements in Ulaanbaatar, Mongolia. In *Proceedings of CORP 2004*, ed. M. Schrenk. Vienna, Austria: Technical University Vienna, pp. 333–339.

Sartori, G., Nembrini, G., and F. Stauffer. 2002. Monitoring of urban growth of informal settlements (IS) and population estimation from aerial photography and satellite imaging. Occasional paper, Geneva Foundation. http://vince.mec.ac.ke/publications/urban-growth-of-informal-settlements-is/ (accessed March 5, 2003).

Shahrokni, A. and R. Feldt. 2013. A systematic review of software robustness. *Information and Software Technology*, 55(1), 1–17.

Sliuzas, R. and M. Kuffer. 2008. Analysing the spatial heterogeneity of poverty using remote sensing: Typology of poverty areas using selected RS based indicators. In *Proceedings of the EARSel Joint Workshop: Remote Sensing New Challenges of High Resolution*, Bochum, Germany, pp. 158–167.

Sliuzas, R., Mboup, G., and de A. Sherbinin. 2008. Report of the expert group meeting on slum identification and mapping. Report by CIESIN, UN-Habitat, ITC. http://www.ciesin.columbia.edu/confluence/download/attachments/39780353/EGM_slum_mapping_report_final.pdf?version=1 (accessed March 7, 2013).

Smith, G. and D. Morton. 2008. Segmentation: The Achilles heel of object-based image analysis? *International Archives of the Photogrammetry, Remote Sensing and Spatial Information Sciences*, XXXVIII-4/C1. CD-ROM.

Subieta, K. 2000. Mapping heterogeneous ontologies through object views. In *Engineering Federal Information Systems: Proceedings of the Third Workshop (EFIS 2000)*, eds. M. Roantree, W. Hasselbring, and S. Conrad. Berlin, Germany: Akad. Verl.-Ges. Aka, pp. 1–10.

Taubenböck, H. and N.J. Kraff. 2013. The physical face of slums: A structural comparison of slums in Mumbai, India, based on remotely sensed data. *Journal of Housing and the Built Environment*. http://link.springer.com/article/10.1007/s10901-013-9333-x/fulltext.html (accessed March 5, 2013).

Trimble. 2012a. *eCognition® Developer 8.8 Reference Book*. Munich, Germany: Trimble Documentation.

Trimble. 2012b. *eCognition® Developer 8.8 User Guide*. Munich, Germany: Trimble Documentation.

Trimble. 2013. *Introducing eCognition 8.8*. http://www.ecognition.com/ (accessed March 2013).

UNESCO. 2012. New research about favelas is launched during international seminar in Rio de Janeiro. http://www.unesco.org/new/en/brasilia/about-this-office/single-view/news/new_research_about_slums_to_be_launched_during_international_seminar_in_rio_de_janeiro/ (accessed March 5, 2013).

UN-HABITAT. 2007. Slum dwellers to double by 2030: Millennium development goal could fall short. http://www.unhabitat.org/downloads/docs/4631_46759_GC%2021%20Slum%20dwellers%20to%20double.pdf (accessed March 5, 2013).

UN-HABITAT. 2011. Cities and climate change. Global report on human settlements 2011. United Nations Human Settlement Programme. Earthscan, London, U.K. http://www.unhabitat.org/documents/SOWC10/R1.pdf (accessed March 5, 2013).

Van Coillie, F.M.B., Pires, P.L.V.M., Van Camp, N.A.F., and S. Gautama. 2008. Quantitative segmentation evaluation for large scale mapping purposes. *International Archives of the Photogrammetry, Remote Sensing and Spatial Information Sciences*, XXXVIII-4/C1. CD-ROM.

Van Rijsbergen, C. 1979. *Information Retrieval*. London, U.K.: Butterworth-Heinemann.

Veljanovski, T., Kanjir, U., Pehani, P., Oštir, K., and P. Kovačič. 2012. Object-based image analysis of VHR satellite imagery for population estimation in informal settlement Kibera-Nairobi, Kenya. In *Remote Sensing Applications*, ed. B. Escalante-Ramirez. DOI: 10.5772/37869. http://www.intechopen.com/books/remote-sensing-applications/object-based-image-analysis-of-vhr-satellite-imagery-for-population-estimation-in-informal-settl (accessed March 6, 2013).

Weber, C. and A. Puissant. 2003. Urbanization pressure and modelling of urban growth: Example of the Tunis Metropolitan Area. *Remote Sensing of Environment*, 86(3), 341–352.

Zhang, H., Fritts, J.E., and S.A. Goldman. 2008. Image segmentation evaluation: A survey of unsupervised methods. *Computer Vision and Image Understanding*, 110(2), 260–280.

17 Automated Techniques for Change Detection Using Combined Edge Segment Texture Analysis, GIS, and 3D Information

Manfred Ehlers, Natalia Sofina,
Yevgeniya Filippovska, and Martin Kada

CONTENTS

17.1 INTRODUCTION

A large number of algorithms for change detection from multitemporal remotely sensed images have been developed and applied. An overview and comparison of different methods can be found, for example, in Coppin et al. (2004), Lu et al. (2003), Mas (1999), Macleod and Congalton (1998), and Singh (1989). In general, change detection methods can be divided into three categories (Mas, 1999): (1) image enhancement methods, (2) multitemporal analysis, and (3) postclassification comparison. Other approaches combine several methods or consist of novel methodologies (an overview can be found in Lu et al., 2003). Image enhancement methods combine data mathematically to enhance image quality (Im et al., 2008). Examples include standards methods such as image difference, image ratio, and principal component and regression analysis.

Multitemporal methods (Coppin et al., 2004) are based on an isochronic analysis of multitemporal image data. This means that n bands of an image taken on date T_1 and n bands of an image of the same area taken on date T_2 are merged to form a multitemporal image with $2n$ bands. This merged image is then used to extract the changed areas (Khorram et al., 1999).

Postclassification analysis is probably the most common change detection technique and allows an assessment of the kind of change from one class to another. It is, however, very sensitive to the achieved classification accuracy. Using pre-event geographic information systems (GIS) information is another way of enhancing change detection reliability and accuracy using object-based analyses (Bovolo, 2009; Chen et al., 2012; Im et al., 2008; Li et al., 2011; Lohmann et al., 2008; Sofina et al., 2012). Recently, inclusion of 3D information has become an additional part of a reliable change detection process (Martha et al., 2010; Tian et al., 2013).

In summary, a wide range of different methods have been developed, displaying different grades of flexibility, robustness, practicability, and significance. Most authors, however, agree that no single best algorithm for change detection exists. Therefore, new methods are still being developed and/or adapted, especially for the detection of damaged buildings and infrastructure in conflict or crisis areas. This chapter is no exception to this, as it describes the development of, and the results for, a set of new change detection algorithms. They were tested with high and very high resolution (VHR) satellite images from QuickBird, GeoEye, and Cartosat sensors. Besides using multitemporal remote sensing images, additional information from GIS and 3D analysis can be incorporated into this cooperative suite of algorithms. The method can be used in catastrophic events or humanitarian crises to show the impact of this particular event.

17.2 MULTITEMPORAL IMAGE CHANGE DETECTION WITH COMBINED EDGE SEGMENT TEXTURE ANALYSIS

In general, simple methods such as image difference or image ratio do not produce reliable change information in complex areas, which means that there is a need to develop a different procedure for automated change detection (Klonus et al., 2012). This procedure is based on several different principles: frequency-based filtering, segmentation, and texture analysis. The frequency domain is used because it allows the direct identification of relevant features such as edges of buildings. If no features are directly visible (such as partial destruction with still standing outside walls), texture parameters are used for debris identification. A segmentation algorithm is used to extract the size and shape of buildings. These methods are combined in a decision tree to improve accuracy. The combination of these processing steps is called combined edge segment texture (CEST) analysis.

17.2.1 Fourier Transform–Based Algorithms

The Fourier transform is defined for single-band or panchromatic images (Cooley and Tukey, 1965). Based on a frequency analysis in the spectral domain, isotropic band pass filters can be designed to highlight selected frequencies and—as such— structures in the images. The design of band pass filters in the frequency domain is based on image size and resolution and the estimated size of buildings and man-made structures where changes are to be detected. The orientation of buildings has no influence due to the use of isotropic band pass filters. The filtered images are then transformed back into the spatial domain for further analysis. Higher frequencies visualize the position of buildings; the highest frequencies, however, contain mostly noise and are not useful for object identification and extraction. Lower frequencies contain mostly general image background, which is also not used for further analysis. To avoid the Gibbs problem, the filter was smoothed with a Hanning window (Brigham, 1997; Ehlers and Klonus, 2004).

After the adaptive band pass filtering, four different methods are possible for extracting changed structures: (1) subtraction in the frequency domain, (2) correlation in the frequency domain, (3) correlation in the spatial domain, and (4) edge detection in the spatial domain. Of these methods, the best results are obtained using the edge detection algorithm (Klonus et al., 2012). Consequently, we incorporated this method as a standard into the CEST analysis. The edge detection in the spatial domain consists of the following steps: The band pass filtered images T_1 and T_2 are first transformed into the spatial domain by an inverse fast Fourier transform (FFT). Thereafter, an edge detection operator is applied to both images (Figure 17.1). The best results are obtained by the Canny edge detector (Canny, 1986). To avoid small registration errors, morphological closing is used before subtracting the scenes from each other; a morphological opening is then applied.

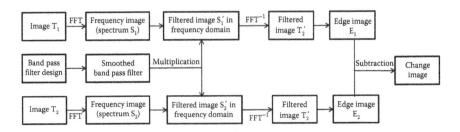

FIGURE 17.1   Change detection with filtering in the Fourier domain and subsequent edge detection.

17.2.2   CHANGE DETECTION BASED ON TEXTURE PARAMETERS

Frequency-based filtering is particularly suited to detect changes in edge structures. If edges remain intact, however, textural features may be used for change analysis. To calculate texture parameters, we make use of the Haralick features. This approach is based on the gray-level co-occurrence matrix (GLCM) (Haralick et al., 1973). The idea is that buildings can have higher texture values than areas without buildings (see, e.g., Ehlers and Tomowski, 2008; Myint, 2007). This is especially true if the surrounding environment is very homogeneous and the buildings are very small or destroyed (with surrounding debris). A GLCM describes the likelihood of the change of the gray value i to the gray value j of two neighboring pixels (Tomowski et al., 2006). To calculate GLCM, the frequency of all possible gray value combinations at two neighbor locations is counted for a defined number of directions (e.g., 0°, 45°, 90°, or 135°). The calculation of the average of these matrices for every element yields a direction-independent symmetric matrix.

Finally, to calculate the likelihood $P_{i,j}$ of a gray value change, every value in this matrix is divided by the maximum number of all possible gray value changes:

$$P_{i,j} = \frac{V_{i,j}}{\sum_{i,j=0}^{N-1} V_{i,j}} \qquad (17.1)$$

where
   V denotes the value in the symmetric GLCM
   i and j are the row and column indexes
   N is the number of rows and columns

The calculation of GLCM for images of high radiometric resolution is very time consuming. To reduce this effect, Haralick et al. (1973) suggest different texture features (the now well-known "Haralick" features), which represent the characteristic of a matrix in one comprehensive value and can be calculated using a moving window technique. Initial tests with several Haralick features showed that "energy" and "inverse difference moment" (IDM, also known as "homogeneity") produced the best results for man-made objects (Klonus et al., 2012). Consequently, these features were used for the CEST approach.

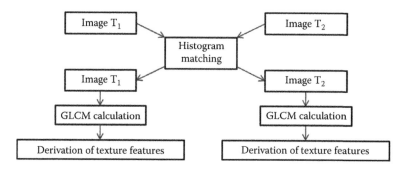

FIGURE 17.2   Steps for derivation of texture features for change detection. Histogram matching and GLCM computation precede the calculation of texture parameters for each image.

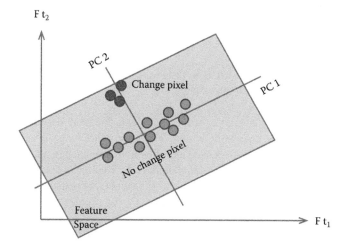

FIGURE 17.3   Change detection through bitemporal selective PCA. Unchanged pixels are clustered along the first principal component, whereas changed pixels are located along the second principal component.

The GLCM (8 bits) for every image is calculated after an initial histogram matching of the multitemporal images (Figure 17.2). Based on the GLCM, the texture features IDM and energy are computed with differently sized windows (ranging from 3×3 to 17×17 pixels). The size of the window depends largely on image resolution and on the size of the man-made structures to be analyzed. The calculated texture images at dates T_1 and T_2 are the input for a selective bitemporal principal component analysis (PCA), which is an excellent tool for the visualization of change (Figure 17.3) (Tomowski et al., 2011).

17.2.3   CHANGE DETECTION BASED ON SEGMENTATION

Object- or segment-based image analysis has gained a lot of interest in the remote sensing community (see, e.g., Baatz and Schäpe, 2000; Blaschke et al., 2008). Segmenting an image seems to be an excellent preanalysis tool, especially for VHR images.

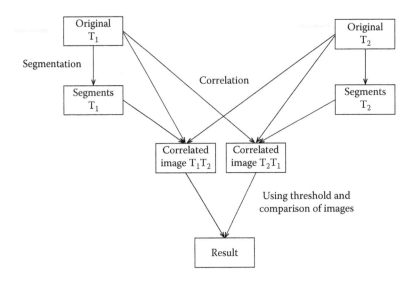

FIGURE 17.4 Change detection based on image segmentation. Segmentation overlay and correlation are calculated for both directions.

Consequently, we developed a gray value Euclidean distance–based segmentation procedure to be used for change detection. The gray value range is calculated and divided by a constant. This result is used as a threshold. For each pixel, the Euclidean distance to each neighboring pixel is calculated. If the Euclidean distance of the gray values is below the threshold, they belong to the same segment. After an independent segmentation of the images at dates T_1 and T_2, the segments of T_1 are selected and overlaid on the T_2 image. For each segment, the T_1–T_2 correlation coefficient is calculated. The result is then assigned to each pixel in the segment. Segments with high correlation represent no changes. Segments with low correlation represent changes. This procedure is repeated with segments from image T_2 overlaid on image T_1 (Figure 17.4). The average correlation value is assigned to each individual pixel. As a final step, thresholds are used to extract the change segments.

17.2.4 Combined Change Detection: The CEST Method

Finally, all three methods are combined in a decision-tree approach (Figure 17.5). The basis for the classification is the result of the change detection algorithm using edge detection based on frequency filtering. If the edge parameter shows "no change," the pixel in the image is classified as "no change." If the edge parameter shows "new building," the pixel is classified as new if the texture feature "energy" is in agreement. If energy shows "change" and one of the features such as "homogeneity" or "segmentation" shows "change," the result is "new." Otherwise, it is classified as unchanged. If the edge parameter shows "change," it is classified as "change" if the texture feature "energy" has a corresponding value. If energy shows "no change," the pixel will be classified as "no change." If energy shows "new" but the segment and homogeneity parameters show "change,"

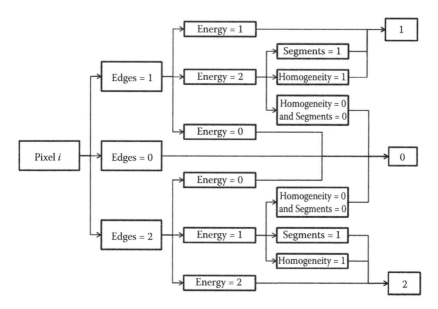

FIGURE 17.5 Decision tree for the combination of change detection methods. Edges = result of edge detection based on filtering in the Fourier domain. Segments = result of change detection using segmentation. Homogeneity and energy = results of Haralick's texture features. Numbers are related to the following classes: class 0 = unchanged buildings, class 1 = changed or destroyed buildings, and class 2 = new buildings.

the pixel is assigned to "change." Otherwise, it is classified as unchanged. The CEST procedure was tested against a number of standard change detection methods (difference, ratio, PCA, multivariate alteration detection [MAD], postclassification analysis).

17.3   CEST RESULTS FOR MULTITEMPORAL IMAGE CHANGE DETECTION

17.3.1   STUDY AREA

The study area is located in Sudan and represents an area that experienced dramatic changes during the Darfur conflict. This conflict is a dispute between different ethnic groups and the Sudanese government. Although the conflict in Sudan has recently been less intense than it was in the past, all sides to the conflict continue to commit violations of international humanitarian law, such as attacks on civilians and on humanitarian convoys. It is estimated that more than 300,000 people have already died in this conflict and more than 2 million people have been displaced (http://www.amnestyusa.org/research/science-for-human-rights).

The study site is located in South Darfur and shows part of the town Abu Suruj in West Darfur. The panchromatic images were taken by QuickBird-2 on March 2, 2006; a subset of the scene is presented in Figure 17.6, before the attack (T_1—Figure 17.6a), and on February 28, 2008, after the attack (T_2—Figure 17.6b). These images

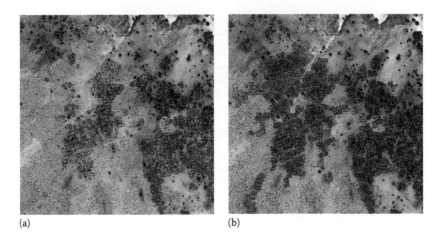

(a)   (b)

FIGURE 17.6 Panchromatic QuickBird-2 images recorded on March 2, 2006 (a) and on February 28, 2008 (b) of the town Abu Suruj (2048 × 2048 pixels). (Images are provided by Amnesty International, courtesy of DigitalGlobe.)

were provided by Amnesty International, courtesy of DigitalGlobe. Because of new settlement areas, this study site is very complex. It contains changes due to destruction and—at the same time—changes due to construction. A change detection procedure should be capable of depicting both types of change. This is demonstrated in Figure 17.7, which shows the manually digitized man-made structures. Black denotes no changes (background), white stands for new buildings (construction), and gray represents changed buildings (destruction). Most changed buildings are located in the east of the image with new buildings in the west. Figures 17.8 and 17.9 show subsets of Figure 17.6. Figures 17.8a and 17.9a present the T_1 image recorded on March 2, 2006, whereas Figures 17.8b and 17.9b show the T_2 image recorded on February 28, 2008. Figure 17.8 displays buildings that are destroyed in T_2 but did exist in T_1. The two existing buildings in Figure 17.9a were destroyed during 2007, but new buildings are constructed at the same place and are visible in T_2 (Figure 17.9b).

A visual comparison and overlay of man-made structures show a high correspondence for both images, so that a new coregistration was not necessary and the problem of possible pseudo change was negligible. These images were used for change analysis. They were preprocessed using a histogram matching procedure. An atmospheric correction was not applied due to missing ground truth data, sparse vegetation, and only one image band.

17.3.2 Results and Accuracy Assessment

In this section, change classification results of the standard methods, the new CEST method, and the achieved accuracies are presented. For assessing the accuracy, three classes were selected:

Class 0 = unchanged buildings
Class 1 = changed or destroyed buildings
Class 2 = new buildings

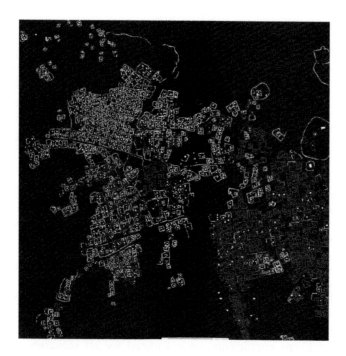

FIGURE 17.7 Manually digitized reference image of the town Abu Suruj (2048 × 2048 pixels). Black denotes no changes (background), white stands for new buildings (construction), and gray represents changed buildings (destruction).

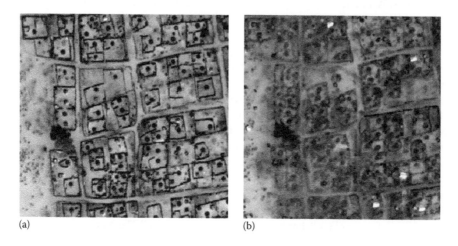

(a) (b)

FIGURE 17.8 Subset of the panchromatic QuickBird-2 images recorded on March 2, 2006 (a), and on February 28, 2008 (b), of the town Abu Suruj. The left image shows intact buildings and the right image shows destroyed buildings.

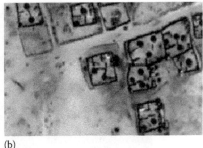

(a)   (b)

FIGURE 17.9   Subset of the panchromatic QuickBird-2 images recorded on March 2, 2006 (a), and on February 28, 2008 (b), of the town Abu Suruj. The image in (a) shows two intact buildings that were destroyed in the image in (b), but new buildings were constructed on the same location.

The reference is a manual digitization of buildings through an independent photointerpreter (see Figure 17.7). Accuracy assessment for classes 1 and 2 is based on 404 randomly chosen digitized objects. Only for class 0, all 404 objects were used. If most of the pixels inside an object are pixels of the correct class, the whole object was considered as correctly detected. Producers' accuracy, users' accuracy, and kappa coefficients are calculated for all scenarios. The CEST method was tested against the standard procedures image difference, image ratio, principle component analysis, MAD, and postclassification analysis. The results for CEST and the image difference and MAD algorithms (Nielsen et al., 1998) are compared in Figure 17.10. Figure 17.11 provides a close-up view of a subsection in the images. All methods, however, were analyzed and visually and quantitatively compared.

17.3.2.1   Visual Analysis

With a simple change detection method such as image difference, it is possible to detect three classes (positive change, negative change, and no change). In this process, however, large areas of pseudo change are detected due to changes in brightness of the sediment in the images. Most of the new buildings that appear in the T_2 image are correctly detected. Buildings that are unchanged are often identified as destroyed or changed buildings. These results are also confirmed by the accuracy assessment (see Figure 17.12). For image ratio, it is difficult to find a threshold between new and changed/destroyed buildings. Therefore, most of the buildings are detected as new buildings. As with image difference, buildings that are unchanged are often detected as destroyed or changed. This leads to the extremely low producers' accuracy of 8.2% for the class "changed or destroyed buildings." The amount of detected pseudo change is relatively low in comparison to image difference. The image processed with the PCA change detection procedure also shows a lot of pseudo change. Similar to the image ratio, most of the buildings are detected as new buildings. Also, nearly 45% of the unchanged buildings are classified as changed/destroyed. On the other hand, 30% of the destroyed or changed buildings are classified as unchanged.

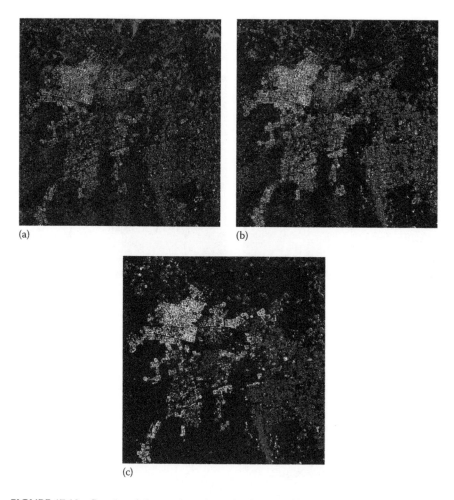

(a)   (b)

(c)

FIGURE 17.10 Results of change detection using image difference (a), MAD algorithm (b), and CEST (c): black denotes no changes (background), white stands for new buildings (construction), and gray represents changed buildings (destruction).

The MAD method produces acceptable results for the classes "unchanged buildings" and "new buildings." More than 60% of the unchanged buildings, however, were detected as changed/destroyed. Additionally, MAD produces a large amount of pseudo change. For the postclassification analysis, we used the isodata algorithm (Jensen, 2005) because no appropriate training areas were available. The postclassification method produces the lowest accuracies. Nevertheless, the producers' accuracy shows that 90% of the changed buildings can be detected but a users' accuracy of 50% means that half of the destroyed buildings are classified as unchanged. Again, pseudo change poses a big problem. For the CEST analysis, it proved possible to identify unchanged areas, new settlements, and destroyed settlements, even single huts and changed walls. As can be seen from the comparison of Figure 17.10 with Figure 17.7,

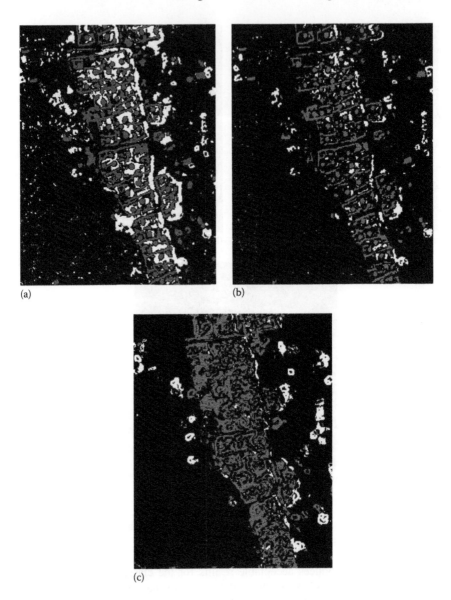

(a)   (b)

(c)

FIGURE 17.11   Blow-up of a subsection of Figure 17.10: image difference (a) shows many false alarms (white color), MAD (b) underestimates the changed (destroyed) buildings, CEST (c) gives a reliable depiction of the destruction area and has just a few false alarms.

CEST has much less noise than the second best algorithm, the MAD. Moreover, misclassification of vegetation as changed buildings is significantly less. In addition, the walls of the buildings are more accurate than those of the other methods. In total, the combination of all three methods generates the most reliable and accurate results for change detection. This is confirmed by the quantitative analysis in the following section.

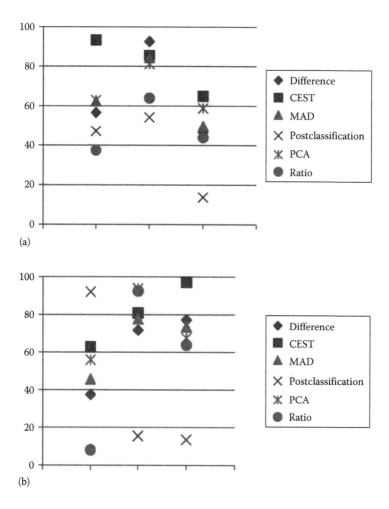

(a)

(b)

FIGURE 17.12   (a) Users' and (b) producers' accuracy for the six change-detection methods (left column "changed or destroyed buildings," center column "new buildings," and right column "unchanged buildings"). For users' accuracy, CEST has always the highest values with one exception: The difference method performs slightly better for the class "new buildings". For producers' accuracy, CEST is always among the top methods. Image ratio and postclassification show the worst results of all methods.

17.3.2.2   Quantitative Analysis

The results of the visual analysis are confirmed by the quantitative accuracy assessment. The accuracies of the CEST method are the best in this study. As much as 97% of the unchanged buildings are correctly detected. Although nearly 35% of the changed or destroyed buildings are identified as unchanged, the CEST result is still acceptable. In comparison to all other algorithms, however, the combined method shows the highest users' and producers' accuracies (Figure 17.12) and also produces less pseudo change. Accuracy figures are based on a per-pixel analysis.

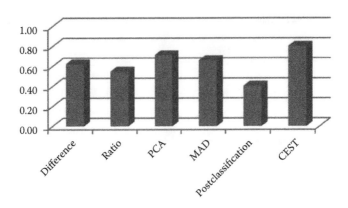

FIGURE 17.13   Overall accuracy for the change detection methods in the study area.

A comparison of overall accuracy is presented in Figure 17.13. As expected, the CEST approach shows the highest accuracy with about 80%. The next best are the PCA and MAD algorithms with 65% and 61%, respectively. The worst results are produced by the postclassification approach with less than 40%. Evidently, the CEST approach provides the best result whereas all other algorithms have lower accuracies and produce a large amount of noise. The superiority of the CEST method is also confirmed by transferring the procedure to other areas where changes have occurred. Details can be found in Klonus et al. (2012).

17.3.2.3 Automated Change Map

The produced change images are to a large degree abstract and hard to interpret. This is particularly true for people not related to remote sensing such as members of official organizations or rescue forces. For planning after a crisis or a catastrophe, the interpretation of change images should be as easy as possible. An algorithm was developed to automatically produce a map that can be easily interpreted. The first step is to generalize the change image. Inside a 20 × 20 pixels window, the amount of change is determined using the information in the change image. The change percentage of this area is calculated and then divided into a number of distinctive general classes. If less than 15% of the area has changed, all pixels are classified as unchanged. Change above 80% marks extensive change and change between 15% and 80% marks low to moderate change. Areas of new buildings with a surface cover of at least 15% are shown as "new areas." The original image of T_2 is used as background for automatically created change maps and for the results. Unchanged areas are transparent, low to moderate change areas are shown as yellow overlay, and areas of strong changes as red overlay. New building areas are shown in green (Figure 17.14). If this technique is applied to areas after catastrophic events, this change map makes it possible to quickly identify the most affected areas or the areas for which high casualties are likely. For the Abu Suruj area, it could be easily depicted that the town has increased, but also that large parts have changed. Buildings were destroyed and new buildings were built on these sites or next to the destroyed buildings.

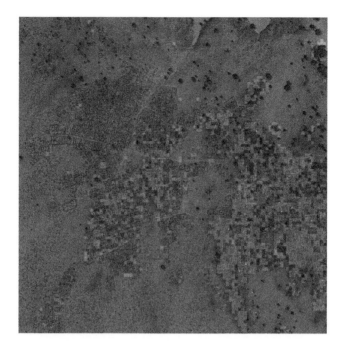

FIGURE 17.14 **(See color insert.)** Generalized change map of Abu Suruj: new buildings (green), low to moderate change (yellow), and extensive change (red).

17.4 CHANGE DETECTION BASED ON GIS AND REMOTE SENSING INTEGRATION

CEST is based on a comparison of remotely sensed images of the same scene obtained at different times. However, for many areas of the Earth, reliable information about buildings and infrastructure exists that may be more accurate and up to date than a remote sensing image that may have been taken a long time ago. Moreover, to achieve successful analysis, it is desirable to take images acquired by the same sensor at the same time of a season, at the same time of the day, and—for electro-optical sensors—in cloudless conditions (Hall et al., 1991). The accuracy of change detection analysis is also adversely affected by variation of acquisition angles. This situation can be improved by taking additional information into account. A comparative analysis of two different data types (vector map and remotely sensed image) is generally performed by extracting spectral, textural, and structural measurements from the image for each individual vector object.

In this method, we employ a single postevent remotely sensed image and GIS vector data with an original urban layout for the detection of building destructions caused by a catastrophic event. A general idea of the proposed approach is the generation of feature sets that characterize the actual state of each individual building and further classification based on the extracted information (Figure 17.15).

To perform a successful classification, the selected features should depict the distinct information that identifies the most effective characteristics specific for all

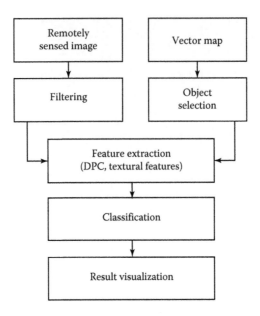

FIGURE 17.15   General scheme of the GIS/remote sensing change detection technique.

objects belonging to the same class. For the determination of building conditions, we focus on building roofs due to their visibility in the images. Thus, two characteristic parameters are employed for developing an integrated change detection system: image homogeneity and integrity of building contour (Sofina et al., 2012).

17.4.1   IMAGE HOMOGENEITY

Obviously, if a building is damaged or destroyed, the structure of its roof image is usually changed, which can be identified by texture analysis. An image area corresponding to the roof of an intact building is usually homogeneous with a low variation of image brightness. Fragments of destruction, on the other hand, adversely affect image homogeneity. To extract this information, the texture feature "IDM" (Haralick et al., 1973) is calculated. Instead of the commonly used average value of four angular feature directions (i.e., 0°, 45°, 90°, 135°), the maximum value (IDM_max) is used as a representative characteristic for the identification of building conditions (Sofina et al., 2012).

17.4.2   BUILDING CONTOUR INTEGRITY

Intact buildings in a remotely sensed image usually display clear contours of the building outline. In the case of damage or destruction, the contour of the building can be partially corrupted (or displaced) or completely absent. The possibility of recognizing the explicit building's contour is used as a representative indicator of its condition. This information is extracted by the "detected part of contour" (DPC) method developed by Sofina et al. (2011). The basic idea behind this method is the correspondence of the building contour in the remotely sensed image with the building footprint of the related vector object. The DPC parameter reaches a maximum value of 100% if the contour of the investigated building can be entirely identified. The first step in the calculation of the DPC involves an edge detection algorithm that is applied to the image for extracting building edges. For our purpose, we again use the Canny edge detector. It yields a raster map with pixel values corresponding to the direction of detected edges. The pixels that do not belong to any edge have a "no data" value. Given that buildings are symbolized as polygons in the vector map, control points are selected along each side of the polygons. For each control point, a search area on the raster map is defined, where pixels with appropriate contour direction are counted. The DPC value is then calculated as the ratio of the detected number of pixels and pixels expected for the intact building. For a detailed description of this method, see Sofina and Ehlers (2012).

17.4.3 EXPERIMENTS

The high potential of the proposed change detection approach is demonstrated on data obtained after the powerful tsunami waves in the wake of the Great East Japan Earthquake on March 11, 2011, which resulted in extensive destruction of roads, railways, dams, and buildings as well as in the loss of thousands of lives. Figure 17.16a presents a pre-earthquake image (April 27, 2005) and Figure 17.16b a post-earthquake image (March 12, 2011) of Kamaishi, provided by Google Earth. As no cadastral information was available to us, we obtained the building footprints by manual digitization. A total of 61 vector objects were digitized from the pre-event remotely sensed image, thus simulating a GIS cadastral map.

The classification of the building conditions was performed by means of a supervised k-nearest neighbor (k-NN) algorithm using both DPC and the maximum of the IDM feature (IDM_max). In our analyses, the IDM_max proved to be the most representative among the texture measurements. For learning the classification algorithm,

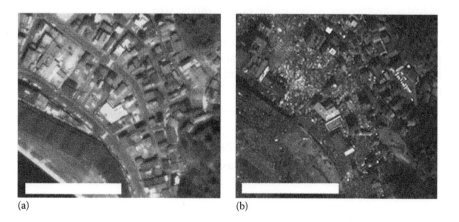

(a)   (b)

FIGURE 17.16   (See color insert.) Test dataset: The pre-event image (© DigitalGlobe 2013) used for digitizing the ground truth information (a) and the postevent image (© GeoEye 2013) used for change detection (b). (Satellite images, courtesy of Google Earth.)

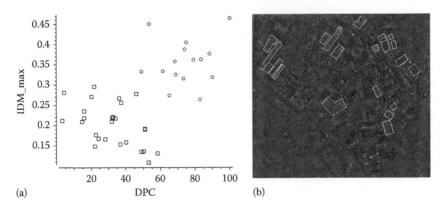

(a)   (b)

FIGURE 17.17 (a) Scatter plot between DPC and IDM_max features. The classes "unchanged buildings" (circles) and "destroyed buildings" (square boxes) can easily be separated; (b) resulting damage map: white objects represent intact buildings, black objects represent destroyed buildings.

30% of the vector dataset (containing 20 objects) was randomly selected as training examples. The scatter plot between the DPC and the IDM_max parameters illustrates the high separability of the objects (Figure 17.17a). Low values of DPC and IDM_max correspond to destroyed buildings (square boxes) and high values to intact ones (circles). It is evident that these two classes can easily be separated. The resulting damage map is presented in Figure 17.17b, where white objects symbolize intact buildings and black objects the destroyed ones.

Accuracy assessment is summarized in the confusion matrix (Table 17.1). An overall classification accuracy of 97.5% proves the validity of the presented integrated GIS/remote sensing approach.

The DPC/IDM approach demonstrates high efficiency for building damage assessment. It is also a robust, fast, and automatic algorithm. Vector maps provide additional information as well as the exact position, size, and shape of each object, which represents a major advantage of this approach. Moreover, due to the object-oriented algorithm, the "map-to-image" strategy enables the extraction of valuable information from the remotely sensed image with regard to an individual vector object. This method constitutes an excellent choice for change detection inside urban areas if GIS information is available.

TABLE 17.1

Confusion Matrix for Accuracy Assessment, Obtained from the Classification

		Prediction		
		Intact	Destroyed	
Actual class	Intact	14	1	15
	Destroyed	0	26	26
		14	27	41

17.5 INCLUSION OF 3D INFORMATION

Over the last few years, integration of 3D information into the task of change detection has gained increasing interest and importance. This is particularly due to the launch of VHR satellite systems like IKONOS, WorldView, and Cartosat that are able to capture stereo images and the development of effective techniques for the automatic processing of this type of data. As an example, semiglobal matching (SGM) has become a widely used approach for generating digital surface models (DSMs) from airborne and more recently also from satellite stereo and multiview imagery (d'Angelo and Reinartz, 2011). The task of change detection can benefit from the additional information provided by these approaches, especially in urban areas.

17.5.1 Change Detection Based on DSM Comparison

Change detection approaches can take advantage of height information directly, for example, by means of DSM differencing, or use it indirectly, for example, as additional contextual knowledge in image classification, and then detect changes in the classification results (Chaabouni-Chouayakh et al., 2010). Subsequent work discusses the effects of denoising the source models by introducing shadow and hole masks into DSM differencing and of combining height and shape information to better differentiate between real changes and false alarms (Chaabouni-Chouayakh and Reinartz, 2011; Tian et al., 2011). In addition, GIS reference data can be incorporated to verify the existence of buildings for one of the time frames (Dini et al., 2012). Champion et al. (2010) extracted 3D primitives from multiple images or DSMs and used them along with 2D contours for detecting changes in a 2D building database.

17.5.2 Building Extraction Using Morphological Segmentation

As far as optical images are concerned, shadows can be used as an alternative source of 3D information, which can be used to generate 3D building models. Shadows are particularly helpful for detecting buildings as they indicate the presence of objects that are higher than the ground. Height is very crucial for differentiating between buildings and spectrally and morphologically similar objects like parking lots, sports fields, and roads. It has been successfully utilized as contextual information for building outline extraction in recent approaches. In this context, it is assumed that buildings are represented in the intensity image by local maxima whereas shadows are local minima. However, as the overall intensity of the two feature types can vary considerably over the image, satisfactory results cannot be achieved by simple intensity thresholding. For reliable identification of local maxima and minima, white top hat and black top hat transformations are used (Serra, 1983). Here, the notion of "top hat" refers to the difference between the original and the morphologically filtered image. The well-known morphological operators opening and closing are often applied as filters. Depending on the shape and size of the chosen structuring element as well as the objects in the image, this procedure can result in artifacts in the filtered image. To avoid these artifacts, opening and closing by reconstruction are preferable (Gonzalez and Woods, 2008; Pesaresi and Benediktsson, 2001). Here, derivative morphological profiles (DMPs) are proposed, which can be calculated as the difference between pairs of morphologically filtered images by using sequentially increasing structuring elements. The idea of multiscale morphological segmentation is based on the assumption that the derivative of profile curves can be considered as some sort of a structural or morphological signature of an object class. Thus, it can be used to discriminate pixels by their morphological characteristics (Fauvel et al., 2005).

In the following section, we will concentrate on the development of a suitable extraction procedure for 3D information, which will later be included in the CEST change detection procedure. As no stereo pair or 3D information is available for the previous study sites, we developed this procedure for a test area with suitable stereo coverage. It can, however, be transferred to other areas with available stereo images.

17.5.3   DATASETS AND STUDY SITE

Although the final aim of our research is the automated 3D change detection for different satellite sensors, the method was developed using Cartosat images. The sensor provides panchromatic stereo images with a spatial resolution of 2.5 m. It is especially interesting as it features high spatial resolution, and a growing number of countries worldwide will have their own satellite systems with similar characteristics in the near future. Thus, a generic change detection approach is required that can be applied to such types of images.

DSMs generated from Cartosat images by means of SGM have lower resolution than the original 5 m stereo pairs. Moreover, they typically have some crucial inaccuracies in comparison to optical images. Hence, the shapes of building footprints are rather distorted, neighboring buildings can hardly be separated and form some sort of 3D composite object, and some buildings are also completely missing. A source for these problems is missing mutual information for some corresponding parts of the stereo images (Tian and Reinartz, 2011). Poor DSM quality is particularly reported at the shadow side of buildings.

The Cartosat images used in this work were acquired in 2008 and depict the Anatolian Coast of Istanbul. This region is especially interesting due to its rapidly developing real estate and new residential areas. In order to clearly illustrate the approach, a smaller subset of this area was chosen, which is located north of the Sabiha Gökçen airport (see Figure 17.18a). As can be noticed, it is often difficult to recognize object boundaries. Some edges of building footprints are blurred and overexposed, so that there is no visible boundary between objects and background. Due to their rectangular shape, it is also difficult to distinguish flat objects like sports fields or parking lots from building roofs. In addition, the basic hypothesis that buildings are usually brighter than their background as assumed in Jin and Davis (2005) and Huang and Zhang (2012) does not always apply. Hence, the availability of shadows is often the only way to detect the presence of 3D objects.

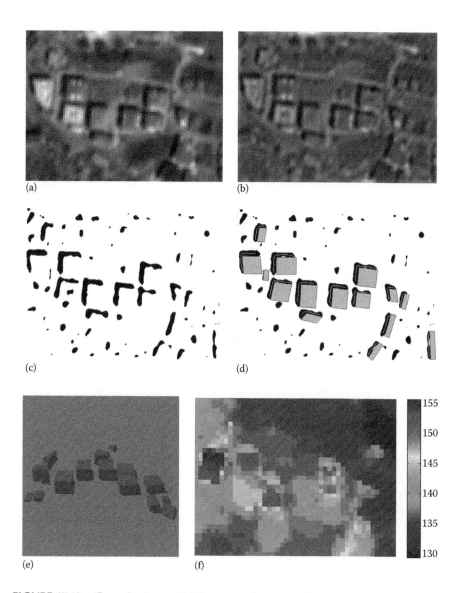

FIGURE 17.18 (See color insert.) (a) Cartosat orthophoto with 2.5 m resolution, (b) filtered orthophoto with 0.5 m resolution, (c) extracted shadows, (d) estimated ground plans overlaid with skeletons, (e) reconstructed 3D building models, and (f) DSM with 5 m resolution generated from the original stereo pair.

17.5.4 PROPOSED 3D CHANGE DETECTION APPROACH

17.5.4.1 Shadow Extraction

Shadows at the used image resolution may just cover 2–3 pixels, which makes comprehensive analysis problematic and complex. For this reason, we increase the image resolution by bilinear interpolation, which avoids abrupt intensity changes between neighboring pixels. Numerous experiments on morphological segmentation by means of closing by reconstruction show that the direct usage of interpolated images does not provide satisfactory results. Important shadow features are missing because of insufficient background contrast. Moreover, the heterogeneity regarding the intensity inside a shadowed region can yield only a partial identification, so that only a small part of a single feature is identified or the feature is divided into two parts. In order to enhance the image features with a controlled frequency response, a filtering in the frequency domain is performed. First, a high-frequency filter is applied in order to sharpen object edges and accentuate smaller features, which may hardly be perceptible in the original resolution. Here, a small constant (an offset) is added to the transfer function of the high pass filter, which preserves image tonality. The second parameter is a filter multiplier, which controls the contribution of high frequencies (Gonzalez and Woods, 2008). The filter image is then smoothed with a low pass filter, which reduces pixel patterns that result from the previous steps. The result is shown in Figure 17.18b. In this example, a 5% threshold of the frequency spectrum is used for high pass filtering with offset and multiplier values 0.5 and 2.0, respectively.

The results of closing by reconstruction with a circular structuring element of pixel size 10 can be seen in Figure 17.18c. It reveals the effectiveness of this method for identifying local minima. It is especially important as shadow features, in some cases, can have even lighter intensities than rooftops. Moreover, even very dark rooftops can also be successfully distinguished from adjacent shadows.

17.5.4.2 Building Outline Generation

In the next step, the extracted shadows are used to generate building outlines. However, depending on the shape and orientation of a building, only those parts of the building that face away from the sun actually cast shadows and those shadow areas might not even be connected. This makes it rather impossible to generate complete building outlines. However, if we decompose the shadow into linearly shaped components, one can assume that each component is bordered by a four-sided building polygon at its sun-facing side. While the length of this polygon can be derived from the length of the adjoining shadow component, the width has to be estimated by using a realistic constant value. If two consecutive components form a convex angle (with regard to the building interior), then both components are bordered by the same four-sided polygon and its length and width can be derived. As concave angles between components cannot be utilized as easily, they are ignored in the process, resulting in the drawback that two small building objects are often extracted instead of a larger one. Keeping these assumptions in mind, one can generate rough building outlines of quadrangular shape as explained in the following paragraph.

The binary shadow image is skeletonized by reducing the boundaries until the shadow areas are only one pixel wide. Then the remaining skeleton pixels are traced according to their 8-neighborhood, generating connected linear components that are subsequently simplified by the Douglas–Peucker algorithm (Douglas and Peucker, 1973). The resulting piecewise linear curves reflect well the topology of the shadows. They are, however, geometrically inaccurate at the vertices, because skeletonizing tends to smoothen the corners. Therefore, each line segment is recalculated from its constituent skeleton pixels, which are identified as lying closer to the currently regarded segment than to any other segment of the curve. An exhaustive search over all the lines going through any two pixels is performed, choosing the one that best fits all pixels. As the number of pixels per line segment is rather small, this approach still performs efficiently. It has the additional advantage that the pixels at the end of the line can be simultaneously filtered out if they start to deviate too far from the line. The line segments are then pulled toward the sun, so that they are located optimally between the boundary pixels of the shadow and their neighboring building pixels. Finally, four-sided building outlines can be synthesized from these linear shadow components (Figure 17.18d).

Once the building outlines are available, their heights are derived by shadow simulation. A large number of 3D building models are generated, all with uniformly increasing heights. At this time, only flat roofs are considered, but the process can be extended to other shapes like saddleback and hipped roofs. Then their shadows are rendered based on the same "sun parameters" as the Cartosat images. We make use of the shadow mapping algorithm (Williams, 1978). The scene is rendered from the sun's position, creating a so-called shadow map. Then the scene is rendered again, but now from the sensor's viewpoint. Each rendered fragment is transformed into light space and checked against the shadow map. If it is further away than the stored value, then it is lying in shadow. The simulated shadows are compared to the extracted shadow image, and the height that results in the best fit is chosen.

17.6 CONCLUSIONS AND FUTURE WORK

In this chapter, a new automated change detection method (CEST) and some possible extensions are presented. CEST consists of (a) adaptive filtering in the frequency domain with edge detection in the spatial domain, (b) calculation of the texture features "homogeneity" and "energy" with a PCA change detection approach, and (c) segment based correlation. This combined method is compared to five standard change detection algorithms (image difference, image ratio, PCA, MAD, and postclassification analysis). Results are visually and quantitatively analyzed. The accuracy assessment shows that the CEST method is far superior to the standard techniques for change detection. Despite the fact that CEST is more complex than the tested standards methods, CEST can be completely automated and transferred to other areas. The combined method yields an overall accuracy of 80% and more than 90% of the unchanged buildings could be correctly identified.

If available, existing building information that is stored in a GIS or a cadastral database can be incorporated into the change detection process. It is shown that the combination of the DPC algorithm and the IDM texture feature is a very reliable option

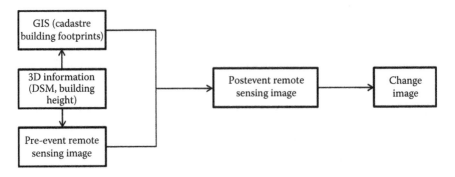

FIGURE 17.19   Extended CEST analysis including GIS and 3D information.

for change analysis with classification accuracies of better than 95%. Consequently, the GIS option will be included in the future CEST procedure (see Figure 17.19).

Cadastral databases do not usually provide 3D information, which is another important input into the change detection process. We therefore developed an algorithm to extract building heights from shadows that are visible in satellite images of very high spatial resolution. The success of this procedure encouraged us to also include the 3D information extraction process in the CEST method. Figure 17.19 presents a general concept for a robust and reliable change analysis using remotely sensed data, GIS, and 3D information.

ACKNOWLEDGMENTS

This work was supported by the German Space Center DLR through the research project "Automated Change Detection Methods for Disaster Applications," the Federal Ministry of Education and Research BMBF through the project "Enhanced Change Detection Accuracy by Integrated 3-D Information and Remote Sensing Data," and a PhD stipend of University of Osnabrueck, Germany. We would like to thank Euromap for providing the Cartosat-1 data of Istanbul. We also very much appreciate the support of Amnesty International, DigitalGlobe, GeoEye, and Google in providing the satellite images. We are grateful to the anonymous reviewers for their valuable comments that helped improve this chapter.

REFERENCES

Baatz, M. and Schäpe, A. (2000). Multiresolution segmentation: An optimization approach for high quality multi-scale image segmentation. *Angewandte Geographische Informationsverarbeitung XII, Beiträge zum AGIT-Symposium*, Salzburg, Germany, 2000, pp. 12–23.

Blaschke, T., Lang, S., and Hay, G. (Eds.) (2008). *Object-Based Image Analysis—Spatial Concepts for Knowledge-Driven Remote Sensing Applications*, Springer Lecture Notes in Geoinformation and Cartography, Heidelberg, Germany.

Bovolo, F. (2009). A multilevel parcel-based approach to change detection in very high resolution multitemporal images. *IEEE Geoscience and Remote Sensing Letters*, 6(1), 33–37.

Brigham, E. O. (1997). *FFT Anwendungen*, Oldenbourg Verlag, München, Germany.

Canny, J. (1986). A computational approach to edge detection. *Transactions on Pattern Analysis and Machine Intelligence*, 8(6), 679–698.

Chaabouni-Chouayakh, H., Krauss, T., d'Angelo, P., and Reinartz, P. (2010). 3D change detection inside urban areas using different digital surface models. *Proceedings, ISPRS Commission III Symposium Photogrammetry Computer Vision and Image Analysis (PCV 2010)*, Paris, France.

Chaabouni-Chouayakh, H. and Reinartz, P. (2011). Towards automatic 3D change detection inside urban areas by combining height and shape information. *Photogrammetrie Fernerkundung Geoinformation*, 205–217.

Champion, N., Boldo, D., Pierrot-Deseilligny, M., and Stamon, G. (2010). 2D building change detection from high resolution satellite imagery: A two-step hierarchical method based on 3D invariant primitives. *Pattern Recognition Letters*, 31, 1138–1147.

Chen, G., Hay, G. J., Carvalho, L. M. T., and Wulder, M. A. (2012). Object-based change detection. *International Journal of Remote Sensing*, 33(14), 4434–4457.

Cooley, J. W. and Tukey, J. W. (1965). Raster algorithm for machine calculation of complex Fourier series. *Mathematics of Computation*, 19, 297–301.

Coppin, P., Jonckheere, I., Nackaerts, K., Muys, B., and Lambin, E. (2004). Digital change detection methods in ecosystem monitoring: A review. *International Journal of Remote Sensing*, 25(9), 1565–1596.

d'Angelo, P. and Reinartz, P. (2011). Semiglobal matching results on the ISPRS stereo matching benchmark. *Proceedings, High-Resolution Earth Imaging for Geospatial Information*, Hannover, Germany.

Dini, G. R., Jacobsen, K., Rottensteiner, F., Al Rajhi, M., and Heipke, C. (2012). 3D building change detection using high resolution stereo images and a GIS database. *International Archives of the Photogrammetry, Remote Sensing and Spatial Information Sciences*, XXXIX-B7, 299–304.

Douglas, D. and Peucker, T. (1973). Algorithms for the reduction of the number of points required to represent a digitized line or its caricature. *The Canadian Cartographer*, 10(2), 112–122.

Ehlers, M. and Klonus, S. (2004). Erhalt der spektralen Charakteristika bei der Bildfusion durch FFT basierte Filterung. *Photogrammetrie Fernerkundung Geoinformation (PFG)*, 6, 495–506.

Ehlers, M. and Tomowski, D. (2008). On segment based image fusion. In: T. Blaschke, S. Lang, and G. Hayes (Eds.), *Object-Based Image Analysis—Spatial Concepts for Knowledge-Driven Remote Sensing Applications*, Springer Lecture Notes in Geoinformation and Cartography, Heidelberg, Germany, pp. 735–754.

Fauvel, M., Palmason, J. A., Benediktsson, J. A., Chanussot, J., and Sveinsson, J. (2005). Classification of remote sensing imagery with high spatial resolution. *Proceedings of SPIE*, 5982, 598201-1–598201-15.

Gonzalez, R. C. and Woods, R. E. (2008). *Digital Image Processing*, 3rd edn., Prentice-Hall, Upper Saddle River, NJ.

Hall, F. G., Strebel, D. E., Nickeson, J. E., and Goetz, S. J. (1991). Radiometric rectification: Toward a common radiometric response among multidate, multisensor images. *Remote Sensing of Environment*, 35, 11–27.

Haralick, R. M., Shanmugam, K., and Dinstein, I. (1973). Textural features for image classification. *IEEE Transactions on Systems, Man, and Cybernetics*, 3, 610–621.

Huang, X. and Zhang, L. (2012). Morphological building/shadow index for building extraction from high-resolution imagery over urban areas. *IEEE Journal of Selected Topics in Applied Earth Observations and Remote Sensing*, 5, 161–172.

Im, J., Jensen, J. R., and Tullis, J. A. (2008). Object-based change detection using correlation image analysis and image segmentation. *International Journal of Remote Sensing*, 29(1–2), 399–423.

Jensen, J. R. (2005). *Introductory Digital Image Processing: A Remote Sensing Perspective*, 3rd edn., Prentice-Hall, Upper Saddle River, NJ.

Jin, X. and Davis, C.H. (2005). Automated building extraction from high-resolution satellite imagery in urban areas using structural, contextual, and spectral information. *EURASIP Journal on Advances in Signal Processing*, 2005, 2196–2206.

Khorram, S., Biging, G. S., Chrisman, N. R., Colby, D. R., Congalton, R. G., and Dobson, J. E. (1999). *Accuracy Assessment of Remote Sensing Derived Change Detection*, American Society for Photogrammetry and Remote Sensing, Bethesda, MD.

Klonus, S., Tomowski, D., Ehlers, M., Michel, U., and Reinartz, P. (2012). Combined edge segment texture analysis for the detection of damaged buildings in crisis areas. *IEEE Journal of Selected Topics in Applied Earth Observations and Remote Sensing*, 5(4), 1118–1128.

Li, P., Haiqing, X., and Song, B. (2011). A novel method for urban road damage detection using very high resolution satellite imagery and road map. *Photogrammetric Engineering & Remote Sensing*, 77(10), 1057–1066.

Lohmann, P., Hoffmann, P., and Müller, S. (2008). Updating GIS by object-based change detection. In: J. Schiewe and U. Michel (Eds.), *gi-reports@igf—Geoinformatics Paves the Highway to Digital Earth: On the occasion of the 60th birthday of Professor Manfred Ehlers*, Band 8, Institut für Geoinformatik und Fernerkundung, Osnabrück, Germany, pp. 81–86. http://www.igf.uni-osnabrueck.de/de/publikationen/gi-reports (last accessed January 22, 2014).

Lu, D., Mausel, P., Brondízio, E., and Moran, E. (2003). Change detection techniques. *International Journal of Remote Sensing*, 25(12), 2365–2407.

Macleod, R. D. and Congalton, R. G. (1998). A quantitative comparison of change-detection algorithms for monitoring eelgrass from remotely sensed data. *Photogrammetric Engineering & Remote Sensing*, 64(3), 207–216.

Martha, T. R., Kerle, N., Jetten, V., van Westen, C. J., and Kumar, K. V. (2010). Landslide volumetric analysis using Cartosat-1 derived DEMs. *IEEE Geosciences and Remote Sensing Letters*, 7(3), 582–586.

Mas, F. J. (1999). Monitoring land-cover changes: A comparison of change detection techniques. *International Journal of Remote Sensing*, 20(1), 139–152.

Myint, S. W. (2007). Urban mapping with geospatial algorithms. In: Q. Weng and D. A. Quattrochi (Eds.), *Urban Remote Sensing*, CRC Press, Boca Raton, FL, pp. 109–136.

Nielsen, A. A., Conradsen, K., and Simpson, J. J. (1998). Multivariate alteration detection (MAD) and MAF postprocessing in multispectral, bitemporal image data: New approaches to change detection studies. *Remote Sensing of Environment*, 64, 1–19.

Pesaresi, M. and Benediktsson, J. A. (2001). A new approach for the morphological segmentation of high-resolution satellite imagery. *IEEE Transactions on Geoscience and Remote Sensing*, 39, 309–320.

Serra, J. (1983). *Image Analysis and Mathematical Morphology*, Academic Press, Orlando, FL.

Singh, A. (1989). Digital change detection techniques using remote-sensed data. *International Journal of Remote Sensing*, 10(10), 989–1003.

Sofina, N. and Ehlers, M. (2012). Object-based change detection using high-resolution remotely sensed data and GIS. *International Archives of the Photogrammetry, Remote Sensing and Spatial Information Sciences*, XXXIX-B7, 345–349.

Sofina, N., Ehlers, M., and Michel, U. (2011). Object-based detection of destroyed buildings based on remotely sensed data and GIS. *Proceedings of SPIE, Earth Resources and Environmental Remote Sensing/GIS Applications II*, Prague, Czech Republic.

Sofina, N., Ehlers, M., and Michel, U. (2012). Integrated data processing of remotely sensed and vector data for building change detection. *Proceedings of SPIE, Earth Resources and Environmental Remote Sensing/GIS Applications III*, Edinburgh, U.K.

Tian, J., Chaabouni-Chouayakh, H., and Reinartz, P. (2011). 3D building change detection from high resolution spaceborne stereo imagery. *Proceedings, International Workshop on Multi-Platform/Multi-Sensor Remote Sensing and Mapping (M2RSM)*, Xiamen, China, pp. 1–7.

Tian, J. and Reinartz, P. (2011). Multitemporal 3D change detection in urban areas using stereo information from different sensors. *Proceedings, International Symposium on Image and Data Fusion (ISIDF)*, Tengchong, Yunnan, China, pp. 1–4.

Tian, J., Reinartz, P., d'Angelo, P., and Ehlers, M. (2013). Region-based automatic building and forest change detection on Cartosat-1 stereo imagery. *ISPRS Journal of Photogrammetry and Remote Sensing*, 79, 226–239.

Tomowski, D., Ehlers, M., and Klonus, S. (2011). Colour and texture based change detection for urban disaster analysis. *Proceedings, Joint Urban Remote Sensing Event (JURSE)*, Munich, Germany, pp. 329–332.

Tomowski, D., Ehlers, M., Michel, U., and Bohmann, G. (2006). *Objektorientierte Klassifikation von Siedlungsflaechen durch multisensorale Fernerkundungsdaten*, gi-reports@igf, Band 3, University of Osnabrueck, Osnabrueck, Germany. http://www.igf.uni-osnabrueck.de/de/publikationen/gi-reports (last accessed January 22, 2014).

Williams, L. (1978). Casting curved shadows on curved surfaces. *Computer Graphics (Proceedings of ACM SIGGRAPH 78)*, 12(3), 270–274. http://www.eyesondarfur.org/villages.html (last access March 20, 2013).

18 Fusion of SAR and Optical Data for Urban Land Cover Mapping and Change Detection

Yifang Ban, Osama Yousif, and Hongtao Hu

CONTENTS

18.1 STATE OF THE ART

18.1.1 FUSION OF SYNTHETIC APERTURE RADAR AND OPTICAL DATA FOR URBAN LAND COVER MAPPING

Urban areas are regarded as a very complex landscape in terms of the diversity of land cover types and the shape and pattern of various urban features. Accurate and up-to-date information on urban land cover is a challenging task but of crucial importance for urban planning, environment protection, and policy-making. Urban extent extraction and land cover mapping have been studied using a range of remotely sensed data and algorithms. Gamba and Herold (2009) summarized the topic with a special focus on global monitoring. High-resolution optical imagery and object-based approaches are often used for urban applications at the local level, while regional or global analysis generally exploits moderate- or coarse-resolution optical images and pixel-based methods (Gamba and Lisini, 2013). In addition to optical data, synthetic aperture radar (SAR) systems have been playing an increasingly important role in urban analysis due to their ability to acquire images day and night in all weathers and to the fact that the number of advanced SAR systems in operation is increasing (Rogan and Chen, 2004). However, the specific imaging characteristics of SAR systems and the existence of speckle noise make the interpretation of SAR images in urban areas generally more difficult compared to the analysis of optical images. Nevertheless, SAR data have been investigated for urban extent extraction and land cover mapping with promising results (Ban et al., 2010, Gamba et al., 2011; Hu and Ban, 2012, Niu and Ban, 2013).

The fusion of optical and SAR data for land cover classification is of increasing interest due to their distinct and complementary features (Ban, 2003; Ban and Jacob, 2013). Optical images contain information on the reflective and emissive characteristics of surface features, while SAR images record the intensities of radar returns from surface features, which are mainly determined by SAR systems properties and surface structure and dielectric properties (Xia and Henderson, 1997; Amarsaikhan and Douglas, 2004). Land cover types that are difficult to discriminate in optical images may be easily separated with SAR images and vice versa because of the complementary information provided by the two datasets (Amarsaikhan and Douglas, 2004). The integration of optical and SAR data for land cover classification has been widely investigated in the literature and this topic was recently summarized by Zhu et al. (2012). All studies indicate that the results from the combined use of optical and SAR data are better than those obtained using an individual data source (Amarsaikhan et al., 2012; Zhu et al., 2012).

Various combinations of optical and SAR images acquired by a range of sensors have been explored for urban extent extraction and land cover mapping. Landsat Thematic Mapper (TM) and Enhanced Thematic Mapper Plus (ETM+), SPOT, and QuickBird are typical optical data sources, while SAR data are often acquired by SAR systems with moderate or high resolutions such as environmental satellite (ENVISAT), advanced synthetic aperture radar (ASAR), RADARSAT SAR, and TerraSAR-X SAR (Alparone et al., 2004; Ban et al., 2010; Leinenkugel et al., 2011).

The simplest way of fusing optical and SAR data for urban extent and land cover mapping is to place all images in a single dataset and then apply certain classifiers to generate a classification map. Zhu et al. (2012) tested the integration of Landsat ETM+ and single-season advanced land-observing satellite (ALOS) phased array type L-band synthetic aperture radar (PALSAR) data for the classification of 17 land cover categories in urban and peri-urban environments. The contribution of different dimensions (spectral, polarimetric, temporal, and spatial) of input data to a random forest classifier was evaluated with map accuracy statistics. The results demonstrated the value of combining multitemporal Landsat imagery, ALOS PALSAR data, and texture variables for land cover classification in urban and peri-urban environments. Griffiths et al. (2010) combined Landsat TM/ETM+ images and SAR images from ERS-1/2 and ASAR by simply stacking them and then employed support vector machines (SVMs) to map urban growth in the Dhaka megacity region between 1990 and 2006. Soria-Ruiz et al. (2010) demonstrated that the combination of Radarsat-1 C-band SAR data and Landsat ETM+ band 5 and 7 images produced more accurate land cover maps than the use of Landsat ETM+ band 2–4 images in a region experiencing rapid urbanization. If either type of dataset contains too many images, principal component analysis (PCA) could be used to decrease the number of inputs of that dataset in order to reduce the training time of the classifier (Pacifici et al., 2008). Haack et al. (2002) compared space-borne radar and optical data for urban delineation and found that the classification of combined radar and TM data produced better results.

Some fusion techniques can be used to derive a new dataset from the original optical and SAR data that is then used for urban extent and land cover mapping. Amarsaikhan et al. (2012) compared the performances of six fusion techniques for the enhancement of urban features and improvement of urban land cover classification. The six fusion techniques include multiplicative method, Brovey transform, PCA, Gram–Schmidt fusion, wavelet-based fusion, and Elhers fusion. Of these methods, wavelet-based fusion of optical and SAR data produced the best results. Lu et al. (2011) investigated the fusion of Landsat TM and radar (i.e., ALOS PALSAR L-band and RADARSAT-2 C-band) data for mapping impervious surfaces. TM and radar data were fused by the wavelet-merging technique to produce a new dataset, which was then unmixed to four fraction images. The impervious surface image was extracted from two fraction images. Their research indicated that the fusion image with 10 m spatial resolution was suitable for such applications and radar data with different wavelengths did not show significant difference in improving the performance of impervious surface mapping. Cao and Jin (2007) applied PCA to fuse ETM+ infrared and SAR images and selected the first three components of PCA as inputs used to train the classifier. The results showed that fusion of the infrared and SAR images improved classification accuracy. Alparone et al. (2004) presented a multisensor image fusion algorithm based on generalized intensity modulation for the integration of multispectral and SAR images. The algorithm was applied to Landsat-7 ETM+ and ERS-2 SAR images of an urban area and the results demonstrated that the algorithm had the advantage of preserving spectral and textural information of both datasets, which can be useful for visual analysis and classification. Henderson et al. (2002) merged SAR data with TM data using four different techniques, namely, concatenation, image addition, image weighting, and PCA fusion, and then used the fused data for land cover mapping in a rapidly urbanizing coastal area. Their results indicated that simple fusion techniques improve classification accuracy more than complex image merge methods do.

Fusion of optical and SAR data can also be conducted at object or feature level for urban land cover classification. Ban and Jacob (2013) investigated the fusion of multitemporal ENVISAT ASAR and Chinese HJ-1B multispectral data for detailed urban land cover mapping using an object-based approach and an SVM classifier. After image segmentation, several features were calculated for each image object from SAR and optical images and these features were used as input into SVM. The results showed that the fusion of SAR and multispectral data improved classification accuracy, and fewer multitemporal SAR images could be used for fusion and classification if multitemporal multiangle dual-look-direction SAR data are carefully selected. Amarsaikhan et al. (2007) found that the rule-based method at feature level achieved higher accuracy than a standard supervised classification method for urban land cover mapping with the integrated use of optical and InSAR data.

Decision level fusion is another approach to the combined utilization of optical and SAR data for urban land cover mapping. Ban et al. (2010) applied an object-based and knowledge-based approach to urban land use/land cover mapping with QuickBird multispectral data and multitemporal RADARSAT Fine-Beam C-HH SAR data. Decision level fusion of QuickBird classification and RADARSAT SAR classification took advantage of the best classifications of both optical and SAR data. Several confused classes from QuickBird images were separated by the integration of SAR classification results.

Urban extent extraction and classification can be carried out using individual datasets, but with the assistance of the other dataset. For example, impervious surface estimation was conducted using SPOT-5 data (Leinenkugel et al., 2011). Single-polarized TerraSAR-X data and the object-oriented classification approach were used in the study to delineate settlement areas in order to exclude irrelevant areas such as natural and undeveloped land from the impervious surface estimation process. Optical images can also be used as ancillary data to improve SAR image classification in urban areas. Normalized Difference Vegetation Index (NDVI) derived from optical data can be utilized to exclude small green urban areas when extracting built-up extent with SAR data (Gamba and Aldrighi, 2012).

Palubinskas (2012) presented a general workflow for optical and SAR data fusion and examined the aspects and parameters that can influence the quality of optical and SAR data fusion and classification accuracy. The aspects and parameters include data acquisition/selection, orthorectification, coregistration, feature extraction, clustering, fusion, classification, and quality assessment. Data selection usually depends on the availability of optical and SAR data and the requirements of the specific application. Improvement of data inputs to the classifiers tends to be more beneficial for urban land cover classification than improvement of classification algorithms (Zhu et al., 2012). The method for fusing optical and SAR data in a specific application should be carefully selected in order to take maximum advantage of the information contained in the data.

18.1.2 FUSION OF SAR AND OPTICAL DATA FOR URBAN CHANGE DETECTION

Similar to mapping urban land cover, monitoring urban land cover change in a timely and accurate manner is of critical importance for urban planning, environmental monitoring, and sustainable management of land resources. Remote sensing data have become a major source used for change detection due to the advances in sensor technology. The ultimate purpose of image fusion is to enhance and improve change detection results for intended applications. Fusion typically involves different types of images, for example, images with different sensors and/or spatial or spectral resolution. In change detection analysis, the fusion of SAR and optical data is important from two perspectives. First, the limited availability of data, often optical data due to cloud cover, forces the generation of a change indicator through the comparison of an image pair acquired over the same area but from a different sensor. Although the images were acquired with sensors that have different modalities, they are two different representations of the same physical reality and consequently can be compared (Inglada and Giros, 2004). Recently, similarity measures have played an essential role in performing such complicated image comparison. Mercier et al. (2008), for example, successfully used the Kullback–Leibler divergence to compare an ERS SAR image with a SPOT image. Second, single-source multitemporal images (optical or radar) are known for their limited capacity to provide exhaustive observation of changes that have occurred on the ground. Change detection analysis can benefit from the complementary nature of the change information represented by SAR and optical multitemporal datasets. In an attempt to improve the quality of the binary change map, for example, Bruzzone and Prieto (2000) proposed an unsupervised change detection approach that uses consensus theory to integrate many change variables. Poulain et al. (2009) fused features extracted from SAR and high-resolution optical images to update cartographic database. Their methodology is not change detection in the traditional sense, as the detection of new buildings is achieved by comparing high-resolution images with the existing older vector database. To achieve this goal, many primitives were extracted from the optical and SAR images, for example, edges, vegetation, and shadows. Evidence constructed based on these features is then fused using the Dempster–Shafer evidence theory to provide a score for each building. This score is then used to decide whether or not to update the vector database.

Although important, only few studies were found in the literature on fusion of SAR and optical data for urban change detection. For example, Onana et al. (2003) fused change information extracted from multitemporal SAR images with the NDVI image extracted from a single SPOT image. Based on multitemporal SAR images, they extracted two change variables, namely, pixel-by-pixel ratio measure and cross-correlation measure. The SPOT-based NDVI image was fused with the change indicators mentioned earlier at the interpretation level, where a set of IF-THEN structures were used. A similar approach was used by Bujor et al. (2001), who extracted distances from each pixel to the nearest road from a topographic map instead of the NDVI image. The problem with these approaches is that intensive prior information is required as the fusion rules were developed by codifying expert knowledge.

Many change detection techniques, however, have been developed in recent decades for the analysis of multitemporal SAR or optical images, respectively, and will be reviewed here. Successful change detection needs to consider characteristics of remote sensing systems and the environment to be monitored, and select a suitable method for change detection. Two steps are essential in change detection: (i) deriving change variables from multitemporal remote sensing images and (ii) generating change maps. In this chapter, the approaches for deriving change variables from multitemporal optical and SAR data are briefly presented first. Then the methods for generating change maps are summarized into four groups: (1) unsupervised approaches, (2) supervised approaches, (3) object-based approaches, and (4) other approaches.

18.1.2.1 Deriving Change Variables from Multitemporal Remote Sensing Images

A comprehensive review was provided by Lu et al. (2004) on the mathematical operators that can be used to compare multitemporal optical remote sensing images, for example, image differencing (ID), image ratioing (IR), and image regression. Berberoglu and Akin (2009) found IR to be effective in reducing topographic effects like variation in illumination and shadowing. However, IR produces relatively poor results compared with ID. Change vector analysis (CVA) is an extension of the concept of ID, which is particularly tailored for comparing multispectral multitemporal images. He et al. (2011) extended the CVA technique to include textural change information. Bovolo and Bruzzone (2007) found that the use of CVA magnitude does not in fact utilize all the information contained in the multitemporal multispectral difference image. They suggested transforming the spectral change vector from the Cartesian to the polar coordinate system, in which they developed rigorous statistical distributions for the magnitude and the direction random variables.

Comparison of multitemporal optical remote sensing images can also be carried out in a new transformed space instead of in the raw data domain defined by the observed multitemporal images. A simple example is differencing multitemporal NDVI images, where the measured intensities in each image—that is, red and near-infrared values—are first transformed to the NDVI space (Lyon et al., 1998). Similarly, Cakir et al. (2006) transformed each individual image in the multitemporal dataset to a component space using correspondence analysis. The difference image is then constructed in this new space.

Comparison of multitemporal SAR images is commonly carried out using the ratio operator as ratioing of the multitemporal radar intensities is shown to be better adapted to the statistical characteristics of SAR data than subtracting (Rignot and van Zyl, 1993; Moser and Serpico, 2006). Ratio-related operators have also been used to compare SAR images, including the log ratio (Bazi et al., 2005), modified ratio (Ban and Yousif, 2012), and the normalized mean ratio (Ma et al., 2012). Bujor et al. (2003) compared different types of parameters to quantify temporal changes based on SAR images and found the ratio operator to be especially suitable for the detection of steplike changes. Hachicha and Chaabane (2010) suggested two different types of change indicators that were developed based on the assumption that SAR amplitudes are Rayleigh-distributed. The first indicator (Rayleigh ratio detector) works per pixel and uses first-order statistics, while the second one, the Rayleigh Kullback–Leibler divergence, utilizes higher-order statistics.

The comparison of multitemporal SAR images can also be carried out using similarity measures. These measures have been used extensively in the field of automatic image-to-image registration as a means of quantifying similarity in the spatial domain. In the context of change detection analysis, given two coregistered images, similarity measures can be used to quantify temporal rather than spatial similarity (Alberga, 2009). The strength of similarity measures lies in the fact that the estimation of the change indicator takes into account the pixel and its neighborhood in contrast to traditional arithmetic operators, which work per pixel and normally ignore the contextual information (Inglada and Mercier, 2007).

18.1.2.2   Unsupervised Change Detection

The advantage of unsupervised change detection algorithms is that no prior knowledge about the study area (in the form of training data) is required. An unsupervised change detection algorithm normally accepts multitemporal images as input and outputs a binary (or ternary) change map that shows changed versus unchanged areas. The main disadvantage of this type of algorithm is therefore the lack of detailed from–to change information. Unsupervised change detection approaches include image algebra (e.g., band differencing, band ratioing), CVA, unsupervised classification of multidate images or change variables, PCA, chi square transformation, minimum-error thresholding algorithms, and contextual-based methods.

Image algebra is relatively easy to implement and interpret, but it cannot provide "from–to" change information and requires careful selection of the "change/no change" threshold. Image differencing subtracts pixel by pixel the same band of images acquired at two times to produce a new image. A threshold is then selected by analyzing the difference image to transform the difference image into a binary "change/no change" map. IR is similar to ID, except that the comparison between images is conducted by computing the ratio. ID can be extended to compare the vegetation index derived from multiple dates of images. For example, researchers computed an NDVI on two dates and then subtracted one from the other to generate a change image (e.g., Townshend and Justice, 1995; Lyon et al., 1998). Bruzzone and Prieto (2000) proposed two automatic techniques (based on the Bayes theory) for the analysis of the difference image. One allows an automatic selection of the decision threshold that minimizes the overall change detection error probability under the assumption that pixels in the difference image are independent of one another. The other analyzes the difference image by considering the spatial–contextual information included in the neighborhood of each pixel. Their experimental results confirmed the effectiveness of both proposed techniques.

In spectral CVA, each pixel at each date is represented by its vector in spectral feature space and the spectral change vector is then computed as the difference between the feature vectors at the two dates. The magnitude of the change vector is used to detect the presence of change using a carefully selected threshold, while the direction of the change vector contains information about the type of change. Kontoes (2008) used the CVA method to detect land cover change in three test sites and achieved high overall accuracies. Sohl (1999) investigated five change detection techniques in the Abu Dhabi Emirate using Landsat TM data, including univariate image differencing, an "enhanced" image differencing, vegetation index differencing, postclassification differencing, and CVA. The enhanced ID technique provided most accurate specific quantitative values of change, while the CVA excelled at providing rich qualitative detail about the nature of a change.

Unsupervised classification of multidate images has been used to detect land cover changes (Jensen, 2005). Images of different dates are first placed together in a single dataset and then unsupervised classification can be performed on this composite dataset to create clusters. The analyst identifies and labels the clusters as "change" or "no change." This method requires only single classification, but it is difficult to label change classes. Mas (1999) tested six change detection techniques including direct multidate unsupervised classification for detecting areas of changes in a coastal zone in Mexico and found that postclassification comparison (PCC) was the most accurate procedure.

The techniques based on PCA can be used to perform change detection by applying the principal component transformation to the multidate composite image. The transformation produces a new, uncorrelated PCA image dataset (Jensen, 2005). The major components of the dataset tend to account for variation in the image data that is not due to land cover change, while the minor components are likely to enhance spectral contrasts between the two dates (Collins and Woodcock, 1996). Another way to apply PCA for change detection is to perform PCA separately and then subtract the second-date component image from the corresponding one (Kwarteng and Chavez, 1998). The difficulty of using PCA for change detection lies in interpreting and labeling each component image. Another disadvantage is that this method cannot provide detailed "from–to" change information. It is possible, however, to classify the PCAs of multitemporal images using supervised classification. Li and Yeh (1998), for example, used the method to monitor rapid land use change and urban expansion in the Pearl River Delta.

Ridd and Liu (1998) introduced a chi square transformation method for change detection. The advantage of the method is that multiple bands are simultaneously considered to produce a single-change image. A potential disadvantage is that change related to specific spectral direction may not be readily identified. Another drawback of this method is that it does not perform well in an area where a large portion of the image is changed.

The minimum-error thresholding algorithm proposed by Kittler and Illingworth (1986) has been used extensively in the field of change detection. This algorithm was developed based on the Bayesian decision theory using the histogram-fitting technique to estimate the unknown probabilities and an optimum threshold for separating the object from the background in the image. The Kittler–Illingworth algorithm is known to be a fast and effective thresholding tool. Melgani et al. (2002) used the algorithm successfully thresholding a change variable derived from multispectral multitemporal images. To deal with the non-Gaussian nature of the SAR images, the minimum-error thresholding algorithm was generalized by Bazi et al. (2005) and Moser and Serpico (2006), whose studies propose different types of density function models suitable for describing the statistics of the changed and unchanged classes in an SAR change image. Ban and Yousif (2012) examined four different density function models for urban change detection using multitemporal SAR data and found that the log normal, the Nakagami ratio, produced better change detection results than the generalized Gaussian model and the Weibull ratio model. Essentially, the algorithm assumes the existence of one object (i.e., one typology of change) and one background. Bazi et al. (2006) successfully applied the algorithm to a case in which more than one typology of change existed in the study area—that is, to a case with more than one threshold. The main drawback of the multithreshold version of the Kittler–Illingworth algorithm is its high computational complexity.

Several contextual-based techniques have been developed for unsupervised change detection using multitemporal images. For example, Bruzzone and Prieto (2000) developed two unsupervised change detection methods. The first method automatically selects the decision threshold that minimizes the probability of error. The second method analyzes the difference image by taking into consideration the spatio–contextual information included in the neighborhood of each pixel. Celik (2010) considered unsupervised change detection to be an intensive search for a change mask that optimizes a minimum mean square (MSE) criterion function. The genetic algorithm is used to search for this optimum mask—that is, the change map. Celik (2009) used the PCA technique to map local neighborhoods in the difference image to a higher-dimensional space defined using nonoverlapping image blocks. The k-means algorithm was then used to automatically separate the changed from the unchanged areas. Moser et al. (2007) developed an unsupervised change detection algorithm based on Markov random field (MRF). In their work, the approximate change map necessary to run the MRF-based algorithm was obtained by applying the Kittler–Illingworth algorithm. Similarly, a context-based unsupervised change detection algorithm was proposed by Bruzzone and Prieto (2002) using a Parzen estimate to model the likelihood function of the observations.

18.1.2.3 Supervised Change Detection

Traditionally, supervised change detection is carried out using PCC logic. It consists of classifying each image in the multitemporal dataset independently using the same classification scheme. The detailed from–to change information can then be extracted by comparing the classified images on a pixel-by-pixel basis. For example, Yuan et al. (2005) applied the method to a series of Landsat images in order to study the dynamics of the land cover change over the Twin Cities Metropolitan Area. Del Frate et al. (2008) used PCC to extract change information from multitemporal SAR images. Instead of using the maximum likelihood classifier, the authors used an artificial neural network for the classification of each SAR image. Walter (2004) and Zhou et al. (2008) extracted the change information by comparing images classified using object-based techniques instead of a per-pixel classification. Castellana et al. (2007) improved the accuracy of the change detection process by combining supervised postclassification logic with an unsupervised change detection algorithm.

Supervised change detection is not restricted to PCC logic. For example, Volpi et al. (2013) investigated supervised change detection using two techniques, namely, multidate classification and analysis of difference image. To address the problem of high intraclass variability, the authors suggested using an SVM classifier. Similarly, Nemmour and Chibani (2006) extracted urban growth in the Algerian capital from Landsat multitemporal images using the SVM classifier.

The main drawback with the supervised change detecting method is the need for high-quality training data to classify each image in the multitemporal dataset. This turns out to be very difficult to achieve, especially for older images. Many semisupervised change detection algorithms have been developed that require only a limited amount of ground information or limited interaction from the analyst (Moser et al., 2002).

18.1.2.4 Object-Based Change Detection

Object-based change detection (OBCD) can be defined as "the process of identifying differences in geographic objects at different moments using object-based image analysis" (Chen et al., 2012). Compared with the traditional pixel-based change detection, OBCD is able to improve the identification of changes for the geographic entities found over a given landscape. There are mainly four ways to perform OBCD: direct comparison, comparison after object-based classification, multitemporal segmentation and comparison, and integration of pixel-based and object-based approaches.

Direct comparison detects changes by directly comparing image objects and applying a threshold. Multitemporal images are usually segmented separately and changes are then analyzed based on objects' spectral information (e.g., averaged band values) and/or associated features of the objects (e.g., texture and geometry). This approach is straightforward, but it needs to search for spatially corresponding objects in multitemporal images and select an appropriate change threshold. Lefebvre et al. (2008) proposed an OBCD method that jointly deals with the analysis of the object contours and the analysis of texture evolution. A geometric change index and a content change index were developed. Their results indicated that both object contour and texture features were effective for change detection purposes in very-high-resolution images. Hall and Hay (2003) introduced an object-based multiscale digital change detection approach, which first segmented panchromatic SPOT images from two dates and then directly applied an ID method to detect object changes at different scales. Their results showed that the proposed approach had the ability to automatically detect changes at multiple scales as well as suppress sensor-related noise.

A direct comparison of image objects cannot easily provide "from–to" change information. Similar to pixel-based postclassification change detection, comparison after object-based classification detects land cover changes by comparing the independently classified objects from multitemporal images. The performance of OBCD is strongly influenced by both segmentation and classification procedures. For example, Laliberte et al. (2004) conducted image segmentation and object-based classification on 11 aerial photos and 1 QuickBird image spanning 67 years to monitor vegetation changes over time and found the usefulness of incorporating both spectral and spatial information in classification. Im et al. (2008) evaluated the performance of incorporating object correlation images and neighborhood correlation images within the classification feature space. Their results showed that object-based change classifications incorporating these new features produced more accurate change detection classes. Blaschke (2005) argued that standard change detection and accuracy assessment techniques, which mainly rely on statistically assessing individual pixels, are not satisfactory for image objects that exhibit shape, boundary, homogeneity, or topological information. Therefore, a Geographic Information

Systems (GIS) conceptual framework for image OBCD was developed and a series of rules were defined by considering object size, shape, and location.

In multitemporal segmentation and comparison approach, multitemporal images are combined and segmented together, producing spatially corresponding change objects. Consequently, the comparison of image objects is straightforward as objects at the same location in multitemporal images are of the same sizes and shapes. However, it is unclear whether this form of change detection is influenced by segmenting multitemporal images together, because different objects may exist at the same geographic location on different dates (Chen et al., 2012).

Integration of pixel-based and object-based approaches is another popular way to detect changes. The preliminary change information can be obtained using pixel-based techniques. Better change results are then generated by applying the object-based paradigm. Several studies have confirmed the effectiveness of integrating pixel-based procedures into OBCD schemes (e.g., Al-Khudhairy et al., 2005; McDermid et al., 2008; Bovolo, 2009). The integration of pixel- and object-based schemes reduces noisy and spurious changes, but it remains unclear how the final results are affected by the different combinations of pixel- and object-based schemes (Chen et al., 2012).

18.1.2.5   Other Approaches

In addition to the change detection techniques discussed earlier, there are also some less frequently used methods developed by researchers. Several studies employed integrated GIS and remote sensing method for change detection. Yang and Lo (2002) used an unsupervised classification approach, a GIS-based image spatial reclassification procedure, and PCC with GIS overlay to map the spatial dynamics of land use and land cover change in the Atlanta metropolitan area. The integration of GIS showed many advantages over traditional change detection methods. Weng (2002) investigated land use change dynamics in the Zhujiang Delta of China by the combined use of satellite remote sensing, GIS, and stochastic modeling technologies and indicated that such integration was an effective approach for analyzing the direction, rate, and spatial pattern of land use change. Durieux et al. (2008) applied an object-based classification methodology to SPOT 5 images with a 2.5 m resolution. The extracted buildings were compared with the existing reference GIS maps to monitor building construction in urban sprawl areas.

Many novel change detection techniques were also developed in the past two decades. For example, Wang (1993) constructed a knowledge-based vision system for detecting land cover changes at urban fringes. Zhang et al. (2002) proposed a new structural method based on road density combined with spectral bands for urban built-up land change detection in Beijing, China. Kasetkasem and Varshney (2002) addressed the problem of image change detection based on MRF models. Nielsen (2007) proposed an iteratively reweighted multivariate alteration detection (IR-MAD) method for change detection in multi- and hyperspectral imagery. Chen et al. (2013) proposed a semisupervised context-sensitive technique for change detection in high-resolution multitemporal remote sensing images by analyzing the posterior probability of a probabilistic Gaussian process classifier within an MRF model.

18.2   CASE STUDIES

18.2.1   Fusion of ENVISAT ASAR and MERIS Data for Urban Land Cover Mapping

18.2.1.1   Introduction

The objective of this research is to evaluate multitemporal, multi-incidence-angle, dual-polarization ENVISAT ASAR, and the synergy of ASAR and MERIS data for urban land cover classification. The study area is located in the Greater Toronto Area (GTA), Ontario, Canada, where rapid urban expansion and sprawl has encroached onto the Oak Ridges Moraine, one of the most distinct and environmentally significant landforms in southern Ontario. The major land use/land cover classes are high-density built-up areas, low-density built-up areas, roads, forests, parks, golf courses, water, and four types of agricultural lands (winter wheat, pasture, corn, and soybeans). These 11 classes, adapted from the United States Geological Survey (USGS) land use/land cover classification scheme, were chosen to characterize the complex landscape and diverse land cover types in the GTA.

Eleven-date ENVISAT ASAR images in alternating polarization mode (C-HH and C-HV) with a spatial resolution of 30 m and a pixel spacing of 12.5 m were acquired from June to October in 2004 (Figure 18.1). The detailed descriptions of these images are given in Table 18.1. These 11-date SAR images can be grouped into two categories based on their beam position. The first group of images was acquired with a steep incidence angle in IS2 beam mode while the second group was acquired with shallow incidence angles.

FIGURE 18.1   (See color insert.) ENVISAR ASAR image over part of Toronto.

TABLE 18.1

ENVISAT C-HH and C-HV ASAR Imagery

Acquisition Date	Beam Position	Incidence Angle Range (Degree)
2004-06-28	IS2	19.2–26.7
2004-07-08	IS6	39.1–42.8
2004-07-24	IS7	42.5–45.2
2004-08-02	IS2	19.2–26.7
2004-08-12	IS6	39.3–42.1
2004-08-31	IS5	35.8–39.4
2004-09-06	IS2	19.2–26.7
2004-09-16	IS6	39.1–42.8
2004-10-02	IS7	42.5–45.2
2004-10-11	IS2	19.2–26.7
2004-10-21	IS6	39.1–42.8

Three 15-band MERIS scenes were acquired based on vegetation phenology, one in early-season scene on June 12, 2004: one in mid-season scene on August 8, 2004 (Figure 18.2) and one in late-season scene on September 18, 2004.

Field data on various land use/land cover types, their roughness and moisture conditions, vegetation heights, and ground coverages were collected during each satellite overpass. Photographs were taken during fieldwork to assist image interpretation and analysis. Other sources of data such as digital topographic data, Landsat ETM+ imagery, and orthophotos were also used to georeference ASAR data and to assist selecting ground reference data for classification, calibration, and validation.

FIGURE 18.2 (See color insert.) ENVISAT MERIS image over Toronto.

18.2.1.2 Methodology

18.2.1.2.1 Geometric Correction of SAR Images

To remove these relief displacements and bring the 11 images from several incidence angles to the same database, multitemporal ENVISAT ASAR imagery was orthorectified to the National Topographic Database (NTDB) using satellite orbital models and a digital elevation model at 30 m resolution.

18.2.2.2.2 Texture Analysis

Texture is one of the significant parameters recognized by the human visual system, besides pixel brightness and color, for identifying objects or regions of interest in an image. Various studies showed that in most cases, texture, not intensity, is the most important source of information in high-resolution radar images (e.g., Ulaby et al., 1986; Dobson et al., 1995). In this study, gray level co-occurrence matrix (GLCM)–based texture measures including mean and standard deviation (SD) were analyzed. Using second-order spatial statistics, GLCM, which is a two-dimensional array, can provide conditional joint probabilities of all pairwise combinations of pixels within a computation window. An 11 × 11 window size was selected based on trials. The mean apply averaging of gray level in the local window and SD calculates the gray level SD in the local window using the following formulas (Jensen, 2005):

For mean,

$$\mu_i = \sum_{i,j=0}^{N-1} i\left(P_{i,j}\right) \quad \mu_i = \sum_{i,j=0}^{N-1} j\left(P_{i,j}\right)$$

SD is the square root of variance in the following equation:

$$\sigma_i^2 = \sum_{i,j=0}^{N-1} P_{i,j}\left(i-\mu_i\right)^2 \quad \sigma_j^2 = \sum_{i,j=0}^{N-1} P_{i,j}\left(j-\mu_j\right)^2$$

18.2.2.2.3 Selection of Calibration and Validation Blocks

For each land use/land cover class, pixel sample blocks were randomly extracted in order to calibrate the classifier. To assess the accuracy of the classifications, validation pixels, independent from the calibration pixels, were randomly selected for each land use/land cover class. The selections of calibration and validation blocks were based on field data, Landsat ETM+ images, NTDB vectors, and maps.

18.2.2.2.4 Image Classification

Artificial neural networks (ANNs) are computer programs designed to simulate the human learning process through establishment and reinforcement of linkages between input data and output data. Presenting a nonparametric and distribution-free approach to image classification, ANN has been increasingly used in remote sensing applications (e.g., Foody, 1995; Jensen et al., 2001; Ban, 2003; Kavzoglu and Mather, 2003; Jensen, 2005). A neural network consists of interconnected processing elements called nodes or neurons. It is an adaptive system that changes its structure during a learning phase.

The most interesting feature in neural networks is the possibility of learning. In remote sensing applications, ANNs are often composed of three elements.

An *input layer* consists of the source data, which in the context of remote sensing are multispectral and/or SAR observations. The *output layer* consists of the classes required by the analyst. With training data, ANN establishes an association between input and output data by the establishment of weights within one or more hidden layers during the training phase. ANNs are trained through the backpropagation algorithm. This can be thought of as a retrospective examination of the links between input and output data in which differences between expected and actual results can be used to adjust weights (Campbell, 2002).

ANN is reported by many authors to be a more robust classifier than the traditional statistical approaches in the classification of multiresolution, multisource remote sensing/geographic data (e.g., Wilkinson et al., 1995, Ban, 2003). Therefore, it is desirable to investigate the effectiveness of ANNs for classifications of spaceborne multiresolution multisensor data.

18.2.2.2.5   Classification Accuracy Assessment

To assess the quality of the image classifications, various measures including overall accuracy and Kappa coefficient of agreement (or Kappa) were analyzed to compare classification results with the validation or reference data in confusion matrices.

18.2.1.3   Results and Discussion

18.2.1.3.1   Classification of ASAR Data

The classification results demonstrated that, for identifying urban land use/land cover classes, ASAR raw data yielded very poor results. The combination of seven-date ASAR raw data in shallow incidence angles performed much better (9% in classification accuracy) than four-date ASAR raw data in steep incidence angles. The best overall validation accuracy and Kappa, however, were very low (Table 18.2). The inaccuracies were, in part, due to speckle in the raw radar images.

Classification of ASAR texture images yielded much better accuracies than that of raw SAR data (Table 18.2). Between the two texture measures, mean texture measure performed much better than SD. Similar to the results of ASAR raw data, the classifications of texture images in shallow incidence angles yielded 12% and 10% better classification accuracies using mean and combined mean and SD, respectively, than that of the steep-angle IS2 texture images. The best classification result of 78.88% (Kappa, 0.77) was achieved with combined mean and SD texture images in both shallow and steep incidence angles. The results indicate that combinations of various texture measures, which can extract unique spatial relationships from the same SAR data, showed improvement over single-set texture measure because of their different, complementary information.

Table 18.3 presents the confusion matrix of the best ASAR classification result. Several land use/land cover classes such as water, winter wheat, and corn achieved very good classification accuracy. Classification accuracies for the majority of land cover classes were in the 70%–80% range. Classification of parks and forests yielded poor results. Forest was confused with low-density built-up areas because low-density built-up areas were dominated by trees and small houses. Parks were confused with roads and golf courses while golf courses were confused with roads due to their similar low backscatter characteristics in SAR images.

TABLE 18.2

ANN Classification Accuracies of ASAR Images

ASAR Data Alone	Overall Accuracy (%)	Kappa
IS2 raw	30.75	0.24
IS5_6_7 raw	39.52	0.33
June 28, mean and SD, C-HH and C-HV	48.34	0.43
July 08 and 24, mean and SD, C-HH and C-HV	57.94	0.54
June 28 and July 08 and July 24, mean and SD, C-HH and C-HV	63.23	0.59
June 28 and August 02, mean and SD, C-HH and C-HV	55.81	0.51
July 08 and July 24 and August 12 and August 31, mean and SD	68.25	0.65
June and July and August, mean and SD, C-HH and C-HV	68.98	0.66
IS2, mean, C-HH and C-HV	51.38	0.46
IS2, mean and SD, C-HH and C-HV	63.49	0.60
IS5_6_7, mean, C-HH and C-HV	63.40	0.60
IS5_6_7, mean and SD, C-HH and C-HV	73.61	0.71
All, mean and SD, C-HV	65.74	0.62
All, mean and SD, C-HH	70.90	0.68
All, mean, C-HH and C-HV	68.93	0.66
All, mean and SD, C-HH and C-HV	78.88	0.77

TABLE 18.3

Confusion Matrix for Classification of IS5_6_7 Mean and SD Texture Images Combined

	Reference Data										
Classified Data	Water	Roads	LB	HB	Golf Courses	Forest	Parks	Corn	Soybeans	Winter Wheat	Pasture
Water	93.3	5.7	0	0	1	0	0	0	0	0	0
Roads	0	74.2	0.1	0	13.2	0	7.6	0	0	2.4	2.6
LB	0	0	74.9	24.8	0	0.2	0	0	0	0	0
HB	0	5	13.5	79.2	0	1.6	0.3	0	0	0.3	0
Golf courses	0.9	16.6	0	0	70.1	0.3	12.2	0	0	0	0
Forest	0	0	25.1	3.2	0.1	67.7	0	0	0.7	0	3.2
Parks	0	18.9	1.2	0	12.7	1.7	61.6	0.9	0	0.4	2.6
Corn	0	0.3	3.1	0	0	0.1	0	88.9	2	2.6	2.9
Soybeans	0	0	0.3	0.4	0	0.2	0	11.1	82.7	0	5.3
Winter wheat	0	0	0	0	0	0	0	3.4	0.3	92.5	3.7
Pasture	0	0	1.9	0	0	7.3	0	0.1	1.6	7.4	81.7

Note: Values in italics highlight the overall accuracy for each land cover class.

18.2.1.3.2 Synergy of ASAR and MERIS Data

The classification accuracies for various ASAR and MERIS data combinations are presented in Table 18.4. The results demonstrate that it is possible to achieve reasonable classification accuracy using early- to mid-season ASAR and MERIS images (70%). The best classification accuracy was achieved using the combination of all ASAR images and the August 8 MERIS images at peak vegetation season (overall 82%, Kappa 0.8, Table 18.5, Figure 18.3). This represents 4% increase in classification accuracy over the best classification of ASAR data alone. With the addition of August 8 MERIS data, confusions between forest and low-density built-up areas, and between parks, roads, and golf courses decreased significantly. Classification accuracies for all classes either improved or remained similar except for winter wheat. Winter wheat was confused with other agricultural classes. The addition of more MERIS images from other dates, however, decreased classification accuracy. This is in part caused by confusion among some land cover classes due to seasonal changes in vegetation that resulted in significant differences in spectral reflectance. The lower overall accuracy could also be caused by the use of more MERIS images in much lower spatial resolution (Figure 18.4).

18.2.1.4 Conclusions

Multitemporal multi-incidence–angle dual polarization ENVISAT ASAR data and synergy of ASAR and MERIS data were evaluated for extracting urban land cover information. Eleven-date ENVISAT ASAR images were acquired from June to October in 2004 for the classification of 11 land cover classes. The results demonstrated that, for identifying landscape/land cover classes, texture images yielded much better results than ASAR raw data. The classification of multitemporal ASAR texture images in shallow incidence angles yielded superior results than that of ASAR texture measures in steep incidence angles. The best classification result of 78.88% (Kappa, 0.766) was achieved with combined mean and SD texture images in both shallow and steep incidence angles.

The synergy of all ASAR texture images and August 8 MERIS images improved the classification accuracy by 4% (overall 82%, Kappa 0.8). The addition of MERIS data was able to resolve confusion between several classes in SAR images such as forest and low-density built-up areas, parks, and roads due to their similar backscatter characteristics. The addition of more MERIS images from other dates, however, decreased classification accuracy. This is in part caused by confusion among some land cover classes due to seasonal changes in vegetation that resulted in significant differences in spectral reflectance. The lower overall accuracy could also be caused by the use of more MERIS images in much lower spatial resolution. The results indicate that attention needs to be focused on the selection of the optical data in appropriate seasons in data fusion.

18.2.2 Fusion of SAR Optical Data for Urban Change Detection

18.2.2.1 Introduction

The objective of this research is to evaluate the fusion of SAR and optical data for urban change detection using the Kittler–Illingworth minimum-error thresholding algorithm. A multitemporal SAR image pair, that is, ERS-2 SAR image acquired on September 7, 1999, and ENVISAT acquired on September 19, 2008,

TABLE 18.4

Classification of ASAR and MERIS Data Combined

MERIS and ASAR	ASAR June, Mean and SD	ASAR July, Mean and SD	ASAR August, Mean and SD	ASAR September and October, Mean and SD	MERIS June 12	MERIS August 8	MERIS September 18	Overall Accuracy (%)	Kappa
A1	✓				✓			56.80	0.52
A2	✓	✓			✓			72.27	0.69
A3	✓	✓	✓		✓			70.79	0.68
A4	✓	✓	✓	✓	✓			80.18	0.78
B1	✓					✓		57.52	0.53
B2	✓	✓				✓		68.28	0.65
B3	✓	✓	✓			✓		72.89	0.70
B4	✓	✓	✓	✓		✓		*82.17*	*0.80*
C1	✓						✓	62.43	0.59
C2	✓	✓					✓	70.93	0.68
C3	✓	✓	✓				✓	70.88	0.68
C4	✓	✓	✓	✓			✓	75.53	0.73

							Accuracy	Kappa
D1	✓				✓	✓		55.20	0.50
D2	✓	✓			✓	✓		67.83	0.64
D3	✓	✓	✓	✓	✓	✓		72.92	0.70
D4	✓	✓	✓		✓	✓		77.07	0.75
E1	✓				✓		✓	59.55	0.55
E2	✓	✓			✓		✓	68.02	0.65
E3	✓	✓	✓		✓		✓	69.12	0.66
E4	✓	✓	✓	✓			✓	73.36	0.71
F1	✓			✓		✓	✓	59.63	0.55
F2	✓	✓				✓	✓	67.31	0.64
F3	✓	✓	✓			✓	✓	66.17	0.63
F4	✓	✓	✓			✓	✓	73.45	0.71
G1	✓			✓	✓	✓	✓	63.45	0.60
G2	✓	✓			✓	✓	✓	65.98	0.62
G3	✓	✓	✓		✓	✓	✓	71.86	0.69
G4	✓	✓	✓		✓	✓	✓	73.91	0.71

Note: Values in italics highlight the best accuracy achieved.

TABLE 18.5

Confusion Matrix for Classification of ASAR and August 8 MERIS Images Combined

Classified Data	Reference Data										
	Water	Roads	LB	HB	Golf Courses	Forest	Parks	Corn	Soybeans	Winter Wheat	Pasture
Water	*92.1*	3.3	0	0	4.5	0	0	0	0	0	0
Roads	0	*78.1*	0	0	2.9	0	18.1	0	0	0.7	0.2
LB	0	0	*70*	30	0	0	0	0	0	0	0
HB	0	4.5	8.1	*85.2*	0	0	1.4	0	0	0.9	0
Golf courses	0	8.5	0	0	*76.8*	0	14.7	0	0	0	0
Forest	0	0	13.9	0.3	0	*75.4*	0.4	0.5	5	0.3	4.3
Parks	0	8.8	0.4	0	0	0	*90.8*	0	0	0	0
Corn	0	0	0	0	0	0.4	0	*89.1*	4.7	5.7	0
Soybeans	0	0	6.4	2.8	0	0	0	0	*87.6*	0	3.2
Winter wheat	0	0	0	0	0	0	0.2	11.4	22.3	*59.9*	6.2
Pasture	0	0	0	0	0	8.3	0	0.3	0	1.5	*89.9*

Note: Values in italics highlight the overall accuracy for each land cover class.

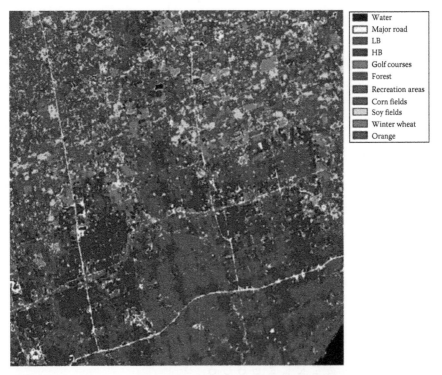

FIGURE 18.3   (**See color insert.**) Urban land cover classification result: ASAR and MERIS.

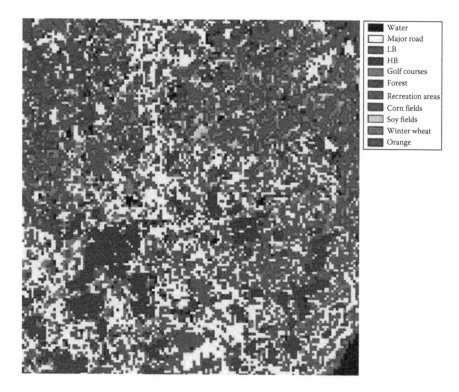

▨	Water
☐	Major road
■	LB
■	HB
■	Golf courses
■	Forest
■	Recreation areas
■	Corn fields
☐	Soy fields
■	Winter wheat
■	Orange

FIGURE 18.4 **(See color insert.)** Urban land cover classification result: MERIS.

and a multitemporal Landsat image pair acquired on November 3, 1999, and July 6, 2008, were selected to investigate the synergy of SAR and optical data for urban change detection in Shanghai.

18.2.2.2 Methodology

In the traditional Kittler–Illingworth minimum-error thresholding algorithm, a single change variable is often used and the spatial dependence between neighboring pixels is deliberately ignored to convert the sophisticated threshold selection process to a simple search in one-dimensional feature space (Bazi et al., 2005, 2006; Inglada and Mercier, 2007; Ban and Yousif, 2012). This one-dimensional feature space is essential to the histogram fitting technique used to estimate the unknown probabilities. The situation is complex when the change variable of interest is represented in a multidimensional feature space. In change detection, the change variable can be a vector consisting of many change indicators derived from multitemporal SAR and multispectral images. In this case, the application of the Kittler–Illingworth algorithm requires the projection or transformation of the multidimensional change variable to a one-dimensional feature space before the application of the Kittler–Illingworth algorithm (Melgani et al., 2002).

Bayesian decision theory is not restricted to scalar feature only. Measured feature can be a vector of random variables represented in multidimensional space that will necessitate a multivariate distribution to describe its likelihood. In this research, multivariate Gaussian will be adopted to model no change/change classes, as it represents the simplest multivariate model. Considering an unsupervised change detection problem with its binary nature, let us assume the change variable r to be an n-dimensional, random variable that is distributed according to multivariate normal distribution with $n \times 1$ mean vector μ_i and $n \times n$ covariance matrix Σ_i, conditioned to change or no change classes. Then, the conditional multivariate normal distribution is given by

$$p\left(\frac{r}{\omega_i}\right) = \frac{1}{(2\pi)^{\frac{n}{2}}|\Sigma_i|^{\frac{1}{2}}} \exp\left[-\frac{(r-\mu_i)^T \Sigma_i^{-1}(r-\mu_i)}{2}\right] \qquad (18.1)$$

where
$|\Sigma_i|$ is the determinant
$i : c$ for change class
$i : u$ for no change class

If the conditional density functions and prior probabilities of change and no change classes are know in advance, it is possible to design a classifier that assigns a measured change vector r to one of the two possible classes as shown in Equation 18.2 (Duda et al., 2001):

$$\text{IF}\left[g_c(r) - g_u(r)\right] > 0 \qquad \text{Then assign } r \text{ to } \omega_c \qquad (18.2)$$
$$\text{Else assign } r \text{ to } \omega_u$$

where

$$g_i(r) = -\frac{1}{2}r^T \sum_i^{-1} r + \mu_i^T \sum_i^{-1} r$$

$$-\frac{1}{2}\mu_i^T \sum_i^{-1} \mu_i - \frac{1}{2}\ln|\Sigma_i| + \ln P(\omega_i)$$

This linear discriminant function is the result of the combination of two individual discriminant functions associated with change and no change classes. This combination does not change the decision surface as explained by Duda et al. (2001).

To construct a multiband change image that accentuates changed areas, let us assume that we have two sets of coregistered multitemporal images. The first dataset is a pair of SAR images from which a change variable is derived based on the modified ratio operator as described in Equation 18.2. The second dataset consists of two Landsat multispectral images, the change variable of which is constructed by taking the absolute value of the differenced bands of interest. It should be understood that the first SAR and first Landsat images were acquired at the same time, and the same also applies to the second SAR and second Landsat images. Equation 18.3 shows the multiband change image that resulted from combining the change variables derived from SAR and optical multitemporal images:

$$r_{nx1} = \begin{bmatrix} MR \\ |\Delta\rho_{Blue}| \\ |\Delta\rho_{Green}| \\ |\Delta\rho_{Red}| \\ \vdots \end{bmatrix} \tag{18.3}$$

where

$$|\Delta\rho_{Band}| = |BV_{Band}^{date1} - BV_{Band}^{date2}|$$

MR is the modified ratio of SAR images

The selection of the multispectral bands to be used in the combined change vector will depend on the nature of the change detection problem; for example, a subset of bands that turn out to be suitable for identifying changes in green cover may not perform well in identifying change associated with urban areas. Sometimes, the high correlation between two individual change variables, for example, may also force the use of one of them in the analysis to reduce unnecessary computation complexity without serious loss of information. Therefore, the selection of the suitable band is dictated by the problem in hand and no general rules can be applied here.

The combined SAR and optical solution for change detection was implemented on Shanghai data (Figure 18.5). The absolute difference of the green and mid-infrared bands of the Landsat multitemporal images was used together with the SAR modified ratio to construct the multidimensional change variable. The combined iterative solution consists of first estimating the class statistics from the multiband change image using the previously generated binary change map as a mask, followed by thresholding the multiband change image and producing a new binary change map. This iterative solution continues until some termination condition is met. As a convergence criterion, the change in the percentage of scene being classified as change between successive iterations should be less than or equal to $10^{-4}\%$. The convergence of the solution signifies the complete separation between two distinctive groups of pixels. In order to test the ability of the algorithm to converge under initial solutions of different qualities, the algorithm was run several times using different initial change maps obtained by thresholding the SAR modified ratio image under the generalized Gaussian (GG), log normal (LN), Nakagami ratio (NR), and Weibull ratio (WR) models (Ban and Yousif, 2012). Starting from an initial change map, the combined SAR and optical algorithm starts to refine the classification using the multidimensional change image. The effect of using both SAR and optical data was introduced

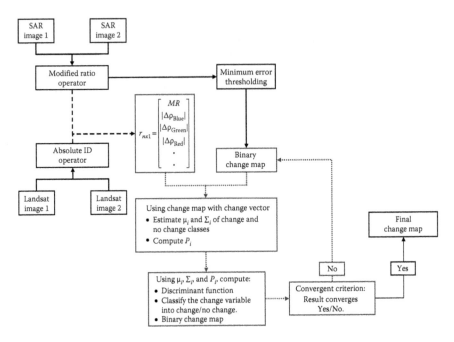

FIGURE 18.5 Flowchart of unsupervised change detection using fusion of SAR and optical data.

gradually during the clustering process. Throughout any single iteration, pixels move between change/no change classes, improving the in-between class discrimination and enhancing the accuracy of the estimation of the parameters of the density model.

18.2.2.3 Results and Discussion

The behaviors of the different measures of accuracy under different initial conditions are plotted in Figure 18.6, showing the variations versus iterations of the positive change detection accuracy, negative change detection accuracy, Kappa coefficient of agreement, overall error rate, and false alarm. The zero in the abscissa of each of the plots in Figure 18.7 corresponds to the value of the measure of accuracy of the solution obtained by using the SAR data alone. Table 18.6 provides the values of the different measures of accuracy of the initial solution and the last result obtained after using the iterative combined solution. The table also shows the number of iteration required for each type of solution to converge. As both Figure 18.6 and Table 18.6 indicate, all measures of accuracy converge to the same final values, though after a different number of iterations, pointing out the differences in the quality of the initial solutions. Considering first the detection accuracy in positive changed areas, originally the result obtained using the modified ratio operator was good for all density models. Therefore, the amount of improvement introduced into this measure is moderate. The largest improvement occurred in the Weibull ratio solution for which the detection accuracy increased from 79.7% to 96.4%. On the other hand, the generalized Gaussian solution received the smallest amount of correction where the accuracy jumped

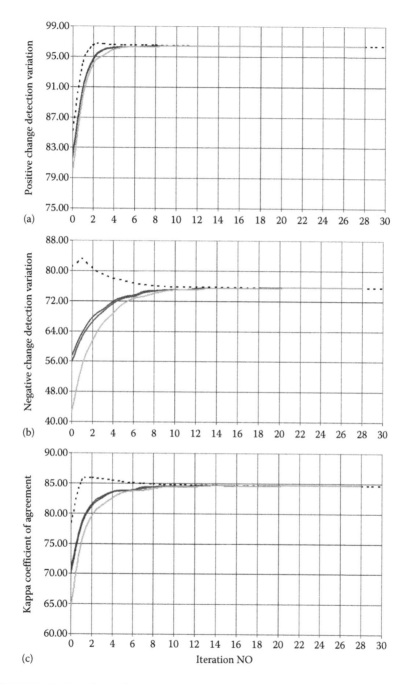

FIGURE 18.6 Variations of measures of accuracy versus iterations of the SAR and optical combined solution: (a) positive change detection accuracy, (b) negative change detection accuracy, and (c) Kappa coefficient.

(*continued*)

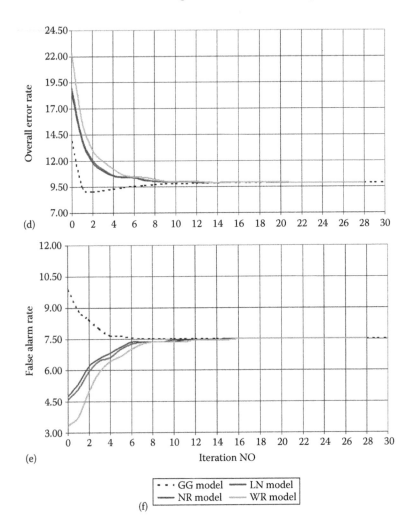

FIGURE 18.6 (continued) Variations of measures of accuracy versus iterations of the SAR and optical combined solution: (d) overall error rate, (e) false alarm rate, and (f) legend.

from 82% to 96%, indicating that this solution was originally the best. Figure 18.6a also shows how the curves of all the models look similar to each other and that the curves of the log normal and Nakagami ratio models are identical, confirming the similarity of the solutions obtained under these two models. For the negative change detection accuracy, the situation is different. As mentioned earlier, the main drawback of using the SAR data alone is the low detection accuracy in negative change areas due to the domination of new built-up areas with their high positive intensity of change. Nevertheless, using the combined solution, the accuracy increased significantly from 55.8%, 57.5%, and 42.5% for the log normal, Nakagami ratio, and Weibull ratio models, respectively, to 75.6%, as shown in Table 18.6 and Figure 18.6b. The Weibull ratio model received the largest correction, confirming that this solution

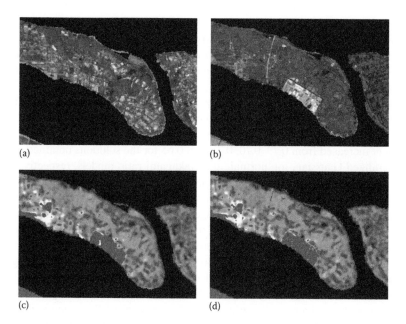

(a)   (b)

(c)   (d)

FIGURE 18.7   **(See color insert.)** Detected changed areas in yellow, overlaid in a false color composite using (a) false color composite of 03.11.1999 Landsat image, (b) false color composite of 06.07.2008 Landsat image, (c) SAR and optical combined solution, and (d) SAR modified ratio with log normal solution.

TABLE 18.6

Accuracy Assessment (%) of the Change Detection Results Obtained Using Combined SAR and Optical Solution with the Algorithm Being Initiated Using GG, LN NR, and WR Models

	SAR GG Model	SAR LN Model	SAR NR Model	SAR WR Model	Optical CVA	SAR and Optical
Positive change detection accuracy (%)	85.27	81.68	82.03	79.69	80.84	96.42
Negative change detection accuracy (%)	80.90	55.80	57.46	42.54	63.22	75.56
Kappa	0.78	0.70	0.71	0.65	0.69	0.85
Overall error (%)	13.87	19.02	18.56	22.36	20.11	9.89
False alarm (%)	9.86	4.60	4.75	3.35	11.04	7.51
Number of iterations	30	27	27	28		

was the worst. In contrast, the detection accuracy decreased slightly for the generalized Gaussian model, which may suggest that the solution under this model slightly overestimated the negative change class.

These results show that changes characterized by an increase in intensity over time are well documented by SAR images, unlike those characterized by intensity decrease. Accordingly, the probability of detecting negative change based on SAR data alone is very low even when using the modified ratio image. This limitation in the capabilities of the SAR images can only be overcome through the use of both SAR and optical images in the combined solution, which significantly enhances the detection accuracy in areas with negative change.

As a result of the improvement that occurred on the detection accuracy in positive and negative change areas, the Kappa coefficient of agreement shows significant enhancements for different initial solutions, as shown in Figure 18.6c. The Kappa coefficient increased from 0.79, 0.70, 0.71, and 0.65 for the solutions obtained, using the generalized Gaussian, log normal, and Nakagami ratio models, respectively, to 0.85 for the combined solution. As it was the worst solution, the Weibull distribution again received the largest enhancement among all other initial solutions.

One of the negative aspects of the change detection solutions based on SAR data alone is the high overall error. However, using the combined solution, the overall error rate has reduced significantly, underlining the positive effects of the optical data. This is particularly true for the solutions obtained using log normal, Nakagami ratio, and Weibull ratio models, which decrease from 19.0%, 18.6%, and 22.4%, respectively, to 9.9%. Regarding the generalized Gaussian model, its original overall error rate was the best among the others; therefore, only a small correction is added as reflected by the relatively smoother overall error variation curve (Figure 18.6d).

The solution obtained by applying the generalized Gaussian model to the SAR images suffered from the exceptionally high false alarm rate. Initializing the combined algorithm with the binary change map obtained using this model, the false alarm decreased from 9.9% to 7.5% after 30 iterations (Table 18.6). On the contrary, the false alarm rate increased slightly for the other three solutions, as shown in Figure 18.6e. This last result is acceptable in light of the improvements connected with the overall error rate, Kappa coefficient, and detection accuracies in areas with intensity decrease/increase that are quite immense compared with the loss associated with the false alarm rate.

The SAR and optical combined solution is in reality a clustering algorithm that restricts the classes to follow a multivariate Gaussian model. An iterative approach is adopted to reach an unambiguous separation between the classes. As such, the number of iterations required can be considered as a measure of the quality of the solution used to initiate the algorithm (training data). In other words, the number of iterations is somehow a measure of the work exerted by the algorithm to achieve a perfect classification. The higher the quality of the initial solution, the smaller is the number of iterations required for convergence. Therefore, the last row in Table 18.6 clearly indicates that the quality of the solution obtained by using the modified ratio operator with the generalized Gaussian model was the worst since the algorithm required 30 iterations to converge. This is mainly due to its very high false alarm rate.

To visually appreciate the enhancement introduced by using both SAR and optical in a combined solution, Figure 18.7 shows (a) the false color composite of the first-date Landsat images, (b) the false color composite of the second-date Landsat image, and the detected change in red overlaid in a false color composite using (c) the SAR data alone and (d) the SAR and optical combined solution. Comparing these two figures, the SAR and optical combined solution is able to detect both new built-up areas and new roads while new roads were missed when using SAR data alone.

18.2.2.4 Conclusions

This case study investigated the fusion of multitemporal SAR and optical data for urban change detection using the Kittler–Illingworth minimum-error thresholding algorithm. The experiment result showed that improvements took place in almost all measures of accuracy, especially in detection accuracy in areas with negative change, which was the main problem when using multitemporal SAR data alone. The Kappa coefficient of agreement increases significantly from 0.78 (the best change detection accuracy using SAR alone) or 0.68 (the CVA result using optical data) to 0.85 using the SAR and optical combined solution.

REFERENCES

Alberga, V. 2009. Similarity measures of remotely sensed multi-sensor images for change detection applications. *Remote Sensing*, 1(3), 122–143.

Al-Khudhairy, D. H. A., I. Caravaggi, and S. Glada. 2005. Structural damage assessments from Ikonos data using change detection, object-oriented segmentation, and classification techniques. *Photogrammetric Engineering & Remote Sensing*, 71, 825–837.

Alparone, L., S. Baronti, A. Garzelli, and F. Nencini. 2004. Landsat ETM+ and SAR image fusion based on generalized intensity modulation. *IEEE Transactions on Geoscience and Remote Sensing*, 42(12), 2832–2839.

Amarsaikhan, D. and T. Douglas. 2004. Data fusion and multisource image classification. *International Journal of Remote Sensing*, 25(17), 3529–3539.

Amarsaikhan, D., M. Ganzorig, P. Ache, and H. Blotevoge. 2007. The integrated use of optical and InSAR data for urban land cover mapping. *International Journal of Remote Sensing*, 28(6), 1161–1171.

Amarsaikhan, D., M. Saandar, M. Ganzorig, H. H. Blotevogel, E. Egshiglen, R. Gantuyal, B. Nergui, and D. Enkhjargal. 2012. Comparison of multisource image fusion methods and land cover classification. *International Journal of Remote Sensing*, 33(8), 2532–2550.

Ban, Y. 2003. Multitemporal ERS-1 SAR and Landsat TM data for agricultural crop classification: Comparison and synergy. *Canadian Journal of Remote Sensing*, 29(4), 518–526.

Ban, Y., H. Hu, and I. M. Rangel. 2010. Fusion of Quickbird MS and RADARSAT SAR data for urban land-cover mapping: Object-based and knowledge-based approach. *International Journal of Remote Sensing*, 31(6), 1391–1410.

Ban, Y. and A. Jacob. 2013. Object-based fusion of multitemporal multiangle ENVISAT ASAR and HJ-1B multispectral data for urban land-cover mapping. *IEEE Transactions on Geoscience and Remote Sensing*, 51(4), 1998–2006.

Ban, Y. and O. A. Yousif. 2012. Multitemporal spaceborne SAR data for urban change detection in China. *IEEE Journal of Selected Topics in Applied Earth Observations and Remote Sensing*, 5(4), 1087–1094.

Bazi, Y., L. Bruzzone, and F. Melgani. 2005. An unsupervised approach based on the generalized Gaussian model to automatic change detection in multitemporal SAR images. *IEEE Transactions on Geoscience and Remote Sensing*, 43(4), 874–887.

Bazi, Y., L. Bruzzone, and F. Melgani. 2006. Automatic identification of the number and values of decision thresholds in the log-ratio image for change detection in SAR images. *IEEE Transactions on Geoscience and Remote Sensing*, 3(3), 3491–353.

Berberoglu, S. and A. Akin. 2009. Assessing different remote sensing techniques to detect land use/cover changes in the eastern Mediterranean. *International Journal of Applied Earth Observation and Geoinformation*, 11, 46–53.

Blaschke, T. 2005. Towards a framework for change detection based on image objects. *Göttinger Geographische Abhandlungen*, 113, 1–9.

Bovolo, F. 2009. A multilevel parcel-based approach to change detection in very high resolution multitemporal images. *IEEE Geoscience and Remote Sensing Letters*, 6, 33–37.

Bovolo, F. and L. Bruzzone. 2007. A theoretical framework for unsupervised change detection based on change vector analysis in the polar domain. *IEEE Transactions on Geoscience and Remote Sensing*, 45 (1), 218–236.

Bruzzone, L. and D. F. Prieto. 2000. A minimum-cost thresholding technique for unsupervised change detection. *International Journal of Remote Sensing*, 21(18), 3539–3544.

Bruzzone, L. and D. F. Prieto. 2002. An adaptive semiparametric and context-based approach to unsupervised change detection in multitemporal remote-sensing images. *IEEE Transactions on Image Processing*, 11, 452–466.

Bujor, F. T., J.-M. Nicolas, E. Trouve, and J.-P. Rudant. 2003. Application of log-cumulants to change detection on multi-temporal SAR images. In *Proceedings of 2003 IEEE International Geoscience and Remote Sensing Symposium (IGARSS'03)*, Toulouse, France, Vol. 2, pp. 1386–1388.

Bujor, F. T., L. Valet, E. Trouvw, G. Mauris, N. Classeau, and J.-P. Rudant. 2001. Data fusion approach for change detection in multi-temporal ERS-SAR images. In *IEEE 2001 International Geoscience and Remote Sensing Symposium (IGARSS'01)*, Sydney, New South Wales, Australia, vol. 6, pp. 2590–2592.

Cakir, I. H., S. Khorram, and S. A. C. Nelson. 2006. Correspondence analysis for detecting land cover change. *Remote Sensing of Environment*, 102, 306–317.

Campbell, J. B. 2002. *Introduction to Remote Sensing*, 3rd edn. New York: The Guilford Press.

Cao, G. and Y. Q. Jin. 2007. A hybrid algorithm of the BP-ANN/GA for classification of urban terrain surfaces with fused data of Landsat ETM+ and ERS-2 SAR. *International Journal of Remote Sensing*, 28(2), 293–305.

Castellana, L., A. D'Addabbo, and G. Pasquariello. 2007. A composed supervised/unsupervised approach to improve change detection from remote sensing. *Pattern Recognition Letters*, 28, 405–413.

Celik, T. 2009. Unsupervised change detection in satellite images using principal component analysis and k-means clustering. *IEEE Geoscience and Remote Sensing Letters*, 6, 772–776.

Celik, T. 2010. Change detection in satellite images using a genetic algorithm approach. *IEEE Geoscience and Remote Sensing Letters*, 7, 386–390.

Chen, G., G. J. Hay, L. M. T. Carvalho, and M. A. Wulder. 2012. Object-based change detection. *International Journal of Remote Sensing*, 33(14), 4434–4457.

Chen, K., Z. Zhou, C. Huo, X. Sun, and K. Fu. 2013. A semisupervised context-sensitive change detection technique via Gaussian process. *IEEE Geoscience and Remote Sensing Letters*, 10(2), 236–240.

Collins, J. B. and C. E. Woodcock. 1996. An assessment of several linear change detection techniques for mapping forest mortality using multitemporal Landsat TM data. *Remote Sensing of Environment*, 56(1), 66–77.

Del Frate, F., F. Pacifici, and D. Solimini. 2008. Monitoring urban land cover in Rome, Italy, and its changes by single-polarization multitemporal SAR images. *IEEE Journal of Selected Topics in Applied Earth Observations and Remote Sensing*, 1, 87–97.

Dobson, M. C., F. T. Ulaby, and L. E. Pierce. 1995. Land-cover classification and estimation of terrain attributes using synthetic aperture radar. *Remote Sensing of Environment*, 51, 199–214.

Duda, R. O., P. E. Hart, and D. G. Stork. 2001. *Pattern Classification*. New York: John Wiley & Sons, Inc.

Durieux, L., E. Lagabrielle, and A. Nelson. 2008. A method for monitoring building construction in urban sprawl areas using object-based analysis of Spot 5 images and existing GIS data. *ISPRS Journal of Photogrammetry and Remote Sensing*, 63, 399–408.

Foody, G. M. 1995. Using prior knowledge in artificial neural network classification with a minimal training set. *International Journal of Remote Sensing*, 16(2), 301–312.

Gamba, P. and Aldrighi, M. 2012. SAR data classification of urban areas by means of segmentation techniques and ancillary optical data. *IEEE Journal of Selected Topics in Applied Earth Observations and Remote Sensing*, 5(4), 1140–1148.

Gamba, P., Aldrighi, M. and Stasolla, M. 2011. Robust extraction of urban area extents in HR and VHR SAR images. *IEEE Journal of Selected Topics in Applied Earth Observations and Remote Sensing*, 4(1), 27–34.

Gamba, P. and M. Herold, Eds. 2009. *Global Mapping of Human Settlement—Experiences, Datasets, and Prospects*. Boca Raton, FL: CRC Press.

Gamba, P. and G. Lisini. 2013. Fast and efficient urban extent extraction using ASAR wide swath mode data. *IEEE Journal of Selected Topics in Applied Earth Observations and Remote Sensing*, 6(5), 2184–2195.

Griffiths, P., P. Hostert, O. Gruebner, and S. van der Linden. 2010. Mapping megacity growth with multi-sensor data. *Remote Sensing of Environment*, 114(2), 426–439.

Haack, B. N., E. K. Solomon, M. A. Bechdol, and N. D. Herold. 2002. Radar and optical data comparison/integration for urban delineation: A case study. *Photogrammetric Engineering & Remote Sensing*, 68(12), 1289–1296.

Hachicha, S. and F. Chaabane. 2010. Comparison of change detection indicators in SAR images. In *Eighth European Conference on Synthetic Aperture Radar (EUSAR)*, Aachen, Germany, pp.1, 4, 7–10.

Hall, O. and G. J. Hay. 2003. A multiscale object-specific approach to digital change detection. *International Journal of Applied Earth Observation and Geoinformation*, 4, 311–327.

He, C., A. Wei, P. Shi, Q. Zhang, and Y. Zhao. 2011. Detecting land-use/land-cover change in rural-urban fringe areas using extended change vector analysis. *International Journal of Applied Earth Observation and Geoinformation*, 13, 572–585.

Henderson, F. M., R. Chasan, J. Portolese, and T. Hart. 2002. Evaluation of SAR-optical imagery synthesis techniques in a complex coastal ecosystem. *Photogrammetric Engineering & Remote Sensing*, 68(8), 839–846.

Hu, H. and Y. Ban. 2012. Multitemporal RADARSAT-2 ultra-fine beam SAR data for urban land cover classification. *Canadian Journal of Remote Sensing*, 38(1), 1–11.

Im, J., J. R. Jensen, and J. A. Tullis. 2008. Object-based change detection using correlation image analysis and image segmentation. *International Journal of Remote Sensing*, 29, 399–423.

Inglada, J. and A. Giros. 2004. On the possibility of automatic multisensor image registration. *IEEE Transactions on Geoscience and Remote Sensing*, 42(10), 2104–2120.

Inglada, J. and G. Mercier. 2007. A new statistical similarity measure for change detection in multitemporal SAR images and its extension to multiscale change analysis. *IEEE Transactions on Geoscience and Remote Sensing*, 45(5), 1432–1445.

Jensen, J. R. 2005. *Introductory Digital Image Processing: A Remote sensing Perspective*, 3rd edn. New York: Prentice-Hall, Inc.

Jensen, J. R., F. Qiu, and K. Patterson. 2001. A neural network image interpretation system to extract rural and urban land use and land cover information for remote sensing data. *Geocarto International: A Multidisciplinary Journal of Remote Sensing & GIS*, 16(1), 19–28.

Kasetkasem, T. and P. K. Varshney. 2002. An image change detection algorithm based on Markov random field models. *IEEE Transactions on Geoscience and Remote Sensing*, 40, 1815–1823.

Kavzoglu, T. and P. M. Mather. 2003. The use of backpropagating artificial neural networks in land cover classification. *International Journal of Remote Sensing*, 24(23), 4907–4938.

Kittler, J. and J. Illingworth. 1986. Minimum error thresholding. *Pattern Recognition*, 19(1), 41–47.

Kontoes, C. C. 2008. Operational land cover change detection using change vector analysis. *International Journal of Remote Sensing*, 29(16), 4757–4779.

Kwarteng, A. Y. and P. S. Chavez. 1998. Change detection study of Kuwait City and environs using multitemporal Landsat Thematic Mapper data. *International Journal of Remote Sensing*, 19, 1651–1662.

Laliberte, A. S., A. Rango, K. M. Havstad, J. F. Paris, R. F. Beck, and R. Mcneely. 2004. Object-oriented image analysis for mapping shrub encroachment from 1937 to 2003 in southern New Mexico. *Remote Sensing of Environment*, 93, 198–210.

Lefebvre, A., T. Corpetti, and L. Hubert-Moy. 2008. Object-oriented approach and texture analysis for change detection in very high resolution images. In *Proceedings of the IEEE International Geoscience and Remote Sensing Symposium (IGARSS'08)*, Boston, MA, vol. 4, pp. IV-663–IV-666.

Leinenkugel, P., T. Esch, and C. Kuenzer. 2011. Settlement detection and impervious surface estimation in the Mekong Delta using optical and SAR remote sensing data. *Remote Sensing of Environment*, 115(12), 3007–3019.

Li, X. and A. G. O. Yeh. 1998. Principal component analysis of stacked multi-temporal images for the monitoring of rapid urban expansion in the Pearl River Delta. *International Journal of Remote Sensing,* 19(8), 1501–1518.

Lu, D., G. Li, E. Moran, M. Batistella, and C. Freitas. 2011. Mapping impervious surfaces with the integrated use of Landsat Thematic Mapper and radar data: A case study in an urban–rural landscape in the Brazilian Amazon. *ISPRS Journal of Photogrammetry and Remote Sensing*, 66(6), 798–808.

Lu, D., P. Mausel, E. Brondízio, and E. Moran. 2004. Change detection techniques. *International Journal of Remote Sensing*, 25(12), 2365–2401.

Lyon, J. G., D. Yuan, R. S, Lunetta, and C. D. Elvidge. 1998. A change detection experiment using vegetation indices. *Photogrammetric Engineering & Remote Sensing*, 64(2), 143–150.

Ma, J., M. Gong, and Z. Zhou. 2012. Wavelet fusion on ratio images for change detection in SAR images, *IEEE Geoscience and Remote Sensing Letters*, 9(6), 1122–1126.

Mas, F. 1999. Monitoring land-cover changes: A comparison of change detection techniques. *International Journal of Remote Sensing*, 20(1), 139–152.

McDermid, G. J., J. Linke, A. Rape, D. N. Laskin, A. J. Mclane, and S. E. Franklin. 2008. Object-based approaches to change analysis and thematic amp update: Challenges and limitation. *Canadian Journal of Remote Sensing*, 34, 462–466.

Melgani, F., G. Moser, and S. B. Serpico. 2002. Unsupervised change-detection methods for remote-sensing images. *Optical Engineering*, 41(12), 3288–3297.

Mercier, G., G. Moser, and S. Serpico. 2008. Conditional copula for change detection on heterogeneous SAR data. *IEEE Transactions on Geoscience and Remote Sensing*, 46(5), 1428–1441.

Moser, G., F. Melgani, S. B. Serpico, and A. Caruso. 2002. Partially supervised detection of changes from remote sensing images. *IEEE International Geoscience and Remote Sensing Symposium*, 1, 299–301.

Moser, G. and S. B. Serpico. 2006. Generalized minimum-error thresholding for unsupervised change detection from SAR amplitude imagery. *IEEE Transactions on Geoscience and Remote Sensing*, 44(10), 2972–2982.

Moser, G., S. Serpico, and G. Vernazza. 2007. Unsupervised change detection from multichannel SAR images. *IEEE Geoscience and Remote Sensing Letters*, 4, 278–282.

Nielsen, A. A. 2007. The regularized iteratively reweighted MAD method for change detection in multi- and hyperspectral data. *IEEE Transactions on Image Processing*, 16(2), 463–478.

Niu, X. and Y. Ban. 2013. Multitemporal RADARSAT-2 polarimetric SAR data for urban land cover classification using object-based support vector machine and rule-based approach. *International Journal of Remote Sensing*, 34(1), 1–26.

Nemmour, H. and Y. Chibani. 2006. Multiple support vector machines for land cover change detection: An application for mapping urban extensions. *ISPRS Journal of Photogrammetry and Remote Sensing*, 61, 125–133.

Onana, V. P., E. Trouve, G. Mauris, J. P. Rudant, and P. L. Frison. 2003. Change detection in urban context with multitemporal ERS-SAR images by using data fusion approach. In *Proceedings of 2003 IEEE International Geoscience and Remote Sensing Symposium (IGARSS'03)*, Toulouse, France, vol. 6, pp. 3650–3652.

Pacifici, F., F. Del Frate, W. J. Emery, P. Gamba, and J. Chanussot. 2008. Urban mapping using coarse SAR and optical data: outcome of the 2007 GRSS data fusion contest. *IEEE Geoscience and Remote Sensing Letters*, 5(3), 331–335.

Palubinskas, G. 2012. How to fuse optical and radar imagery? In *2012 IEEE International Geoscience and Remote Sensing Symposium (IGARSS)*, Munich, Germany, pp. 2171–2174.

Poulain, V., J. Inglada, M. Spigai, J. Y. Tourneret, and P. Marthon. 2009. Fusion of high resolution optical and SAR images with vector data bases for change detection. In *2009 IEEE International Geoscience and Remote Sensing Symposium (IGARSS'09)*, Cape Town, South Africa, vol. 4, pp. IV-956–IV-959.

Ridd, M. K. and J. J. Liu. 1998. A comparison of four algorithms for change detection in an urban environment. *Remote Sensing of Environment*, 63, 95–100.

Rignot, E. J. M. and J. J. van Zyl. 1993. Change detection techniques for ERS-1 SAR data. *IEEE Transactions on Geoscience and Remote Sensing*, 31(4), 896–906.

Rogan, J. and D. Chen. 2004. Remote sensing technology for mapping and monitoring landcover and land-use change. *Progress in Planning*, 61(4), 301–325.

Sohl, T. L. 1999. Change analysis in the United Arab Emirates: An investigation of techniques. *Photogrammetric Engineering & Remote Sensing*, 65(4), 475–484.

Soria-Ruiz, J., Y. Fernandez-Ordoñez, and I. H. Woodhouse. 2010. Land-cover classification using radar and optical images: A case study in Central Mexico. *International Journal of Remote Sensing*, 31(12), 3291–3305.

Townshend, J. R. G. and C. O. Justice. 1995. Spatial variability of images and the monitoring of changes in the normalized difference vegetation index. *International Journal of Remote Sensing*, 16, 2187–2195.

Ulaby, F. T., R. K. Moore, and A. K. Fung. 1986. *Microwave Remote Sensing: Active and Passive, Vol. III: Volume Scattering and Emission Theory, Advanced Systems and Applications*. Dedham, MA: Artech House, Inc., 1100pp.

Volpi, M., D. Tuia, F. Bovolo, M. Kanevski, and L. Bruzzone. 2013. Supervised change detection in VHR images using contextual information and support vector machines. *International Journal of Applied Earth Observation and Geoinformation*, 20, 77–85.

Walter, V. 2004. Object-based classification of remote sensing data for change detection. *ISPRS Journal of Photogrammetry and Remote Sensing*, 58, 225–238.

Wang, F. 1993. A knowledge-based vision system for detecting land changes at urban fringes. *IEEE Transactions on Geoscience and Remote Sensing*, 31(1), 136–145.

Weng, Q. 2002. Land use change analysis in the Zhujiang Delta of China using satellite remote sensing, GIS and stochastic modelling. *Journal of Environmental Management*, 64, 273–284.

Wilkinson, G. G., S. Folving, I. Kanellopoulos, N. McCormick, K. Fullerton, and J. Megier. 1995. Forest mapping from multi-source satellite data using neural network classifiers: An experiment in Portugal. *Remote Sensing Reviews*, 12, 83–106.

Xia, Z. and F. M. Henderson. 1997. Understanding the relationships between radar response patterns and the bio- and geophysical parameters of urban areas. *IEEE Transactions on Geoscience and Remote Sensing*, 35(1), 93–101.

Yang, X. and C. P. Lo. 2002. Using a time series of satellite imagery to detect land use and land cover changes in the Atlanta, Georgia metropolitan area. *International Journal of Remote Sensing*, 23(9), 1775–1798.

Yuan, F., K. E. Sawaya, B. C. Loeffelholz, and M. E, Bauer. 2005. Land cover classification and change analysis of the Twin Cities (Minnesota) Metropolitan Area by multitemporal Landsat remote sensing. *Remote Sensing of Environment*, 98, 317–328.

Zhang, Q., J. Wang, X. Peng, P. Gong, and P. Shi. 2002. Urban built-up land change detection with road density and spectral information from multi-temporal Landsat TM data. *International Journal of Remote Sensing*, 23(15), 3057–3078.

Zhou, W., A. Troy, and M. Grove. 2008. A comparison of object-based with pixel-based land cover change detection in the Baltimore Metropolitan Area using multitemporal high resolution remote sensing data. In *IEEE International Geoscience and Remote Sensing Symposium*, Boston, MA, pp. IV-683–IV-686.

Zhu, Z., C. E. Woodcock, J. Rogan, and J. Kellndorfer. 2012. Assessment of spectral, polarimetric, temporal, and spatial dimensions for urban and peri-urban land cover classification using Landsat and SAR data. *Remote Sensing of Environment*, 117, 72–82.

Index

Printed and bound by CPI Group (UK) Ltd, Croydon, CR0 4YY

01/11/2024

01782625-0007